ECOLOGICAL ECONOMICS FOR THE ANTHROPOCENE

Ecological Economics
for the Anthropocene

An Emerging Paradigm

EDITED BY PETER G. BROWN AND
PETER TIMMERMAN

 COLUMBIA UNIVERSITY PRESS *NEW YORK*

Columbia University Press
Publishers Since 1893
New York   Chichester, West Sussex
cup.columbia.edu

Library of Congress Cataloging-in-Publication Data
Ecological economics for the anthropocene : an emerging paradigm /
    edited by Peter G. Brown and Peter Timmerman.
        pages cm
    Includes bibliographical references and index.
    ISBN 978-0-231-17342-1 (cloth : alk. paper) —
    ISBN 978-0-231-17343-8 (pbk. : alk. paper) —
    ISBN 978-0-231-54042-1 (e-book)
    1. Ecology—Economic aspects. 2. Environmental economics.
I. Brown, Peter G. II. Timmerman, Peter.

    HC79.E5E25253 2015
    333.7—dc23

                                                    2015009578

Columbia University Press books are printed on permanent
and durable acid-free paper.
This book is printed on paper with recycled content.
Printed in the United States of America

c 10 9 8 7 6 5 4 3 2 1
p 10 9 8 7 6 5 4 3 2 1

COVER IMAGE: Courtesy of NASA
COVER DESIGN: Milenda Nan Ok Lee

References to websites (URLs) were accurate at the time of writing.
Neither the author nor Columbia University Press is responsible for URLs
that may have expired or changed since the manuscript was prepared.

CONTENTS

Foreword
JON D. ERICKSON   ix

Acknowledgments   xv

INTRODUCTION
The Unfinished Journey of Ecological Economics
PETER G. BROWN AND PETER TIMMERMAN   1

PART I Proposed Ethical Foundations of Ecological Economics
Introduction and Chapter Summaries   15

CHAPTER ONE
The Ethics of Re-Embedding Economics in the Real: Case Studies
PETER TIMMERMAN   21

CHAPTER TWO
Ethics for Economics in the Anthropocene
PETER G. BROWN   66

CHAPTER THREE
Justice Claims Underpinning Ecological Economics
RICHARD JANDA AND RICHARD LEHUN   89

PART II Measurements: Understanding and Mapping Where We Are
Introduction and Chapter Summaries   119

CHAPTER FOUR
Measurement of Essential Indicators in Ecological Economics
MARK S. GOLDBERG AND GEOFFREY GARVER   125

CHAPTER FIVE
Boundaries and Indicators
*Conceptualizing and Measuring Progress Toward an Economy of Right
Relationship Constrained by Global Ecological Limits*
GEOFFREY GARVER AND MARK S. GOLDBERG   149

CHAPTER SIX
Revisiting the Metaphor of Human Health for Assessing Ecological
Systems and Its Application to Ecological Economics
MARK S. GOLDBERG, GEOFFREY GARVER, AND NANCY E. MAYO   190

CHAPTER SEVEN
Following in Aldo Leopold's Footsteps
Humans-in-Ecosystem and Implications for Ecosystem Health
QI FENG LIN AND JAMES W. FYLES   208

PART III Implications: Steps Toward Realizing an Ecological Economy
Introduction and Chapter Summaries   233

CHAPTER EIGHT
Toward an Ecological Macroeconomics
PETER A. VICTOR AND TIM JACKSON   237

CHAPTER NINE
New Corporations for an Ecological Economy: A Case Study
RICHARD JANDA, PHILIP DUGUAY, AND RICHARD LEHUN   260

CHAPTER TEN
Ecological Political Economy and Liberty
BRUCE JENNINGS   272

CHAPTER ELEVEN
A New Ethos, a New Discourse, a New Economy
Change Dynamics Toward an Ecological Political Economy
JANICE E. HARVEY   318

CONCLUSION
Continuing the Journey of Ecological Economics
Reorientation and Research   357

Contributors   371

Index   377

FOREWORD

The Unfinished Journey of Ecological Economics

JON D. ERICKSON

Ecological economics began with a rather audacious promise to change the world. It was a promise to ground the study and application of economics within the biophysical realities of a finite world and the moral obligations of a just society; a charge to search for truth across disciplines and begin to erase artificial boundaries between C. P. Snow's "two cultures" of the sciences and humanities; and an agenda for political action emanating from the United Nation's Earth Summit in 1992 that billed ecological economics as the science of sustainable development.

I was in graduate school in 1992 and wholeheartedly bought into the vision and course of ecological economics. I devoted my professional life to the development of this transdisciplinary lens on the study and management of human communities embedded in our social and biophysical environments. In 1997, with Ph.D. in hand, I landed one of the first jobs advertising for an ecological economist at Rensselaer Polytechnic Institute. We set out to build the first doctoral program in ecological economics, and in subsequent years I helped found the U.S. Society for Ecological Economics; served on the board of our international society; authored and coauthored the requisite number of ecological economic papers and books to get tenure and then a full professorship in the field of ecological economics; worked with some of the pioneers in our field, including Herman Daly, John Gowdy, and Bob Costanza; and went on to manage the Gund Institute for Ecological Economics at the University of Vermont, which is today one of the main hubs of research, application, and education in ecological economics in North America.

I share my journey as an ecological economist to make a point. When the field was formalized in the late 1980s and early 1990s through the creation of a professional society and journal, people like me were supposed to be incubated, indoctrinated, and infiltrated into society. I did not discover ecological economics midcareer, nor did I transform myself into an ecological economist after tenure. I defined myself as an ecological economist in graduate school. For more than twenty years, I have believed that ecological economics would bring about a major paradigm shift in economics and perhaps the social sciences and humanities more broadly.

The premise of this book is that the revolution got a bit sidetracked. I certainly agree. There are scholars, activists, policy makers, professionals, and citizens from all walks of life that identify strongly with ecological economics (and many even call themselves ecological economists as I do). We are all on an unfinished journey, and it is high time that we take stock of where we are, make course corrections, and get on with changing the world.

This will take some soul-searching. For me, it starts with evaluating the field against some core foundational aspects of ecological economics. Ecological economics was initially framed as the study of the economy grounded in the principles of ecology—what Herman Daly called for in his 1968 essay "On Economics as a Life Science." Today, I would argue, in practice and in perception, ecological economics has largely become the application of mainstream economics (economics as an orthodox social science) to the existing agenda of ecologists and environmentalists, thus in practice facilitating a growth agenda. Much of what gets published in our journal, presented at our conferences, and picked up by the press is what we called in graduate school "environmental economics"—a subdiscipline of economics applied to environmental problems.

This book is first and foremost about re-embedding the study of the economy within the hard-won physical principles of natural science and through a discourse built more on ethical argument than mathematical formalism. In graduate school, I was convinced that my newfound charge as an ecological economist was to expose the faulty assumptions upon which the house of cards called neoclassical economics was built—particularly the all-too-convenient myths of the rational-actor model and the market-efficiency hypothesis that it was designed to support. Then, with the growing ranks of ecological economists, we were to build an economics that transcended disciplinary boundaries—an economics built on biophysical realities and real human beings ("irrational" emotions and all) as decision makers; an economics that does not always and everywhere assume that more is better; an honest economics built on scientific integrity and democratic discourse.

Instead, I fear at times we practice a hypocritical economics. We too often expose the inconsistencies between our behavioral assumptions and hard-won, testable facts from other disciplines while we kneel at the altar of the market as the one true path to sustainability. This has particularly been the case in the North American expressions of ecological economics, where the discussion has largely focused on market failure, missing the broader critique of the failure of markets, and the role of nonmarket institutions in moving away from what David Korten has called "suicide economics." Institutions matter, and the European brand of ecological economics seems to distinguish itself on at least this point. For example, we have learned from the first generation of payments for ecosystem-services schemes, largely in the tropics, that the role of institutions is critical to successful conservation. Quantifying the economic benefit of nature means little without the institutions in place to assure sustainable and equitable outcomes.

I fear that many who practice ecological economics have taken the road most traveled: to double-down on market efficiency as the primary goal of the economy through an exercise of "getting the prices right." However, by putting the free market and the overarching goal of efficiency ahead of sustainable scale and just distribution, this bandwagon version of ecological economics has simply become a prescription for fairer, cleaner growth—but growth nonetheless. The "unfinished journey" has much to do with reaffirming a vision of economic development that embraces ethics, affirms life, and argues for well-defined limits to human economies. Natural science must help the ecological economist define the boundary conditions of sustainable scale. Ethical debate and public process must negotiate just distribution. Sustainability and justice then frame the design of well-regulated markets to achieve genuine economic efficiency.

Ecological economics also has some catching up to do with other inter-disciplinary pursuits. If Milton Friedman were alive today, my guess is that rather than "Keynesians," he would say, "We're all behavioral economists now." All the analysis in the world about energy return on investment, thermodynamics, and the folly of infinite growth will continue to have a limited effect on policy and planning in absence of a grounding in human nature. The Madison Avenue advertising firms threw out their Economics 101 textbooks many years ago and aligned themselves with the behavioral sciences; so have the political-campaign consultancies, mass media, and news agencies.

The biological underpinnings of our decisions are as much a constraint to sustainability as the energy and ecological limits to growth. Much progress has been made on illuminating the proximate causes of economic choice through the neurosciences, addressing questions about how we make decisions. More work remains to be done on ultimate cause: questions about why we make the decisions we do. Ecological economics as a life science should have much to contribute to investigating the evolutionary roots of our behavior, the relationship between resource scarcity and systems of governance, and the creation of adaptive strategies to get our species through current and future resource bottlenecks. With a few notable exceptions, we have contributed little to this leading edge of science.

Finally, my own soul-searching leads to questions about the status of ecological economics as a social movement. Was I naïve to believe that our charge to reform economics based on principles of sustainability and justice could also reform the economy? What role should the ecological economist play in not just pointing out what is wrong with our current system but also advocating for change? At professional meetings of ecological economists, it is easy to find like-minded people who are asking all the right questions but are perhaps uneasy with the honest answers. We have grown content to talk to one another but are unwilling to step outside our own comfort zone to articulate and lobby for change in policy circles. Do we have an alternative to the mainstream or not? Can we provide the credibility of an academic field of study and still offer the much-needed guidance that is necessary to lead the transformation our economy so desperately needs?

This book provides a roadmap to this unfinished journey of ecological economics. The authors are not afraid to throw away other maps that no longer serve us, differentiating between consumer choice and citizen obligations; cautioning against an economic model that values only what is priced; and searching for the right relationships between humans and all of life on Earth. Ecological economists must return to telling the truth about the economy. We must stick to principles of economics based on biophysical reality. Also, we should catch up to the major influence the behavioral sciences are having on a broader revolution in the social sciences. Then perhaps we can return to challenging the status quo instead of complaining about it in one breath and serving it in the next.

We live in an incredible time, with a responsibility to tell the truth, educate this generation on the ecological realities of our economic decisions, and use our privileged position of scientific knowledge to make a difference. I found my inspiration in the essays of Herman Daly over two decades ago. You might just find yours in these pages and join us in the unfinished journey of ecological economics.

ACKNOWLEDGMENTS

We wish to acknowledge those who helped in the overall development, criticism, and review of the manuscript: Julie Ann Ames, Rachel Bruner, Herman Daly, Kate Darling, Brett Dolter, Jon Erickson, Robert Godin, Richard Howarth, Brenda Lee, Robert Nadeau, Richard Norgaard, Alex Poisson, and Ray Rogers. We are especially grateful to Jon Erickson for a foreword that holds ecological economics to its revolutionary agenda.

We also wish to acknowledge funding by the Social Science and Humanities Research Council of Canada; McGill University for its hospitality in hosting the project; Qi Feng Lin for assistance throughout the editorial process; and Margaret Forrest for diligent work in preparing the manuscript.

ECOLOGICAL ECONOMICS FOR THE ANTHROPOCENE

INTRODUCTION

The Unfinished Journey of Ecological Economics

PETER G. BROWN AND PETER TIMMERMAN

1. THE UNFINISHED JOURNEY

A specter is haunting the Earth—the living ghost of an economic theory that, no matter how much it is assaulted or how much damage it causes, refuses to die. The economic order that is based on the premises of this theory is grinding itself into the physical face of the planet. Many indicators suggest that we are witnessing a rapid decline in the richness of life processes, including accelerating climate change, increasing loss of natural diversity, changing and expanding disease vectors, and the spreading of an unsustainable growth and consumption model of what constitutes human well-being and happiness across the globe. The spectral nature of standard economics is reflected in its inability to halt, or even recognize, our seemingly inexorable movement toward some critical boundary conditions necessary for the flourishing of life on Earth.

Why does this economic order hold us so captive? Two important elements of the current economistic approach are often overlooked: (1) it not only provides a explanation of how markets, transactions, and so on function, but it also contains (in spite of its value-neutral rhetoric) a powerful ethical formulation of what it is to be a human being in search of well-being; and (2) at its heart is an abstract, ideal model—a set of quasi-scientific claims about the operations of a social system. Apart from any other issues that these elements of economics may involve, they help protect it from proof or refutation. If the facts or dynamics of the actual system in operation are different, then either the facts are not yet captured by the ideal model or the human beings are being perverse or irrational, not living up to their designated role as rational agents.

Among the attempts to bring down this toxic mix of quasi-science and social psychology, ecological economics is perhaps the best equipped. In its short history, the main strength of ecological economics has been its focus on the shortcomings of the quasi-scientific model upon which standard economics has been based (Costanza 1991, 2003; Daly 2005; Georgescu-Roegen 1971, 1986). The model been shown to be based on ill-grounded nineteenth-century models and rhetoric concerning science and scientific processes (De Marchi 1993; Mirowski 1989, 1994; Nadeau 2006). These assumptions render it incapable of dealing with the actual physical dimensions of the situation that it is helping to spawn around the world. Because it dominates public policy and discourse alike, this model holds life's prospects in a death grip.

The fundamental, original premise of ecological economics is to insist on seeing the human economy as embedded in and part of Earth's biogeochemical systems. Energy, matter, entropy, and evolution, among others, have been neglected by standard economics. This neglect imperils the present and future well-being of humanity and the other creatures with whom we share Earth's heritage and destiny. To be faithful to its fundamental insights, ecological economics must address this situation—first by developing this physical understanding and underpinning, then by drawing out its implications for human economic, social, and political experience in this critical moment in Earth's history.

However, we believe that ecological economics has only just begun to consider the radical implications of its original promise—that its "journey is unfinished." Part of what is unfinished is the consideration of how the new physical understanding and the human experience together demand some form of ethical foundation for their mutual enhancement. As mentioned, standard economics, in spite of its pretensions to being value neutral, in fact contains a wealth (or illth) of ethical assumptions and implications drawn from its eighteenth- and nineteenth-century underpinnings in emergent Western individualistic capitalism. Ecological economics has operated with many of the same assumptions as the model from which it is trying to get away.

This volume is an attempt to revisit, reconfigure, and challenge some of these assumptions. We begin with the working assumptions already in place in ecological economics concerning a more necessary and appropriate understanding of the relationships between economics and the earth's biogeochemical processes. We use this as a beachhead in the unfolding battle for the heart and soul of Earth. Our particular point of entry is in developing the ethical, social, and normative foundations of ecological economics, which so far consist mainly of borrowings from the neoclassical dustbin.

This book forms a link between the new understandings of thermodynamics, evolutionary biology, cosmology, and ecology, as well as a better understanding of economics and its fundamental and operational constructs. In this sense, we want to explicitly broaden the mandate of ecological economics beyond its concern with scale, which is its centerpiece in the work of Nicolas Georgescu-Roegen and Herman Daly. To accomplish this, we must understand the heterodox nature of ecological economics itself and the conflicting agendas that have emerged in its short history.

2. THE THREE AGENDAS OF ECOLOGICAL ECONOMICS

In our view, ecological economics poses a simple but very broad question: what would economics entail if it rested on a worldview based on current science? Ecological economists have an existing agenda that requires making important adjustments to the frameworks of macroeconomics and public finance. These changes can be made while leaving the bulk of the frameworks of economics, in particular, and contemporary culture, in general, intact. This we call the *explicit agenda*.

At the same time, ecological economics can be thought of as something that is radically revolutionary, stronger, and more fundamental, offering a clarion call for a thorough rethinking of the human relationship with life and the world. This is accomplished simply by asking the same questions of many other disciplines that make up the edifice of contemporary thought in addition to economics. These disciplines or frameworks include law, governance, finance, ethics, and religion; like economics, each offers norms of conduct—they tell us what we should do. As we will see, we are in a situation of "all fall down," as in the nursery rhyme. Like a row of dominos, once one of these structures falls, so do the others. This we call the *implicit agenda*.

Given the fulfillment of these agendas, a different and brighter prospect for life's future comes into view. Scientific developments of the last two hundred years, particularly in the post–World War II period, have challenged mainstream Western culture's assumptions about the place of humans on Earth and in the universe. Developments in evolutionary, organismal, and molecular biology; equilibrium and far-from-equilibrium thermodynamics; quantum theory; complex-systems science; astrophysics; cosmology; neuroscience; and certain branches of theology are just some of the elements of a new understanding of who we are, where we came from, and where we may be headed. The implications of *this* burgeoning narrative are profound and far-reaching for developments not only in science but also in economics, finance, law, governance/political science, religion, and ethics—in short, for all of human culture. This we call the *reconstruction agenda*.

2.1. The Explicit Agenda

With roots in the early-twentieth-century work of Frederick Soddy, ecological economics emerged in the 1980s as a seemingly natural and unthreatening subject of study. Much of its justification came from *The Limits to Growth* report (Meadows et al. 1972) and the work of Nicolas Georgescu-Roegen in the early 1970s (Georgescu-Roegen 1971). Drawing on the work of John Stuart Mill, Herman Daly's *Steady-State Economics* (1977) provided further justification for the idea of placing limits on the economy. Ecological economics gained force at about the same time as the Brundtland Report (World Commission on Environment and Development 1987) and seemed to accord well with the idea of sustainable development.

One could describe the explicit objectives of ecological economics as being concerned with three issues: scale, distribution, and efficiency. Scale refers to how the economy can be regarded as a subset of local and global biogeochemical processes (which both determine its content and set limits to its growth). Distribution or fairness is embodied in a commitment to sustainable development with both intra- and intergenerational dimensions (e.g., Müller 2007). Efficient allocation, the central goal of the neoclassical school, is retained. However, it must be constrained by the considerations of scale and equity. In addition, there is a methodological commitment to use material and energy flows in conceptualizing and measuring the performance of the economy.

Concerns with scale were important precursors in classical economics in the work of Thomas Malthus, and distribution is important in the works of David Ricardo and Karl Marx. Even in the 1950s, Richard Musgrave (1959) had included distribution as one of the key branches of public finance, along with stabilization and allocation. Thus, the emphasis on distribution within the frame of ecological economics offers adjustments to the neoclassical framework that are restorative (rather than revolutionary) of certain classical economic insights and concerns.

Indeed, it seemed to some ecological economists that some tools of the neoclassical model could be extended to assess the relationship between the economy and the natural world in which it is embedded. Many ecological economists thought they could retain the ethical perspective that underpinned the neoclassical model that they had rejected. One of the forms this takes is the movement to assign monetary values to ecosystem services. This is, in a way, a turn back toward the neoclassical idea of internalizing externalities. It is thought by many, including these authors, to constitute a regrettable retreat to the subset of neoclassical economics known as environmental economics. It has served to blur the meaning of ecological economics, to hobble its mission, and even to endanger its existence. Partly for this reason, ecological economics is in danger of going extinct, although it might continue to exist in name only.

2.2. The Implicit Agenda

Taken at its word, ecological economics is radical compared to the status quo—yet it is an essential framework, especially given the current ecological crisis. It both is grounded in a scientific understanding of the world and also has the potential to guide the behavior of one of our principal normative structures (economics)—one that happens to be laying waste to the earth and that has a goal to digest the biosphere. As Robert Nadeau (2006) has convincingly argued, neoclassical economics is *intrinsically* incapable of addressing the crises it has created; it urges acceleration toward the precipice. Seeing this, the founders of ecological economics have issued a clarion call: abandon the fantasies of the neoclassical vision and live in the world as it is currently understood. To be consistent and visionary, we must take advantage of the insights of the current scientific revolution. This is apparent not only for economics but also for other disciplines founded on dated and unrevised metaphysical and prescientific visions. They may be thought of as orphans whose intellectual parents have perished while they live on.

Here is a preview of the sweep of the revolution: current conceptions of property, which assume boundaries and severability, underlie the law in many countries. Yet contemporary science emphasizes the interpenetrating character of Earth's natural systems. The evolutionary worldview dethrones humanity and undercuts the presumption that human ownership is morally justified as a gift from God (Brown 2004; Cullinan 2011). "Ownership" is, at best, a diminished concept. Western liberal political systems rest on the idea that human actions can be independent of one another—an idea sharply at variance with the law of the conservation of matter and energy, which emphasizes that there are no actions that affect only the actor. Traffic jams in Dallas may affect the flooding of fields near Dakha. The gasoline burned in Dallas may deplete the supply and inflate the price of fossil fuels needed to produce the inexpensive fertilizers on which billions depend for their food supply. The world's natural systems live under the shadow of the guillotine of finance. Being designated as the source of a lucrative commodity or an attractive "emerging market" can be a death sentence for forests and the flora and fauna within them, not to mention the peoples who have depended on them from time immemorial. However, leading finance textbooks do not contain a single word about the relationships between money and the fate of these forests and their peoples or of the imbalances in the carbon, nitrogen, and hydrological cycles, which are massively perturbed by money (Bodie and Merton 1998). Mammon appears to have escaped this world. Our ethics are the residue of the crumbled foundations of metaphysics past. For those who consider us to be free of these dusty encumbrances, a fantasy that one way of behaving is as good as another is espoused. The result is moral and conceptual chaos that eviscerates public discourse and blocks the development of the collective responses needed to avert catastrophe. Ironically, many of the "faithful" use their energies in quarrels of doctrine and retain and pronounce views that are sharply at variance with well-grounded empirical beliefs. Yet at the same time, strong leadership from the Catholic, Jewish, Protestant, Baha'i, Quaker, and other faiths, as well as in the scientific community, is emerging to respond to the decline in life's prospects. Certain movements emerging in these faiths are reasserting humanity's place in the universe based on a modern scientific synthesis.

Simultaneously, these traditions and others are reconsidering our relationship to the rest of the ecological community with which we share the planet. Human beings need other species and beings to flourish independently of our needs, wishes, and desires. This has little or nothing to do with instrumental concepts, such as "ecosystem services"—it has to do with fundamental understandings of our place in Nature. Part of our need stems from what Wilson (1984) called "biophilia"—our long history with other species with their own worlds and autonomies (Kahn and Kellert 2002); part is due to what Louv (2005) called "nature-deficit disorder"—that we are ever more removed from nature and natural processes. However, more profoundly, animals inhabit other and different worlds from ours—they are "other" to us, even though we share genetic and environmental histories with them. We need to respect and assist in maintaining and enhancing their differences—communing with them in a "communion of subjects" (Berry 1999:82). Because of our current influence over the planetary ecosystems, our main aim should be to manage—and reduce—that influence so as to provide our fellow beings with their own room to live, breathe, flourish, and even die on their own terms. Instead, at the moment we appear to be heading for a planet where there will be little other than versions of ourselves and our goals grinning (or weeping) back at us as we survey the world about us.

2.3. The Reconstruction Agenda

Scientific understanding has advanced markedly in the twentieth century. It provides a broad evolutionary narrative that Thomas Berry coined as the "New Story" (1999). From this vantage point, a sweeping agenda emerges for informing and shaping the human prospect as we recognize that we have entered the Anthropocene. The narratives from which we currently take our bearings are simply not true to our circumstances.

The idea of freeing ourselves from nature, myth, and "the primitive" runs strong and deep in Western culture. Mainstream Western conceptions of ethics and metaphysics begin with ideas of separation. Humanity, created in the image of God, is different in kind than the rest of creation. The biblical account of humankind's early acts sets us apart from God, which has contributed to the fall of humanity and nature alike. We are awarded dominion over a profane Earth. As Carolyn Merchant and Hugh Brody have independently pointed out, an imperative to transform Earth finds its roots in the idea that humanity must retake its rightful place in paradise (Brody 2000; Merchant 2013). Merchant argued that the quest for restoration to paradise forms an underlying narrative that has legitimated the global transformation of nature. This works hand in glove with the "divine mandate" to convert the non-Christian people of the world who are heathen by definition—a mandate that often resulted in the domination, enslavement, and extirpation of people and myriad other species (Pagden 2001).

The Greek tradition that forms the other main root of the Western tradition is a culture that, even before the time of Plato (424–348 BCE), had wasted much of its own ecological base. Although Greek "primitive religion" had strong naturalist roots, the philosophical traditions stemming from it did not take their nourishment from them. These traditions have heavily influenced historical and contemporary mainstream Western culture. The traditional Greek hero is Ulysses, who outwitted the gods, outmaneuvered his primitive adversaries, and mastered his own temptations by being tied to the mast (Horkheimer and Adorno [1947] 2002). In this rationalist stream, humanity is regarded as inherently superior to the other animals. In the convergence of the biblical and Greek traditions, the world in which we live, move, and have our being is something to be owned and used, not loved and respected. Regardless, forms of naturalism still persisted on the margins of the dominant culture (Thomas 1996), at least until the European Enlightenment of the seventeenth and eighteenth centuries, and they continue in various forms in indigenous and even mainstream environmentalism—particularly in the legacy of John Muir. Despite the rebellions of movements such as romanticism, we continue to imagine a world in which we are the principal agents—masters of self and world alike.

For these reasons, the Western tradition (in its main manifestation) now struggles with difficulty to formulate an ethic of respect and relationship with the nonhuman world. These distant foundational trends set the stage for the scientific and technical revolutions of the last five centuries. These revolutions have radically transformed large parts of Earth's surface and massively altered the chemistry of the atmosphere and oceans. These trends enabled a vast expansion in human population and consumption that now overwhelm Earth's biophysical life-support systems. We are embarked on a tragic course. More fateful still is our resistance to envisioning and articulating an alternative relationship with life and the world.

3. THE PIVOTAL SIGNIFICANCE OF ECOLOGICAL ECONOMICS

"In for a penny, in for a pound" describes the situation of ecological economics. Once you insist that economics must be grounded in thermodynamics, you have ushered into the conversation the whole of contemporary science. This is not to say that contemporary science is flawless, complete, final, or anything of the sort. Nor is it to say that deep mysteries do not remain, such as the nature of dark energy and matter, the relationship between quantum mechanics and the theory of relativity, or even the nature of consciousness. However, it is to say that, at this point in history, scientific knowledge represents a converging synthesis built around the idea of evolution and complex-system theory (Kauffman 1995).

Here, for those of us who live in the "tower of Babel," is an offer of salvation. Here is a rock on which the edifice of a civilization worthy of respect can be constructed. As Albert Schweitzer argued in *Philosophy of Civilization*, ethics is arbitrary unless grounded in a worldview—a theory of the universe and the human place in it (Schweitzer [1949] 1987). Contemporary science, subject to the limitations noted above, offers just this.

What this volume aims to do is to use this perspective to illuminate the ethical dimensions of ecological economics and to trace their implications for its theory and practice. It will turn out that the move to an ecological economics framework is fundamentally about justice and that accepting this implication will transform our relationship with life and the world.

The project upon which this volume is based thus seeks to extend and enrich ecological economics in four main directions. *First*, it proposes an analysis of *what the most appropriate ethics or ethical systems might be for ecological economics in general*, given our deteriorating situation with regard to the physical systems within which we are deeply embedded and upon which we are fully dependent, as well as the new understanding of nature and natural processes characteristic of the twentieth and twenty-first centuries. By "ethics," we do not simply mean rules or norms for individual behavior; we include issues of environmental justice, ecological politics, and social concern more broadly. Three chapters make up this section. The first, by Peter Timmerman, is a lively and quick-moving overview of the myriad ways in which cosmology, culture, ethics, and exchange differ and intertwine. It is meant to loosen up the mind to enable escape from the conceit, attributed to Margaret Thatcher, that "there is no alternative" to the current standard model of economics. The second chapter, by Peter Brown, argues that the new understanding of the limits to knowledge and the nature of the embedded person requires an ethics grounded in respect and reciprocity. The third, by Richard Janda and Richard Lehun, seeks to reposition the centrality of our own pursuit of freedom. It insists on wider justice considerations that bear upon both our stewardship of the conditions for the possibility of life itself and the incontrovertible contemporary reality that each of our free choices impinges upon every other being.

Second, the project proposes to develop *credible indicators of an ecological economy*. Despite decades of work and controversy on the adequacy of indicators such as gross domestic product, the standard model of economics continues to use them—thus legitimating and masking the process of civilizational self-liquidation that is gathering increasing momentum. Rather than joining this conversation, we offer four chapters on measurement.

First, Mark Goldberg and Geoffrey Garver propose a methodological framework for the development of indicators relevant to ecological economics. This consists of five elements: contextual considerations, considerations of scale and dimension, considerations of scope, considerations of commensurability, and considerations of interacting systems in order to identify and target paths leading to the indicator. Second, Geoffrey Garver and Mark Goldberg address the ecological boundaries and related parameters of the space in which the human economy must function to avoid overshoot, injustice, and collapse in interdependent ecological, economic, and social systems. They set out the policy-oriented indicators that can accurately and reliably show the extent to which human enterprise respects those boundaries and parameters. In the third chapter on measurement, Mark Goldberg, Geoffrey Garver, and Nancy Mayo point out the difficulties in using the metaphor of health in judging the success of an economic system by stressing the many difficulties in defining the seemingly familiar idea of human health. The fourth and final chapter of the section, by Qi Feng Lin and James Fyles, explores two related changes in our current thinking on the relationship between humans and the environment that are crucial to successful measurement. The first is the idea of humans as being part of the ecosystem as opposed to the common assumption that humans are separate from it. The second is a reinterpretation of the concept of ecosystem health in view of this humans-in-ecosystem perspective.

Third, we consider the implications of ethical measurement considerations for both microeconomics and macroeconomics. At the macro level, Peter Victor and Tim Jackson argue that we are missing a convincing ecological macroeconomics—that is, a conceptual framework within which macroeconomic stability is consistent with the ecological limits of a finite planet. Despite promising developments in this direction (e.g., Jackson 2009; Victor 2008), there is an urgent need for a much more fully developed ecological macroeconomics to avert immanent and massive disaster. In the second chapter of this part, Richard Janda argues that an ecological-economics perspective requires the rethinking of individual behavior and corporate charters alike.

In the third chapter, Bruce Jennings critically scrutinizes received conceptions of liberty that are too individualistic and atomistic to realistically and rationally guide human norms and self-understanding in the coming era. He seeks to formulate an understanding of liberty that is consonant with the demands of the new era, the Anthropocene. Finally, Janice Harvey draws on two theoretical perspectives on how change could be facilitated. From a critical perspective, discourse theory holds that the process of institutional-cultural change is discursive and dialectical (Fairclough, Mulderrig, and Wodak 2011; van Dijk 2011). World-systems theory, on the other hand, reminds us that such social constructions occur within a unique historical context that either constrains or propels but ultimately shapes systemic change (Wallerstein 1999, 2004). The publication of this book is also a discursive event, but it occurs in a context of greater and greater appreciation that we are in a global crisis. While intended to help ecological economics on its journey, it aspires to change our relationship with life and the world by grounding our self-understanding in an evolutionary worldview. In this task, it joins an ever-expanding corpus of ecological discourse with similar aspirations. The objective of this chapter is to sketch how we could change course.

Last, the project discusses how *we can fashion a way forward* toward a more fruitful path than the one that lies along the current fateful trajectory.

REFERENCES

Berry, Thomas. 1999. *The Great Work: Our Way Into the Future*. New York: Three Rivers Press.

Bodie, Zvi, and Robert C. Merton. 1998. *Finance*. Upper Saddle River, NJ: Prentice-Hall.

Brody, Hugh. 2000. *The Other Side of Eden: Hunters, Farmers and the Shaping of the World*. Vancouver: Douglas & McIntyre.

Brown, Peter G. 2004. "Are There Any Natural Resources?" *Politics and the Life Sciences* 23 (1): 12–21. doi: 10.2307/4236728.

Costanza, Robert, ed. 1991. *Ecological Economics: The Science and Management of Sustainability*. New York: Columbia University Press.

Costanza, Robert. 2003. "Ecological Economics Is Post-Autistic." *Post-Autistic Economics Review* 20 (June 3): 2. http://www.paecon.net/PAEReview/issue20/Costanza20.htm.

Cullinan, Cormac. 2011. *Wild Law: A Manifesto for Earth Justice*. 2nd ed. White River Junction, VT: Chelsea Green Publishing.

Daly, Herman. 1977. *Steady-State Economics: The Economics of Biophysical Equilibrium and Moral Growth*. San Francisco: W.H. Freeman.

Daly, Herman E. 2005. "Economics in a Full World." *Scientific American* 293: 100–107.

De Marchi, Neil, ed. 1993. *Non-Natural Social Science: Reflecting on the Enterprise of More Heat Than Light*. Durham, NC: Duke University Press.

Fairclough, N., H. Mulderrig, and R. Wodak. 2011. "Critical Discourse Analysis." In *Discourse Studies: A Multidisciplinary Introduction*, edited by Teun Adrianus van Dijk, 357–378. London: Sage Publications.

Georgescu-Roegen, Nicholas. 1971. *The Entropy Law and the Economic Process*. Cambridge, MA: Harvard University Press.

Georgescu-Roegen, Nicholas. 1986. "The Entropy Law and the Economic Process in Retrospect." *Eastern Economic Journal* 12 (1): 3–25. doi: 10.2307/40357380.

Horkheimer, Max, and Theodor W. Adorno. 2002. *Dialectic of Enlightenment: Philosophical Fragments*. Edited by Gunzelin Schmid Noerr. Translated by Edmund Jephcott. Stanford, CA: Stanford University Press.

Jackson, Tim. 2009. *Prosperity Without Growth: Economics for a Finite Planet*. London: Earthscan.

Kahn, Peter H., and Stephen R. Kellert. 2002. *Children and Nature: Psychological, Sociocultural, and Evolutionary Investigations*. Cambridge, MA: MIT Press.

Kauffman, Stuart A. 1995. *At Home in the Universe: The Search for Laws of Self-Organization and Complexity*. Oxford: Oxford University Press.

Louv, Richard. 2005. *Last Child in the Woods: Saving Our Children from Nature-Deficit Disorder*. Chapel Hill, NC: Algonquin Books of Chapel Hill.

Meadows, Donella H., Dennis L. Meadows, Jørgen Randers, and William W. Behrens. 1972. *The Limits to Growth: A Report for the Club of Rome's Project on the Predicament of Mankind*. New York: Universe Books.

Merchant, Carolyn. 2013. *Reinventing Eden: The Fate of Nature in Western Culture*. 2nd ed. New York: Routledge.

Mirowski, Philip. 1989. *More Heat Than Light: Economics as Social Physics, Physics as Nature's Economics*. Cambridge, UK: Cambridge University Press.

Mirowski, Philip, ed. 1994. *Natural Images in Economic Thought: "Markets Read in Tooth and Claw."* Cambridge, UK: Cambridge University Press.

Müller, Frank G. 2007. "Ecological Economics as a Basis for Distributive Justice." In *Frontiers in Ecological Economic Theory and Application*, edited by Jon D. Erickson and John M. Gowdy, 72–90. Cheltenham, UK: Edward Elgar.

Musgrave, Richard. 1959. *The Theory of Public Finance: A Study in Public Economy*. New York: McGraw-Hill.

Nadeau, Robert. 2006. *The Environmental Endgame: Mainstream Economics, Ecological Disaster, and Human Survival*. New Brunswick, NJ: Rutgers University Press.

Pagden, Anthony. 2001. *Peoples and Empires: A Short History of European Migration, Exploration, and Conquest, from Greece to the Present*. New York: Modern Library.

Schweitzer, Albert. (1949) 1987. *The Philosophy of Civilization*. Translated by C. T. Campion. Amherst, NY: Prometheus Books.

Thomas, Keith. 1996. *Man and the Natural World: Changing Attitudes in England 1500–1800*. New York: Oxford University Press.

van Dijk, Teun Adrianus, ed. 2011. *Discourse Studies: A Multidisciplinary Introduction*. 2nd ed. London: Sage Publications.

Victor, Peter A. 2008. *Managing without Growth: Slower by Design, Not Disaster*. Cheltenham, UK: Edward Elgar.

Wallerstein, Immanuel Maurice. 1999. *The End of the World as We Know It: Social Science for the Twenty-First Century*. Minneapolis: University of Minnesota Press.

Wallerstein, Immanuel Maurice. 2004. *World-Systems Analysis: An Introduction*. Durham, NC: Duke University Press.

Wilson, Edward O. 1984. *Biophilia*. Cambridge, MA: Harvard University Press.

World Commission on Environment and Development. 1987. *Our Common Future*. Oxford: Oxford University Press.

PART I

PROPOSED ETHICAL FOUNDATIONS OF ECOLOGICAL ECONOMICS

INTRODUCTION AND CHAPTER SUMMARIES

The first section of the book explores a reconsideration of the ethical foundations of ecological economics, beginning with a summary statement outlining what we consider to be the necessary basic elements of such foundations. It then proceeds through different articulations of the consequences of embracing these elements.

Summary Statement

We start from the belief that ecological economics has only begun to consider the radical implications of its original promise—that its journey is unfinished. Part of what is unfinished is the consideration of a new ethical foundation based on the insights of ecological economics, and this foundation cannot be laid without considering the *implications* of those insights. Among the reasons why the journey is unfinished are the following: (1) standard economics continues to mesmerize its possible critics; (2) there is appeal in tinkering around the edges of standard economics through various forms of what is usually referred to as "environmental economics"; and (3) it is immensely difficult to come to terms with the urgency and true dimensions of the planetary crisis—a difficulty shared with most people who have tried to face it.

The fundamental, original premise of ecological economics is to consider the human economy as embedded in and part of natural ecology—that is (among other things), the dynamics of the physical world—energy, matter, entropy, evolution, etc.—have been neglected by standard economics.

Crucially, we now believe that this neglect imperils the present and future well-being of humanity and the other creatures with whom we share heritage and destiny. The extraordinary ethical shift that ecological economics needs to incorporate into its workings is that these physical demands have ethical implications; that is, ecological facts are values, and "is" has become "ought." The German philosopher Hans Jonas made the fundamental point that there "ought" to be a continuation of "is," otherwise there will no longer be any ethics or anything else; his "ontological imperative" therefore is that nothing should be done to threaten the continued flourishing of life on earth. These are ethical claims about the implications of certain qualities of the physical world.

Concerning the ethical foundations of ecological economics, we propose three interlocking postulates:

(1) *Membership:* Humans are members, not masters, of the community of life.
(2) *Householding:* The earth and the living systems on and in it should not be seen as merely "natural resources." They are worthy of respect and care in their own right.
(3) *Entropic Thrift:* Low-entropy sources and sink capacities, the things that undergird life's possibilities and flourishing, must be used with care and shared fairly. Ecological economics is inexorably and fundamentally about justice.

Taken together, these postulates can be understood—and should be lived— as foundational to an ethos of right relationship with life and the world. Versions of these postulates and their accompanying ethos can be found in indigenous and traditional economic systems, past and present, and we have much to learn from them.

SUMMARY OF CHAPTERS

"The Ethics of Re-Embedding Economics in the Real: Case Studies," by Peter Timmerman

Peter Timmerman's chapter presents a series of case studies or examples of economic systems that operate according to nonstandard economic theory as a way of illuminating economic elements, practices, and concepts that do not "fit" the theory. These case studies were also chosen for a number of other reasons, of which the most important is to highlight the fact that standard economic theory is anomalous in human history. It is based on a very specific range of assumptions drawn from a very specific period in modern history—a historical artifact of a particular period of thought and time rather than a universal truth. The alternative cases sketched out in this chapter range from hunter-gatherer economic practices (the Nayaka of India) through complex sustainable agricultural systems (Bali) to Western economic approaches that were consigned to the so-called dustbin of history (Aristotelean, medieval, etc.). They highlight social, cultural, physical, and ethical factors in actual economic practices throughout history and in contemporary life that have been eliminated, downplayed, or too narrowly construed by standard theory. These alternatives provide resources for reconsidering and recasting economics within a more "embedded" ecological and ethical world.

"Ethics for Economics in the Anthropocene," by Peter G. Brown

Peter Brown argues that ecological economics has profound ethical implications. He points out that ethical systems typically have at least five features, although they can be weighted and may function extremely differently. These features are: a foundation or justification, postulates, structured principles or rules, virtues, and a guiding metaphor or ethos. First, to work toward an appropriate foundation, he grounds his argument in an affirmation of human life; however, he places the understanding of what life is in broader contexts (e.g., what it means to be a person). Then, as already outlined, Brown wishes us to accept the three interlocking postulates of membership, householding, and entropic thrift. Following on from these and moving forward to proposed principles or rules, Brown notes that ecological economics has an implicit structure of concern. These are matters of scale or size of the economy relative to the earth's capacities, fair distribution of these capacities, and efficiency in allocation. Brown argues that each of these, although valid, has stronger ethical implications for the unfinished journey of ecological economics than has hitherto been explored. For example, although "efficiency" is a core concept in ecological economics, it remains constricted by the neoclassical conception. Once scale and distribution concerns have been met, a more complex and highly modified version of the neoclassical conception of efficiency comes into play. Brown further argues that in an enriched ethical perspective, notions like "virtue" become important as part of the emerging ethos of ecological economics. We require such virtues as courage, epistemological humility, and a sense of (and quest for) atonement for what has already been wrought by our carelessness and attempts at domination. Ultimately, however, we cannot take our first ethical step upon presumed ideas about what we ought to do. Rather, it must emerge from an empirical understanding of our place in a broader community of life.

"Justice Claims Underpinning Ecological Economics," by Richard Janda and Richard Lehun

Richard Janda and Richard Lehun link the insights drawn from the previous two chapters by contrasting the constellation of justice claims recognized in the existing market economy with those that would be recognized in an ecological economy. The existing market economy has "flattened" possible conceptions of justice so as to include only the sum of individual preferences. Thus, it excludes any claims upon us from the world or from outside of ourselves, which can be called metaphysical claims. Absent an accounting for metaphysical claims, there is no exogenous principle that might serve to correct or adjust the outcomes of our preferences. Metaphysical justice claims, characteristic of premodern societies, were displaced and repressed by the Enlightenment project in the name of emancipation. Emancipation now confronts an impasse it created. Rather than the free will willing itself to be free, to use Hegel's formula, the free will that is not other-regarding wills its own destruction in appropriating and depleting all life-support capacity. Ecological economics cannot maintain a justice conception that relies on a self-regarding calculus of utility functions. Rather, it must begin to model the interaction of self-regarding and other-regarding justice conceptions as it plots a trajectory toward fair and reproducible shares of Earth's life-support capacity. Thus, it will expand its conception of the household and of its members and will shift attention to rendering justice with entropic thrift for present and future generations.

An ecological economy would thus not only be an economy of taking "resources" but also an economy of safeguarding the prospects of life into the future. This has practical, substantive implications. The governance of economic actors would have to be redesigned to ensure that they are accountable for producing environmental and social benefits, not only to produce return for shareholders. Economic transactions would have to be restructured so as to send both a price signal reflecting private utility and a social score reflecting the burden of the transaction on public goods. The tax regime would have to be reconceived so as to channel the resources of the gift economy into the environmental and social goods that are now depleted through collective action. Also, the basic policy framework for the deployment of economic instruments—cost-benefit analysis—would have to be replaced by a full life-cycle accounting scheme to ensure the mitigation of any effects of a project or activity that tend toward transgressing planetary boundaries. Nonetheless, any new justice model will entail new sets of unrecognized and unanticipated justice claims. Ecological economics should already begin to model the inadequacy of its emerging justice conception, particularly in its application to existing norms.

A *picture* held us captive.
—Ludwig Wittgenstein, *Philosophical Investigations*

Americans are not addicted to oil, Americans are addicted to freedom—
the freedom to move freely and independently where and when we want.
—Neoconservative former Virginia governor George Allen, 2008

The good man orders himself in relation to the whole, and the wicked man
orders the whole in relation to himself. The latter makes himself the
centre of all things, the former measures his radius and keeps to the
circumference. Then he is oriented to the common centre, which is God,
and in relation to all the concentric circles, which are the creatures.
—Jean-Jacques Rousseau, *Emile*

CHAPTER ONE

The Ethics of Re-Embedding Economics in the Real: Case Studies

PETER TIMMERMAN

1. INTRODUCTION

This chapter presents a series of case studies or examples of economic systems that operate according to nonstandard economic theory as a way of
illuminating economic practices and concepts that do not "fit" standard
theory. These case studies were chosen for this book to help explore themes
in ecological economics, but their most important function is to underscore
the fact that standard economic theory is anomalous in human history. It
is based on a specific range of assumptions drawn from a specific period in
modern history—a historical artifact of a particular period of thought and
time rather than a universal truth. Claims to universal theoretical purity
are a recurring feature of various historical periods and movements and
were never more attractive, perhaps, than to nineteenth-century Western
societies, faced with the emergence of physics and mathematics as world-
defining powers.

If we intend to replace standard economic theory, it is important to
understand why this theory has been so seductive and so powerful for so

long. Central to the appeal of standard or mainstream economics has been its ideal quality: it describes a modeled world that enables, among other things, mathematizable descriptions of its operations once one adopts its basic assumptions. It is immensely powerful in its operations, providing for most people the most readily available model of a self-organizing system that appears to work out its complexities all by itself (the infamous "invisible hand"). It encapsulates and appears to explain a wide variety of human experience, from ordinary transactions in daily life to global movements of goods, services, finances, etc. It also embodies a plausible if debased model of social interaction. As a kind of bonus, it also has ethical and political implications; for instance, it refuses to make moral judgments about what people wish to spend their money on, and it is severe on external interference with transactions. Furthermore, and perhaps most prominently, it has until recently been associated with the rise of liberal democracies, capitalism, and "progress" generally. The elision of "freedom" with "free, frictionless motion" underpins the godlike power of the market that punishes and rewards from a passionless, apolitical standpoint—and claims to being universal and apolitical (markets, nature) are among the most powerful political moves there are.

The difficulty with ideal models, of course, is their relationship to reality, which is full of particulars and details and is messier and more "abundant" (Feyerabend 1999) than any model. This is not to dispute the value of models and modeling. However, the further an ideal model strays from reality, the less heuristic it becomes if the understanding of reality is the object (the neoconservative economist Milton Friedman [1953] famously disagreed with this). Also, as will be discussed shortly, there is a difference between ideal models in physical systems and ideal models in social systems (although many people confuse the two for good and bad reasons). In the case of standard economics, a variety of challenges have been made over the years, to which theoreticians have responded with various forms of improving, tinkering, or simply saving face. These challenges have included concerns over basic elements of the model, including the assumptions of perfect information, the rationality of the actors, the mysterious origins of the "desires" of the actors, the misappropriation of physical analogies, and so on. In recent years, the number of challenges has increased to the point where—to the unfriendly observer—economics is in a Ptolemaic state; that is, it is attempting to protect an outdated vision of an earth-circling cosmos through the endless addition of epicycle after epicycle so as to save the foundational explanation of increasingly awkward phenomena.

Part of the contention of ecological economics is that it has brought forward the one element in the current theoretical economic system that cannot be persuaded or ultimately coerced into its fortress walls: the physical realm. As noted already, there are differences between models of physical systems and social systems, and one clear difference is that physical systems are unpersuadable by rhetoric: gravity operates, whatever one may think or complain about it; by contrast, people can be persuaded to believe in a seductive model of a social system and then proceed to try and alter their world accordingly. The great danger of standard economic theory is not that it models the way people act; rather, it takes elements of how people might act and then does what it can to prove that that is how they do in fact act—or if they do not act in that way, they are somehow in the wrong or misguided as to the true nature of what they are doing "deep down." As this theory increasingly pervades the social world, more and more people become convinced by it—and in some cases benefit mightily from it; thus, people's behavior, and the social and cultural worlds that support it, alters accordingly. People proceed to see the world through the lens of that theory and so change the world accordingly—as much as they are able. It may be per contra that their daily intimacies, their own instincts, and various social processes to which they are witness contradict the theory: if so, so much the worse for them. They are sinners against the model and will be punished in due course.

The threatening signs of planetary distress are, however, for ecological economists, prima facie physical evidence that the standard model is flawed to the point of refutation. For others who are not immediately concerned with ecology, there are other signs of distress, such as in various failures of the financial system, inequities, and so on. However, these other signs can—and have—been "saved" by various internal tinkerings with or obfuscations of the model for a hundred or more years. But physical deterioration on a planetary scale is not subject to this kind of "saving the appearances,"—calling into question the rest of the dogma. Ecological economics poses this question by bringing the ideal model back down into gritty reality, out of the realm where it can move in its own frictionless orbit—a realm whose abstractness contributes to its potential for fostering intoxicating dreams of infinite progress, infinite space, and the fulfillment of infinite desires. Ironically enough, one of the great reasons this world-view has lasted so long was the discovery and promotion of oil in the late nineteenth century, which "oiled" the world and removed (for a time) the friction of reality and the previous historic limits to perpetual economic growth (Wrigley 2010). Fossil-fuel use essentially encouraged the delusion that human progress could ignore time and space and greased the spread of the machine worldwide in association with physical "freedom" (see this chapter's epigraph from Governor Allen). The end of the fossil-fuel era undermines the plausibility of the standard economic model, and not just for economic reasons.

Ecological economics, by focusing initially on the physical system and our role in it, thus calls into question the rest of the dogmas of standard economics—and loosens their grip on us. This important physical challenge to economics has been the primary focus of ecological economics to date. However, having pried open the crystal box, as it were, we, throughout this volume, are considering what else has been mislaid, gone missing or simply misguided.

This chapter approaches this array of missing or misguided items through a variety of sketches of cases that "re-embed" the economy in specific places and times, in human relations, and in the environment. The opposite notion, "disembedding," was characterized initially by Anthony Giddens (1990) as fundamental to modernity: the pulling of people, objects, landscapes, etc., out of their roots and intrinsic connections and thereby making them available for commodification—another form of frictionless-ness. This disembedding draws on standard economic theory (indeed, it is hard to determine which created the other) and is essential to the removal of brakes on the exploitation of societies and the earth. The oddness of this, historically, will be underlined in the case studies to follow, but the following quotation from an anthropologist, Ernst Gellner, is as appropriate an introduction as one could ask for:

A man making a purchase [in our society] is simply interested in buying the best commodity, at the least price. Not so in a multi-stranded social context: a man buying something from a village neighbour in a tribal community is dealing not only with a seller, but also with a kinsman, collaborator, ally or rival, potential supplier of a bride for his son, fellow juryman, ritual participant, fellow defender of the village, fellow

council member . . . all these multiple relations will enter into the economic operation, and restrain either party from looking only to the gain and loss involved in that operation, taken in isolation. In such a many-stranded context, there can be no question of "rational" economic conduct, governed by the single-minded pursuit of maximum gain.
(Gellner 1989; similar material can be found in Polanyi [1944])

The following sections of this chapter echo and reinforce this. They include economic systems and ways of thinking that involve rich, alternative considerations of what constitutes the role of economic activity in people's lives. These alternatives provide resources for recasting economics in a more "embedded" ecological and ethical world.

When considering these cases, it is important to be reminded that, historically and in non-Western traditions, rather than starting from physical properties as ecological economics tends to do, the understanding, articulation, and ethical assessment of economics have been socially or religiously construed—only part of which involved physical elements. These assessments have included such themes as basic needs, equitable distribution of God's abundance, appropriate husbandry, and so on. However, one characteristic of these other traditions that is of significant potential for ecological economics is that the traditions operate from within something barely captured by the term "embeddedness": they operate from within deep webs of relationships that are reinforced and illuminated by the flow of goods and services. Exchange outside of these webs—the neutral exchange process of the market—is (if it exists at all) a task for those outside the family, the community, and the tribe. The encroachment of the market into these deep webs has historically been the source of agonies of a kind that can be exemplified by the seemingly strange recurring struggle in the pages that follow over such processes as "usury" and "interest." Usury—charging extra for the return of borrowed funds—is a betrayal of what was considered (in the Old Testament, the Koran, and elsewhere) to be an act of relational love among brethren and not an act of commerce.

As briefly discussed by Bateson ([1972] 2000b) and more recently, and somewhat more fully, by Berkes (2012), the embeddedness of these traditions is also cultural and often spiritual. The cultural and spiritual traditions intimate through story, myth, and symbol that the local system within which economics plays itself out is only a smaller subsystem within a much larger system of meaning. Failure of the members of that smaller subsystem to understand or abide by that larger system is dangerous and potentially catastrophic. This threatening possibility underpins the concerns of many traditional ontologies and epistemologies that on the surface appear to be trivial or absurd. These concerns link threats of a physical kind ("runaway" processes) to threats of a personal and psychological kind; that is, one is warned not only of the potential unknown material consequences of transgressing various limits but also of the potential for anyone who loses track of their niche (Livingston 2002) or "place in the world" to become disoriented, misguided, and easily led astray (e.g., Schelling [1809] 2006; Timmerman 2010; Wittgenstein [1953] 2009).

I have made a deliberate choice in this chapter of cases that very roughly shadow the historical sequence—from hunter-gatherers, through various forms of agriculture, and toward modern capitalism—but I have avoided making a great deal of the implications of this trajectory here for lack of space; of course, these cases are simply sketches of what is available in much more detail in a myriad of studies from social, anthropological, and historical perspectives. As the chapter goes on, I eventually focus on the comparisons with and the transitions to market economies. However, looking historically in terms of fundamental concepts and practices, it is the earlier transition from hunter-gatherer economic systems to agriculturally based systems that is perhaps the most important of all transitions in human history.

For the purposes of this chapter, the recognition of the finite nature of the planet may not be as important for the ethics of ecological economics as the recognition that this finitude regrounds the nature and bonds of our mutual relationship as a bounded interdependent community (thus recasting justice issues, as Janda and Lehun discuss in chapter 3). This raises significant questions about the nature of the individual self in a postromantic world.

2. CASE STUDIES

Of the case studies discussed below, some were influenced directly or indirectly by historic Western economic theory (i.e., Plato, Aristotle, and the medieval Scholastics), so I have begun with setting that stage.

The examples that then follow are Franciscan economics and a curiously related hunter-gatherer economy, the Nayaka—which can also be seen as a kind of "ground zero" for economic systems. This is followed by Buddhist and Gandhian economics (which is somewhat different than Buddhist economics, although it depends in part on the former). After this, we return to a "Western" economic system—Islamic economics, which was heavily influenced by Aristotelian ideas and developed its own version of the "usury" debate and which has had a resurgence in recent years. The two last cases discussed are of sophisticated cultures (unlike the Nayaka) that have survived and been sustainable for many centuries: the Hopi of the southwestern United States and Bali. Bali in particular has evolved a sophisticated ecological system whose cultural underpinning has been in place for nearly 1,000 years (see Bateson 2000a). As will be seen, a case like Bali provides a "worked-through" example of sustainable "steady-state" ecological systems based on a completely alternative ethical matrix. The Hopi and the agricultural realm of Bali are perhaps the most intriguingly and obviously "3E" (i.e., possessing ethics, economics, and ecology) of the case studies provided, and they make a fitting conclusion to this chapter.

3. ALTERNATIVE WESTERN TRADITION: ARISTOTLE/SCHOLASTICISM

We begin with a quote from Schumpeter on Marx:[1]

> He was under the same delusion as Aristotle, viz., that value, though a factor in the determination of relative prices, is yet something that is different from, and exists independently of, relative prices or exchange relations. The proposition that the value of a commodity is the amount of labour embodied in it, can hardly mean anything else. If so, then there is a difference between Ricardo and Marx, since Ricardo's values are simply exchange values or relative prices. It is worthwhile to mention this, because if we could accept this view of value [sc. as a common property intrinsic to commodities], much of Marx's theory that seems to us untenable or even meaningless would cease to be so. Of course we cannot.
>
> (Schumpeter 1950)

Vilfredo Pareto was even more emphatic:

> In a recently published book, it is said that price is "the concrete mani-
> festation of value." We have had incarnation of Buddha, here we have
> incarnation of value. What indeed can this mysterious entity be? . . . It
> is useless to entangle ourselves with these metaphysical entities, and we
> can stick to the prices.
> (cited in Meikle 1995)

These examples of scorn could be multiplied a thousand times. The recurring theme is that previous generations were confused or primitive or
naive; only now does clarity reign, given that the search for a "theory of
value" was abandoned with the rise of neoclassical economics. Part of this
abandonment meant that one of the critical ethical dynamics of the drive
toward a theory of value—which was to determine if prices were "just"
and wealth was distributed equitably—disappears behind the veil of prices.
Marx's famous search for a "labour theory of value" was only one version of
a long alternative tradition that sought to evaluate the fairness and equity
of economic practice under a variety of themes. Because of space considerations, this chapter will only deal with two of these: use/exchange value
considerations and usury, primarily through a look at Aristotle.

Aristotle is the first to set out the immensely influential distinction
between use and exchange value in his very brief discussion of economics
in *Politics* (1.9.1257, 6–13):

> With every article of property there is a double way of using it: both uses
> are related to the article itself, but not related to it in the same manner—
> one is peculiar to the thing and the other is not peculiar to it. Take for
> example a shoe—there is its wear as a shoe and there is its use as an
> article of exchange; for both are ways of using a shoe, inasmuch as even
> he that exchanges a shoe for money or food with the customer that wants
> a shoe uses it as a shoe, though not for the use peculiar to a shoe, since
> shoes have not come into existence for the purpose of exchange.

In chapters 3 and 5 of *Nicomachean Ethics*, Aristotle deals with the just
exchange issue (i.e., exchanging a shoe for a house is not just) by arguing
that appropriate exchange is "reciprocity . . . on the basis of proportion, not
on the basis of exact equality" (1132b32–33).

Aristotle's subsequent attempts to determine what "proportion" means moved into metaphysical terrain. Although he grounded some proportionality on "need" (*chreia*, not the same as demand), in the end he was unable to come up with a measure of commensurability. He was simply clear that whatever it was it was certainly not money, which fulfills a different role (see Langholm 1979, 1984; Meikle 1995). Money is merely a convenient means of exchange (and there was exchange before there was money).

The important practical distinction Aristotle goes on to discuss in *Politics* (again, in book 1) is between *oikonomike*—which will later become "economics" but in Aristotle's terms meant household management, including "such things necessary to life, and useful for the community of the family or the state" (1256b27–30)—as well as something called *chrematistike*, which is "wealth getting, and that is so called with justice; and to this kind it is due that there is thought to be no limit to riches and property." (156b27).

Modern economics (in Aristotelian language) ought to be chrematistics, not economics. It is interesting that he stressed the unlimited nature of its processes, because wealth getting has no limits. In a famous shorthand, borrowed here from Marx's Capital: *A Critique of Political Economy* ([1867] 1990, chaps. 3–4), Aristotle compared an economic system of C-M-C (commodity-money-commodity), where money is in its rightful place, with one that is M-C-M (money-commodity-money), "unnatural," and (one can infer) subject to a kind of cancerous runaway growth (Kaye 1998).

This cancerous runaway growth is associated with usury (asking interest for money), and medieval Scholastics used the preceding material from Aristotle as part of their suspiciousness toward, and denunciation of, usury. Apart from the assaults on usury in the Bible (and later in Islam), Aristotle remains a general source for these arguments (although his influence is not total because *Nicomachean Ethics* was "lost" for most of the early medieval period and only recovered in the thirteenth century; see Wood [2002]). Aristotle argued that the *telos* (the goal or end purpose) of money is as a medium of exchange and that to pile up money purely as money is a perversion of money's role. Money should not "grow" interest—it is intrinsically sterile; in Marx's shorthand, moneylenders are engaged in: M-M+. Natural things have a natural growth toward their endpoint or goal (their *telos*), but money has no goal as such and therefore has no stopping point: "And this term 'interest' [*tokos*], which means the birth of money from money, is applied to the breeding of money because the offspring resembles the parent. Wherefore of all modes of getting wealth this is the most unnatural" (1258b1–8).

Aristotle and later medieval commentators were troubled by new economic growth, which can be seen from the attempt to apply biological metaphors to such processes (although there appears to be no crossover between the more rigorous discussion of biological growth in Aristotle and economic growth; e.g., his pathbreaking work on embryology, a crossover which did happen in the eighteenth and nineteenth centuries). It is also worth noting that much of the discussion on this topic in the Middle Ages was completely hypocritical: while the church denounced usury, it simultaneously took out loans and in some cases lent money. There were a variety of theological escape hatches (perhaps the most famous being that merchants could wait until they died to make retribution for their usurious behavior through donations to the church—thus funding all those glorious churches and paintings through to the end of the Renaissance).

Discussion and Implications

Perhaps the most interesting aspect of the Aristotelian/medieval approaches is that they took place in what was assumed to be a stationary state. "Just prices" and the inappropriateness of usury take place against the backdrop of a (supposedly) unchanging universe where everything is already in its place. One of the outcomes of a growth economy is that it undermines the "naturalness" of these ideas of an unchanging state. (That this is very unsettling psychologically is clear when serious bouts of inflation affect the familiar costs of basic foodstuffs or "touchstone" commodities). It is interesting that Schumpeter (1934) put forward the hypothesis that in a stationary state there would be zero interest and that interest is somehow intrinsic to capitalism—which is another version of the anti-usury argument in a stationary state. It is intriguing to ask if a return to a no-growth or stationary state would bring about calls for zero interest and a return to a theory of just pricing.

Concerning use versus exchange, it does seem that ecological economics already considers that the materiality/usefulness of things should carry some weight, which would suggest at least that the border between "use value" and "exchange value" would have to be reconsidered. It might be worth also considering in detail the opposition cited above between economics and chrematistics: ecologists are fond of arguing that economics stole *oikos-nomos* (the laws of the house) from its proper home in ecology; perhaps ecological economists could argue that economics also stole it from the original *oikonomike*?

4. FRANCISCAN ETHICS/ECONOMICS/ECOLOGY: THE ETHICS OF ABUNDANCE

St. Francis (1182–1226) is widely regarded as the patron saint of ecology in the Christian world.[2] Francis's radical teachings and still more radical lifestyle continue to pose fundamental challenges to our ways of thinking and acting in the world.

Francis was born in the little hill town of Assisi, the son of a cloth merchant who spent his time traveling from market to market. It is believed that Francis got his name from the fact that his father was in France when he was born. His father had hopes that Francis would go into merchant banking, like his father. However, as a teenager, Francis had a series of personal crises of faith and began to disappear into local caves and obscure places to meditate and pray. At some point, he was struck by lines in the Bible that emphasized the poverty of Jesus—particularly one of Jesus's sayings that one should take no thought for tomorrow, that "God would provide" if one trusted him.

Francis made a commitment to a life of "Holy Poverty," which, among other things, infuriated his father, who sought to have him arrested and locked up for lunacy. In a dramatic confrontation in the Assisi city square, Francis stripped off his wealthy clothes and repudiated his father. From then on, Francis threw himself on the mercy of God and swore allegiance to "Lady Poverty." Lady Poverty was a romantic phrase, and it signified that Francis was taking the tradition of the troubadour, or wandering singer, as one of his models—and there are a number of stories of him and his disciples singing and dancing through the highways and byways of Italy.

For the next few years, Francis lived in absolute poverty, hand to mouth, begging on the streets. He gradually acquired a group of disciples who did the same. The basic idea, as already stated, was to trust completely in the love of God, that he would provide out of his abundance for Francis and his disciples, just as he provided for the lilies of the field and the birds of the air.

Francis and his disciples developed a set of working principles—such as taking only so much as would feed them for the day—to ensure that they trusted only in God. Voluntary poverty was the entry into an ontological vision of an abundant universe.

Interestingly enough, it worked completely. There were daily "miracles" in which people suddenly appeared to give him food or shelter. As Francis continued, his followers increased, and (ironically enough) more and more people began to donate to him. This was in spite of his concern over greed and money. One story tells of a rich banker who came to Francis and his disciples and said: "I respect you very much. Can you tell me what I can do for you?" Francis and his disciples huddled together, and one of them came up with a penny that somehow had not yet been spent. Francis took the penny, and gave it to the banker saying: "Yes, this is what you can do for us—take this away."

Of even more interest is the fact that as time passed, Francis began to talk and sing about the change that the arrival of complete poverty had made in his way of seeing the world. In particular, as the stories and legends of St. Francis show, there is a deep connection between his absolute poverty and the way he became more and more connected to animals and birds and the whole of creation. His famous canticle is a song about the way in which, through the love of God, Francis was able to see all the beings of the world as friends and relations, sisters and brothers. It is as if the absence of property—or perhaps the aura of absolute trust that emanated from him—broke down the barriers between human and animal. Whether this actually happened or is merely legend is unclear, but the symbolism is powerful and important.

The overwhelming success of Francis's ethics of abundance turned into a serious management problem. The refusal of riches or property—living like Jesus—threatened the power of the medieval Catholic Church, which received great sums of money, built churches all over Europe, and kept legions of priests. Francis's life overlapped with the Crusade against the Cathars and Waldensians in the south of France, many of whom were also involved in preaching poverty and assaulting the wealth of the Church. A series of popes found that they had to deal with this extraordinary person within the bosom of the church, and it remains somewhat astonishing that Francis was never excommunicated. Many people accused him of threatening the church; however, Francis seems to have not accused anyone directly but chose instead to be personally exemplary. After a complicated internal struggle, the church agreed to a series of rules for the practice of the Franciscan way of life, which were composed by Francis before his death. Many of these rules were designed to cope with the refusal of Francis to deal with the extraordinary abundance that people kept trying to donate to him and his movement.

Following Francis's early death in 1226, the growth of the Franciscan movement skyrocketed, the huge basilica that still looms over Assisi was built, and money continued to pour into the Franciscan community from people all over Europe. This caused significant problems over the years. Among the most challenging results of Francis's legacy was a movement at the end of the thirteenth century called the "Spiritual Franciscans," which demanded that the church repudiate its wealth. A part of this movement was declared heretical, and the papacy was forced to articulate its views on the sanctity of private property (e.g., did Jesus have property?) and to make its initial declaration on "alienability" and "inalienability" in *Cum inter nonnullus* (1322) (the ultimate source of this language in the rights debate). This, in turn, provided legitimacy to the burgeoning early Renaissance explosion in banking and merchandizing that Francis's father had been involved in and that Francis had walked, nakedly, away from.

Discussion and Implications

Curiously enough, the other economic system to which Franciscan economics seems closest is the "immediate return" of Nayaka hunter-gatherers (see section 5), who have a similar belief in an economics of abundance. Perhaps begging is a form of hunting-gathering in the jungle of modern life.

It is possible to make a larger argument concerning the relationship between an "ontology of abundance" and an "ontology of scarcity." One of the characteristics of many societies is that they operate with an ontology of abundance—that is, that the cosmos is fundamentally abundant. Given that we did not create ourselves, the earth, or the things within it, the evidence suggests that all of this is a "gift" of some kind. In ontologies of abundance, the things of the world are given to us generously, and it is our role to give thanks, celebrate, and imitate that abundant generosity in our own relationships and rituals. This sets up, as only one element, rituals designed to mimic and celebrate the generosity of, for example, the salmon runs along the coast of British Columbia and to ensure that they will continue, year after year, by returning the first of the catch to the run and not wasting the gift. Power comes, in part, by mimicking the generosity of the gods (e.g., potlatches).

When scarcities arise in an ontology of abundance, it is because the abundance has been temporarily withdrawn as a punishment for something human beings have done wrong (the Nayaka operate slightly differently). There is no sense that the scarcities are permanent—when human beings repent, the abundance returns. A classic example is the Northern Cree hunting practices, where it has been shown that the animals run the hunt. Strict rules that show respect for the animals must be followed; when hunters are careless or disrespectful, the animals withdraw themselves and make themselves "scarce" (Berkes 2012). To summarize, in these cases there is primary abundance and secondary scarcity.

In Western society, arguably since the eighteenth century, we have switched ontologies to an ontology of scarcity. Modern economics is based on an assumption of scarcity, which implies that we must be in competition for scarce goods and resources, and this generates (among other things) the market. In a world of scarcity, the natural world appears to be stingy and withholding and must be "developed" to provide us with the abundance we now need. Scarcity compels us into production. In this approach, there is primary scarcity and secondary abundance. The irony, of course, is that our drive to create this secondary abundance is in fact creating ecological scarcity.

As will be discussed again in the section on the Nayaka, this primary scarcity/secondary abundance view is generated by—and generates—a lack of trust in the fundamental gift, or abundance of things. The result is the need for private property, insurance, and storage of various kinds to buffer ourselves against the expected scarcity. Francis's radical claim was that trust in God would make us see that these needs were unnecessary.

It is further worth noting that there have been similar movements of absolute poverty that have also been successful. Ittoen in Japan is a comparable modern movement, based on a similar commitment by a wandering saint to absolute poverty, with similar kinds of subsequent "miraculous" success and the creation of a community and philosophy (see Nishida [1969]):

> When human beings are born, we come into the world with no posses-
> sions, and our very life has been granted to us. There is nothing that we
> can truly claim is our own, and we are in no position to really insist on
> rights. Thus, having nothing, and no property ownership, is the original
> state of human beings.
>
> "Roto", being on the roadside and serving, outside of home, and even
> without a home, is also the foundation and true picture of this state of
> having nothing and of non-possession. At the root of all conflicts among
> men is desire (greed) and self-centeredness. And human beings become
> attached to properties, to social position, to fame, and so on. It is neces-
> sary to cut off this kind of attachment and possessiveness, to cleanse one's
> heart of selfishness, and get back to our starting point. This is called the
> practice of "roto", that is, going back on the roadside. We need to dispose
> of all the things that have gathered around us, to return them to the Light
> [the Original Abundance] from which we received them, and get back
> out on the road having nothing—not to merely go adrift, but to get back
> to our original mode and Source.
>
> (Ittoen n.d.)

5. THE NAYAKA: THE ETHICS OF THE GIFT

The Nayaka are a tribe of hunter-gatherers who live in a deep jungle valley in Tamil Nadu, in southern India. They represent the ground zero of
economic history, living and working in an environment that they perceive
to be fully abundant (providing them with wild yams, fruit, berries, fish,
animals). Their 3Es reflect this perception. They are an example of what
anthropologists call an "immediate-return" society (in that they have no
storage; another example would be the Mbuti pygmies of the Congo; see
Turnbull [1983]). In discussing them, some background on the theme of
the "gift economy" is required.

The "gift economy" has received significant recent attention as an alternative to the market economy. There is significant confusion about this
because of ambiguities (or alternative interpretations) of what a "gift" and

a "gift economy" entail. The first confusion dates back to the beginning of the discussion. Marcel Mauss, a French anthropologist, drew on anthropological investigations in various parts of the world (significantly in the Pacific) to explore the ways in which gift giving cemented or greased or ran various societies. He assimilated the processes of these "gift" economies to the market economy through the notion of reciprocity—"gifts" were a form of exchange. Mauss (1925) and others (e.g., Mary Douglas) thus argued that the gift was a disguised or proto-market because the gift had "hidden" in it the expectation of a reciprocal gift of equivalent value at some time in the future. (An example from contemporary circumstances would be the complicated decision-making people undertake in ensuring that one gives gifts at Christmas "at the same level" as the gifts one expects from others.)

While still based on a model of the gift as disguised reciprocity, gift giving was made more complicated later on by Marshall Sahlins, author of *Stone Age Economics* (1972), who pointed out that there were different forms of reciprocity implied in gift giving in different societies: "negative reciprocity" (i.e., cheating), "balanced reciprocity" (equal exchange in some form or other), and "generalized reciprocity" (a diffused gift).

The first two reciprocities are fairly obvious. However, the diffused gift complicates the enterprise in a number of ways and also introduces the possibility of the "nonreciprocal gift." The nonreciprocal gift is the gift without expectations of reciprocity. We have seen this kind of gift already in the case of St. Francis—giving everything away as a mirror of God's giving away. Mauss noted in puzzlement that there seemed to be in Indian rituals examples of nonreciprocal gift giving, but he never developed this insight. In these rituals (particularly in the Buddhist tradition [Ohnuma 2005]), the nonreciprocal gift is stipulated as essential: no expectation of reward or reciprocity should be hoped for from the gods at all. This is part of the salvific power of the gift—otherwise the gift is "tainted." In spite of this, various commentators (disbelieving in any form of nonreciprocal gift economy and attempting once again to assimilate gifts to an exchange system) have suggested that God or the gods do expect prayer and worship as exchange. Without going into the theology of the mind of God, this does seem to be contrary to the expressed spirit of the gift in this situation. An extensive postmodern discussion of the impossibility of the "pure gift"—that is, the nonreciprocal gift—was triggered by Jacques Derrida (1992), who argued that all gift relationships involved "forgetting" the looming shadow of reciprocity (if only for a moment). For the classic discussion of nonreciprocity in various domains, including artistic or creative "gifts," see Hyde (1983).

The essence of the nonreciprocal-gift model is of a lack of expectations and control over the response. In a reciprocal model, X gives to Y, expecting a return from Y over which there is some kind of social or individual control (e.g., if nothing is forthcoming, then there are one-to-one consequences). A nonreciprocal model, rather than being linear (X to Y, then Y back to X) is at least triangular: X gives to Y, who may give the thing given by X now to Z, over whom X has no control. The triangularity is the first step in a broader circulation model: X to Y to Z to A to B to C, and maybe then to X. (This is related to what Sahlins meant by a "diffused" model—that is, one that benefits the system as a whole.) Giving without expectation of immediate or even delayed one-to-one return can be an affirmation of a basic trust in the abundance of the system as a whole (I benefit from everyone benefiting, or some of the circulation will come back to me eventually). Pierre Bourdieu (1997) noted that some of this discussion is a function of time and control. Immediate return is under some control (the parties are face to face; expectations can be immediately fulfilled) while delayed return is subject to the vagaries of time. This leads into further explorations by various theorists into the relationship between the gift and trust (see, among many other recent discussions, Caputo and Scanlon [1999] and Guenther [2006]).

This brings us back to the Nayaka. The Nayaka believe in nonreciprocal gift giving, based on a view of their environment—the forest—as their parents ("big mother" and "big father"). Giving is assumed to be normal—either things are given outright or given upon being asked—as if it was being shared by a family. The environment is a "player"—part of the sharing. The gods are invited to festivals (through shamanism), and these festivals (to which all Nayaka are invited) reinforce what is called the "Way of the Parents."

One example of the Way of the Parents would be the sharing of an animal brought back from a hunt. The animal is shared by all without preferential treatment (this contrasts with other societies based on reciprocity where a hierarchy of return is normal). Similarly, it does not matter who puts in the initial work. Bird-David (1990) gave the example of catching fish through upstream poisoning of the river, which takes a substantial amount of initial work; however, anyone can come and take fish once the work is done.

Even when Nayaka go to work for pay (there are plantations nearby), all wages are spent immediately or given away upon request. Ownership is temporary and is based on first discovery or long association (as in keepsakes from the dead)—possession is a function of the ability to give permission to take. So, in the case of X giving something to Y and Y then giving it to Z, if W then wants it, W has to go back to X and see if it is acceptable (Bird-David calls this "the wheel of giving around X" [1990]). There are wrinkles: if, for example, someone does not want to give something away, they hide until the request(er) goes away. There are also "intergifts"—gifts that are between couples and designated as such that others would not request. However, by and large, the sharing ethic works reasonably smoothly.

Discussion and Implications

The case of the Nayaka is obviously not one that immediately applies to a modern society, but it might be interesting to work through in more detail the metaphor of the family and the parent. After all, there is at least global rhetoric about humanity being one family (the "family of man"). If we are being enjoined to share our responsibilities, what is the ethical frame underlying it all? (It is also worth noting that there are complexities involved the different ethics that metaphorically underpin family, community, clan, ancestors, etc.)

It is important to stress that, in this section, I have not dealt in any significant way with the complex discussions among anthropologists and others about the relationship between hunter-gatherer ethics/economics and agricultural ethics/economics (see, among others, Brody [2000]). Woodburn (1982) and others have argued that the "immediate return" economy of hunters promotes a different kind of social structure (e.g., a radical egalitarianism) than the "delayed return" of agriculturalists (though there are complications involving "gatherers" and transitional societies or hierarchical hunter-gatherers, such as certain famous "potlatching" of tribes in British Columbia).

6. GANDHIAN ETHICS/ECONOMICS/ECOLOGY: THE ETHICS OF DHARMA

To enter into the Gandhian 3Es requires a brief look at his two main sources: John Ruskin's "Unto This Last" ([1860] 2004), which Gandhi translated into

Gujarati, and the Hindu *Bhagavad Gita*, which he hybridized for his own purposes (Gandhi [1926] 1980; Zaehner 1969).[3]

Ruskin's "Unto This Last" ([1860] 2004) was an attack on economics as he saw it unfolding (mostly through Mill). Ruskin argued that the model of the human as a self-interested maximizer was false, and he used a variety of examples to show this, including examples of sacrifice for others—a mother sacrificing everything for her children, a soldier dying for his country, etc. He also used the parable of the overseer giving the same wages to someone who signed on at the end of the day to work as to those who had signed on, and been working, from daybreak (thus giving "unto this last" the same wages for much less work, contrary to all economic principles). Ruskin's position was that economics as it was taught was "catallactics" (a word coined by Richard Whately in 1831, who objected to the term "political economy" and preferred a term that would mean "the science of exchange") and as a result was mostly irrelevant. What mattered to Ruskin was the role people played in life: a shopkeeper's role was to provide good products at a good price; nobility was fulfilling your role in life, whatever it was. His ideas formed a kind of nineteenth-century updating of feudalism. Indeed, Ruskin argued that the aristocracy had abandoned its role as the protector of the poor, and it was hardly a surprise that the new ascendant class of liberal merchants should have produced an economic theory for its own benefit. Ruskin variously called himself an "old Tory" and a communist.

Gandhi hybridized this with his other main source, the *Bhagavad Gita* (probably first century CE). The Gita is an insert into the vast epic *Mahabharata* and takes place just before a battle. A warrior, Arjuna, is unsure about whether he ought to fight, given that millions will die; many of whom are his close relatives on the other side. His charioteer, the disguised god Krishna, shows him that it is his role (his *svadharma*—personal truth/law) to fight, just as it is Krishna's role to keep the universe going (fulfilling the larger *dharma*—universal truth/law). What makes the role—and its tasks— ethical is that Arjuna must act in such a way as not to desire or be attached to the outcomes: it is just what he is supposed to do. As long as he does what he is supposed to do, he can leave the outcomes to God (Krishna).

Gandhi reinterpreted this quite warlike poem (in its own way) as a metaphor: as the warfare within the self as it struggles to find its own *svadharma*. Work is part of that struggle to find oneself. Everyone has a role in life, all of which are dignified, including (and especially) manual labor. The struggle to find oneself is part of the larger struggle for the truth, which Gandhi again sees as something to be carried out selflessly—the self is in the way of such a search. Furthermore, this search must be carried out nonviolently because violence is a sign of impatience, greed, and self-protectiveness. Violence is fundamentally a failure to trust in the truth of one's cause. Gandhi called his nonviolent search *Satyagraha*—the struggle for truth. He argued that part of the power of nonviolence was that the nonviolent activist shows his or her commitment to the search for truth as a common venture by being prepared to submit to the violence of the unenlightened other.

Gandhi's economic policies followed suit. Personally, he embraced voluntary poverty as another form of nonviolence:

> I found that if I kept anything as my own, I had to defend it against the whole world. I found also that there were many people who did not have the thing, although they wanted it, and I would have to seek police assistance also if hungry, famine-stricken people, finding me in a lonely place, wanted not merely to divide the thing with me but to dispossess me . . . possession seems to me to be a crime.
> (Gandhi 2008:79)

Gandhi harked back to St. Francis: "A seeker after truth, a follower of the law of love, cannot hold anything against tomorrow. God never stores for the morrow" (Gandhi 2008:91). Because not everyone can achieve this, equality of shares is perhaps more plausible. Gandhi said:

> First of all, in order to translate this idea into our lives we should minimize our needs, keeping in mind the poorest of the poor in India. One should earn just enough to support oneself and one's family. . . . Strict restraint should be kept over small matters in our lives. Even if a single individual enforces this ideal in his life, he is bound to influence others.
> (Gandhi 2008:101)

Interestingly enough, however, as a follower of Ruskin, Gandhi considered that the rich should not be forced to give up their wealth. They should hold their superfluous wealth in trusteeship for the benefit of the remainder of society. This definitely separated him from communism and certainly endeared him to a number of Indian capitalists who were among his firmest supporters.

Gandhi's solution for the larger Indian population was what he called *svaraj* (self-rule) based on self-reliance and handwork (*svadeshi*) leading to the welfare of all (*sarvodaya*). This grassroots approach was literally based on a series of expanding concentric circles, beginning with the self-ruling individual, then the self-ruling village, then the self-ruling nation. Gandhi believed that on the day that the self-ruling nation came into being, there would no longer be any perceived need for British colonial rule (of course, the end came before that).

Gandhi's ecology comes out of his nonviolence but also from his Buddhist, Jain, and Hindu backgrounds. For example, concerning the sacredness of the cow, he remarked, "It takes the human being beyond his species. The cow to me means the entire sub-human world. Man, through the cow, is enjoined to realize his identity with all that lives." Similarly, he stated: "If our sense of right and wrong had not become blunt, we would recognise that animals had rights, no less than men" (cited in Weber 1999). In a summary of his entire philosophy, Gandhi stated: "The purpose of life is undoubtedly to know oneself. We cannot do it unless we learn to identify ourselves with all that lives. The sum total of that life is God. Hence the necessity of realizing God living within every one of us. The instrument of this knowledge is boundless selfless service" (Gandhi 2008:41).

Discussion and Implications

Gandhian principles of self-realization through nonviolence have arguably been the most powerful political force for good in the last fifty years (witness the fall of the Soviet Union, the democratization of the Philippines, the Iranian struggle, etc.). In environmental terms, Gandhism has had a direct influence on such movements as the Chipko movement in northern India and the fight over the Narmada dam, as well as many other similar fights. One of the showcase developing-country nongovernmental organizations, Sarvodaya in Sri Lanka, is a Buddhist adaptation of Gandhi's Swaraj. Gandhism has also had a strong influence on the development of deep ecology. It would seem to have implications for an appropriate ethic for ecological economics.

In 1960, a year before he went to India, E. F. Schumacher wrote:

A way of life that ever more rapidly depletes the power of earth to sustain it and piles up ever more insoluble problems for each succeeding generation can only be called "violent." . . . In short, man's urgent task is to discover a non-violent way in his economics as well as in his political life. . . . Non-violence must permeate the whole of man's activities, if mankind is to be secure against a war of annihilation. . . . Present day economics, while claiming to be ethically neutral, in fact propagates a philosophy of unlimited expansionism without any regard to the true and genuine needs of man which are limited.
(Schumacher 1960)

7. BUDDHIST ETHICS/ECONOMICS/ECOLOGY: THE ETHICS OF INTERDEPENDENT IMPERMANENCE

I am a Marxist monk.
(Dalai Lama 2009)

The Buddhist tradition stems from the awakening of the Buddha (approx. 500 BCE)—"buddha" means the awakened one—to the nature of reality. In contradistinction to the then-pervasive early Hindu tradition, the Buddha argued that there was no such thing as a permanent kernel of self or soul and that perceiving this was only the first step is realizing that there was no such thing as a permanent anything, self or world. Human desires and the unhappiness associated with them are the result of attachment to the illusion of something that can be insulated from change. For the Buddha, the dharma (i.e., the rules of the universe) proved on examination to be what is called "*paticca-samutpadda*"—that is, everything is mutually arising and mutually disappearing. Their temporary intermeshing is what we see around us. Among the implications of this worldview is the notion of constantly emerging interdependence (nothing can be insulated from the web of things), and this has led to the recent interest in Buddhism among physicists and ecologists. The abandonment of a fixed sense of self (Harvey 1995) has also interested philosophers and ethicists in the West, in part because of a perception that we are living in a postmodern, post-Christian world where a fixed sense of self has been eroded (Collins 1982).

Most people who connect Buddhism to economics refer to Schumacher's *Small Is Beautiful* (1973) essay (see next section, "Schumacher: 'Buddhist Economics'"). Schumacher's essay has actually little to do with Buddhist theory or practice—influential as it has been, it is more Gandhian and sociological than anything (reflecting his witnessing of Buddhist societies in action; e.g., Burma before the recent junta). It focuses on what Buddhism (and Hinduism) refers to as the rules for the life of the householder or layperson, which are mostly general pieties. The link between the householder life and the monastic life is usually considered to be the transference of merit (good karma) from the monk to the layperson. The layperson does his or her job, and the monk does the same—the meditative practices of the monks as they struggle to achieve enlightenment generate well-being for the whole community (note that this transference is possible because of interdependence). But again, the gifts (called *dana*) from the householder to the monk are clearly stated to be nonreciprocal.

All this is not to discount the fact that the Buddhist tradition had a very strong ecosocial impulse as well: the digging of wells and the planting and protection of trees and animals are longstanding Buddhist practices. The early Buddhist tradition recognized the lesser folk gods of the surrounding culture, and there are rules against the harming of other creatures. Heroic efforts have been made to extrapolate from these practices a theory of "Buddhist ecology," but the results have been mostly superficial theoretically. (Practice is different: for example, Buddhist monks in Thailand have ordained tree spirits as a way of protecting forests from destruction.) Later Buddhist teachings are more radical in their teachings about human responsibility for other creatures: the bodhisattva is a Buddhist who forgoes individual salvation in favor of assisting all other beings down to the lowliest blade of grass (one implication being that there is no such thing as individual salvation if all beings are interconnected).

The core of Buddhist practice is mindfulness training, and this covers the whole range of life experiences, lay and monastic alike. Mindfulness is essentially paying extremely close, slow attention to how one experiences the world. It is assumed that not only will adept practitioners come to see, as the Buddha did, that the things of the world are impermanent, but they will also see that their daily thoughts and actions are saturated with grasping, desire, greed, and fear. These are the ingredients for suffering. Practice is designed to peel these away over time, relieving suffering and leading to insight or "awakening" to the reality of things as they truly are. This careful interrogation of our perceived reality is at the heart of anything that could be called Buddhist ethics/economics/ecology. Buddhist economics, if it is anything other than general rules for monks, nuns, merchants and householders, including strictures against Buddhists being employed in certain trades (e.g., hunting, leather-work, etc.), is grounded in the deep interrogation of the web of impermanence out of which appear the objects and transactions in our lives. Do they lead to attachment and suffering or assist us in awakening?

As might be expected from the foregoing, Buddhist teachings strongly attack self-interest and desire and argue for a lessening of these. Nevertheless, there is also a recognition that not everyone can go completely into monastic practice in this life, and there are a number of Buddhist sutras that discuss the life of the merchant and householder. Certain sociologists (e.g., Ling 1976) have pointed out that the Buddha spent his teaching life during the changeover from a rural to an urban society in the Ganges Valley in India and that his early monasteries were situated on the edge of the emerging cities. The teachings require householders to follow in their way various precepts, underlying which are the four "abidings" or mental stances: friendliness, compassion, empathy, and equanimity (Digha Nikaya 3:223), and these are in turn part of what is called "right livelihood" (positive as well as negative injunctions about jobs one can do). These provide and pervade attitudes toward daily life in Buddhist countries.

Discussion and Implications

There are two strong elements of the Buddhist tradition that are worth further consideration in an ecological economics perspective. The first, already mentioned, is that Buddhism takes what some people might call "process theorizing" to radical lengths. Virtually all of the discussion about interdependence in the current conversations is not really interdependence at a Buddhist level: the basic image is still of fairly fixed individuals being affected by outside "environments." A later Buddhist tradition (Hua-Yen) spends a certain amount of time dealing with the problematic side of real interdependence (e.g., as I tell my students, total cell-phone interconnection is a good thing, but do you really want your parents to know where you are all the time?). If we are in many ways "semipermeable," then there are important questions about what we allow to influence us directly, indirectly, and so on (Tucker and Williams [1997] collected a range of ecological responses to this challenge).

Another aspect of the mindfulness tradition that has perhaps been of more influence on deep ecology is the way in which the careful perception of things reveals that part of interdependence is our inability to declare where we leave off and the environment begins. For example, in the classic meditative practice of breathing, part of the learning is to experience the shift back and forth between the involuntary and the voluntary in the breath, to follow the breath in its pathways in and out, and to do the same with foodstuffs and so on. In that sense, deep ecologists and others consider Buddhism (and other meditative practices) to be demonstrating local ecological ethics that should be "scaled up."

A variety of attempts have been made to develop contemporary versions of Buddhist economics, none very convincing in my opinion (see Golden, 2009, for a review). The most famous, discussed in the next section, was that of E. F. Schumacher.

Schumacher: "Buddhist Economics"

E. F. Schumacher (1911–1977) was an economist, policy advisor, and essayist. His most famous work, *Small Is Beautiful* (1973), has sold millions of copies and helped spark the "intermediate" or "appropriate technology" movement. Schumacher was born in Bonn in 1911. He studied economics from an early age, becoming a Rhodes Scholar, and made his first trip to England to study at New College, Oxford. Throughout this period—the late 1920s and early 1930s—Schumacher's life and work were overshadowed by the struggles in his native Germany to cope with a series of economic and political crises that led to the rise of Adolf Hitler and the Nazis. At the outbreak of World War II, Schumacher was interned in England as an enemy alien and subsequently was forced to work as an agricultural laborer. Ironically, this experience fostered interests in farming, soil management, and Marxist socialism that were to influence his future activities.

At the end of the war, Schumacher—now a British citizen—was made a member of the Strategic Bombing Survey (which examined the effects of bombing on Germany), followed by membership in the British Control Commission in Germany. Upon his return to England, he became the economic advisor to the National Coal Board, where he stayed until his official retirement in 1971. His early interests in the postwar period revolved around reconstructing Germany and the efforts to establish an international financial and trading system (e.g., the World Bank and the International Monetary Fund).

In 1955, Schumacher was invited to go to Burma for a short term as an economic advisor, an event that changed his life and resulted in an essay published in 1955 entitled "Economics in a Buddhist Country," later revised as "Buddhist Economics." The essay drew upon his experience of a country whose core way of life was completely different than the worlds he had hitherto inhabited. He was especially astonished at the happiness of a people who were living in poverty. In this essay and subsequently, he ascribed this to a series of approaches to life: minimizing wants, work as a means of enhancing life, and a spiritual understanding of human beings. The enrichment of life was seen as the whole purpose of economics. Schumacher's new interest in the problems of developing countries—and Gandhism—next led him (as mentioned at the end of the previous section) to India. The vast distance between the high-technology future being dreamed of by government assisted by international economic advisors and the direst poverty of the masses of the population led Schumacher to push for "intermediate technologies," which would enhance the work people were already doing through the application of appropriately considered improvements.

Small Is Beautiful: Economics as If People Mattered (1973) is a series of overlapping essays of extraordinary prophetic power and insight. Only a few examples can be cited. The essay "Nuclear Power—Salvation or Damnation" not only attacks the economics of nuclear power but also points out in great detail the intransigence of the waste problem. "Development" articulates the emerging problem of rich elites in developing countries as enclaves of Westernization in a sea of poverty. "Social and Economic Problems Calling for the Development of Intermediate Technology" puts forward a diagnosis and a work agenda that would soon be followed by the Intermediate Technology Development Group (ITDG). The most famous chapter, "Buddhist Economics," sets out the essentials of ecological economics long before anyone had given it a name: Schumacher makes clear the distinctions between renewable and nonrenewable resources, the folly of living off capital rather than interest, and so on. "How to obtain given ends"—the dignity of the human—"with minimum means" is Schumacher's version of what a Buddhist economics ought to be.

Good Work (1979) is a further complementary series of speeches and short essays, published after Schumacher's death, including (among other things) an indictment of the world's increasing dependence on oil. *A Guide for the Perplexed* (1977) is a not very successful attempt by Schumacher after a late religious conversion to Catholicism to put forward a grander philosophy and theology. It is essentially an updated version of medieval Christianity, based primarily on St. Thomas Aquinas, which argues that contemporary philosophies are "horizontal" (only concerned with the material world), whereas a "vertical" approach is more appropriate to beings who ascend from mere physical life, the mineral, through plants, to consciousness, and then to self-awareness. The general point of the book is that we need an adequate level of response to the problems provided by each level.

A more interesting insight into spiritual economics is perhaps provided by a discussion in his final essay, "Education for Good Work," where Schumacher points out that the poor are frugal with ephemeral goods but are abundantly generous in those things that are meant for eternity.

8. ISLAMIC ECONOMICS

In Abbasid times, the Caliph asked the great Islamic scholar Muhammad Ibn Idris al-Shafi'i why God created flies, which were such annoying creatures, and he replied: "In my humble opinion the purpose is to show those in power their own helplessness."
(quoted in Foltz 2006)

The Islamic ethics/economics/ecology perspective echoes a number of the themes in some of the other examples: a fundamental sense of the universe as an abundant gift, human beings being responsible for each other and for creation, a sense of a static society, a suspicion of usury, and so on.[4] In many ways, Islamic ethics/economics is the most influential alternative viewpoint in action today. It has affected, among other things, the economic and political systems of a variety of Middle Eastern countries and various attempts to create a "guardian sphere" wherein Western capitalism is to be prevented, such as the era of the Taliban rule in Afghanistan.

Islam is driven by a foundational belief that all things are God's, and that the worship of God is the basic role of human beings on Earth. Given that there is only one source to be reverenced and worshipped, "all being is a unity" (Nomani and Rahnema 1994); and in some sense, all are equal under God. However, Islam takes far more seriously than the other monotheisms from which it draws some of its theology the notion that human beings are God's "vice-regents" on Earth—that they are stewards of the Earth, and that they will be called to account for it on the last day.

The specific community of Islam (the umma) is bound by the *Koran*—assumed to be direct messages from God—heavily influenced by the hadith (collected sayings of the prophet Muhammad), guided by shari'a, which is the traditional judicial system that evolved over the historic period, and in a vague sense subject to mutual consultation (shura). These teachings reflect in many ways the social ecology of a community made up of a mixture of urban and desert dwellers and traders of the seventh century. For example, shari'a, the law, originally meant "a path to the water"—and Islam strongly forbids denying people access to water wells (thus providing a foundation for common property law in Islam—the Prophet said, "People are partners in three things: water, fire, and pastures" [hadith cited in Nomani and Rahnema 1994:67]) and its public parks are notable for the provision of fountains and flowing water. There is a strong sensitivity to the well-being of animals in Islam (the prophet Muhammad was appalled at the mistreatment of horses), about which a great deal has been written (Foltz 2006; Masri 1989).

Because Muhammad came from a trading background, Islam is not opposed to people making money—it can be a sign of God's favor—but it must always be considered as a gift from God, who occasionally takes it away as a warning (28:71–81 [all Koranic quotations refer to chapter and verse]). Duties to God take precedence over business. Israf (waste/excess) is condemned (Nomani and Rahnema 1994). There is a similarly powerful strain of egalitarianism in Islam: the Prophet is reported to have said, "Anyone of you with cloth in excess, should return it to the person who needs it, and anyone who possesses food in excess of his needs should return it to the person who needs it" (Afzal-ur-Rahman 1974:222). There is a strong commitment to the well-being (*maslahah*) of all members of the community, at least at the level of basic subsistence.

There are three special aspects of Islam that have attracted the attention of ethicists and economists: the ban on usury (riba), the zakat (required offerings to the poor [2:1777]), and the Islamic banking system. I focus here briefly on the first and the third: zakat controversy is mostly about how much of one's income should be devoted to the poor, what should be subject to zakat, and whether it is as efficient as a straightforward tax (e.g., while zakat used to be individually assessed, countries like Saudi Arabia now require it of corporations).

The ban on usury has a complicated interpretative history in Islam (see section 3 for the related alternative Western history). Some Islamic historians believe that riba refers to a pre-Islamic loan system where defaulters on loans were penalized by a doubling of the principal and then a redoubling—thus reducing them to penury. Others consider it to refer, similarly, to any kind of extortionate interest. However, the mainstream interpretation is that it is, in fact, a ban on all interest. A familiar theme is trying to prevent borrowers from borrowing under conditions of (what was considered to be) duress.

Islamic financing and banking (some of which are twentieth-century inventions) have worked with this in a number of ways. First (as in medieval Europe), certain aspects of banking have remained untouched: putting in deposits for safekeeping (wadiah) and taking them out again later, subject to a small "gift" reward for doing so, and money-changing (again for a fee). Other aspects are different, with a focus on sharing risk and loss between both lender and borrower. The ethical argument (Kuran 2004) is that it is unjust to earn money without assuming some risk; that is, it is assumed that it is unjust that the borrower should be assuming the risk of the venture and also paying interest. (This, of course, ignores the other varieties of risks that banks undertake, including the potential loss of all the original capital [Kuran 2004].) In any case, in this interpretation of Islam, the lender should be more involved and share in the risk more directly (a Western example might be venture investing, where the investor is in fact a strong upfront partner). Many Islamic versions are designed so that what would otherwise be open-ended "interest" is turned into agreed-upon sharing of profits at the end of a specific time. Variations include straightforward profit sharing (mudharabah), joint ventures with multiple partners (musharakah), rent-to-own agreements (murabahah), and leasing (ijarah)—the bank owns the operation/thing and leases it back to the entrepreneur. A recent example of innovation in this sphere is interest-free mortgages on homes (the homeowner gets equity from the bank and pays rent back to the bank). The "bank" in this case, for example, could be a pool of Islamic lenders associated with a mosque or a real established bank. The rent is not fixed but is reassessed up or down, depending on changes in the value of the home (e.g., the wider real-estate market). The bank is thus involved in the risk. If the owner defaults, the ownership shares are divided up, with the owner usually retaining some level of accumulated equity (Kuran 2004). Finally, it should be said that there is a widespread, well-organized, informal system for avoiding banks completely: the fund transfer hawalah (entrusting note) used by migrant workers and others (Tripp 2006).

In his sardonic treatment of the history and contemporary practice of Islamic economics, Islam and Mammon, Timur Kuran summarized his findings:

> There are vast incongruities . . . between the rhetoric of Islamic economics and its practice. Specifically I demonstrate that the impact of Islamic banking has been anything but revolutionary, that obligatory zakat has nowhere become a significant vehicle for reducing inequality, and, last, that the renewed emphasis on economic morality has had no appreciable effect on economic behaviour. By its own lofty yardstick, then Islamic economics is a failure. This assessment needs to be qualified by the fact that the strictly economic impact of Islamic economics is not the only measure of its achievements [compared to the political and cultural consequences].
>
> (Kuran 2004:7)

The political and cultural elements of recent Islamic economics have been described as a critique of the individualism and acquisitive drive of Western capitalism, as well as an idealization of a presumed ideal community at the time of the Prophet (whose life is generally regarded as an ideal model for human conduct; Tripp 2006); however, critics like the Iranian writer Ali Shari'ati have argued that many Koranic injunctions were local to the seventh century, and not for all time and place (Shari'ati 1971). Anticapitalist Islamic figures have come up with various alternatives across the political spectrum, from calls for Islamic socialism to a reliance on individual spiritual resistance.

The ecological links to all this are varied. It is very important to recognize that, while the Koran does paint a hierarchical world, it also acknowledges that there are other umma (communities) who obey their own shari'a (e.g., bees [6:38]), and a hadith tells of a wantonly destroyed ant colony that praised God in its own way (Masri 1989). These communities are not only intrinsically valuable; they are "signs" for humans to read. Among the most important of these signs is the fact that the rich and diverse natural world is nevertheless ordered, measured, and proportionate. This order, measure, and proportion form the strong connection to an ecological ethic: they are part of a "balanced" environment that human beings must recognize and "may not upset"; indeed, God goes on to link this with equity generally: "Keep the balance with equity, and fall not short in it" [55–3-13].

A number of points concerning the relationships between ethical and economic views and the wider natural ecology were nicely drawn by a famous Islamic theologian of the twentieth century, Sayyid Abu A'la Mawdidi (1903–1979), who wrote concerning creation, invoking israf:

> Islam says that all the creation has certain rights upon man. They are: he should not waste them on fruitless ventures nor should he unnecessarily hurt or harm them. . . . Regarding the beasts of burden and animals used for riding and transport, Islam distinctly forbids man to keep them hungry, to take hard and intolerable work from them and to beat them cruelly. To catch birds and imprison them in cages without any special purpose is considered abominable. . . . Islam does not approve even of the useless cutting of trees and bushes. Man can use their fruit and other produce, but he has no right to destroy them. Vegetables after all possess life, but Islam does not allow the waste of even lifeless things; so much so that it disapproves of the wasteful flow of too much water. Its avowed purpose is to avoid waste in every conceivable form and to make the best use of all resources—living and lifeless.
> (cited in Masri 1989:30–31)

9. THE HOPI

> For whites it is almost axiomatic that not to know the past is to be condemned to repeat it; or Hopi not to know their past is to be condemned not to repeat it—to have, in fact, no idea who they are.
> (Page and Page [1970] 2009)

The Hopi People are famous for their intensely practical and ritualized spirituality rooted in their environment.[5] They have lived for at least nine hundred years in a very harsh desert/mesa region (tusqua, "the land") and have always affirmed that it is their ritual practices and worldview that have sustained them in a natural ecology that is unforgiving of mistakes. Their entire way of life—their myths, histories, lives, economies, rituals, and stories—is tightly and deeply interwoven with the cycles and demands of the natural world around them. Among their claims to a wider interest by a vast multitude of historians and researchers over the last hundred years (Loftin [2003] suggested that more than three thousand papers have been written on the Hopi) has been the fact that the Hopi have been able, by their intense privacy and the forbidding nature of their situation, to keep their culture mostly intact, despite being in the midst of the United States.

Three aspects of their society deserve our attention:

1. They have a fully localized Earth-centered mythology and a related ethic.
2. Their ethic is simultaneously ecological and social.
3. They have a powerful gift economy that extends throughout their cosmology.

The Hopi reservation is located within a larger Navajo reservation, itself situated at what is known as the "Four Corners," where the states of Utah, Arizona, Colorado, and New Mexico meet. There are roughly five thousand Hopi living in small villages, strung out along the tops of three mesas. The area is famous for its archeological history of failed cultures and abandoned cities, such as the cliffside Anazazi ruins; Hopi history and mythology recount the wanderings and failures of many other tribes. The Hopi historically migrated to their homeland after many trials; now they survive on dryland farming, which is labor intensive and revolves around the care and feeding of maize corn and a few other crops (e.g., beans, squash). Thus, there is great concern for the provision of water, which is in very short supply; precipitation averages 10 inches per year, often coming inconveniently in winter snow and flash floods (Page and Page [1970] 2009). The rest of their water is provided by scattered springs and pump wells.

Hopi society is bound by ritual to an extraordinary degree. Although many Hopi rituals are secret and off limits (there are serious restrictions even now on filming and obtaining information about these rituals), a number of their practices have been described in substantial detail (Courlander 1971; Loftin 2003; Waters 1963). The entire society revolves around—and indeed survives by—these rituals, which take up vast amounts of Hopi time: there are at least nine major rituals in the course of a year that each take many days, as well as a variety of social dances that bind the various clans and village communities together. These rituals revolve around times of planting and harvesting of their crops. Furthermore, it is no exaggeration to say that the whole arc of Hopi life is ritualized. The rituals are dynamic embodiments of their working mythology that operate from earliest child-hood to death and beyond.

Earth Story

The central myth of the Hopi is based on the mystery of plant growth (corn, beans, squash, etc.). When the Hopi arrived in this Fourth World, they were met by the god of fire and death (Maasaw) who gave them (along with other tribes) a choice of different corn to grow (yellow, white, red, purple, etc.). The Hopi chose the smallest, blue corn, because it was humble yet persistent. They value humility and hard work, and they are suspicious of big talkers (Talayesva 1942).

The central guiding images or metaphors for the Hopi are of the secret shaping of a plant underground before it breaks the surface of the earth and of the plant stalk poking straight up through the soil (Geertz 1984). These images saturate Hopi life. For example, when a baby is born, it is hidden from light for twenty days, and then baptized with cornmeal, as described by a Hopi elder:

> A baby is like a plant that has started to grow from seed. It must be protected in just the same way. It needs the ten days that were required while the seed was sending the little plant up to the surface of the ground [while it is being shaped], and the other ten days while the right leaves were forming. For twenty days the sun must not shine on it. Then on the twenty-first day its mother takes it in her arms, and along with her mother, or some other woman, carries it out of the house to the edge of the mesa and there prays for its health and happiness.
> (quoted in O'Kane 1950)

This ritual echoes the original story of the Hopi as a people. Their origin story is a "deep earth" story—it is based on a series of worlds emerging from the interior of the earth, one above the other "breaking the surface." Like seeds or embryos, as they rise from world to world through holes in the sky ceiling, living beings are shaped and brought to life. There are various versions of how and who did the shaping, such as Tawa, the Sun Spirit or the Spider Grandmother. When the Hopi entered our current world (the Fourth), climbing up on a plant stalk, they also came through a hole in the ceiling of the previous world (sipapuni, navel). This fundamental movement is echoed in the structure of their sacred houses (kivas), which have a hole in the floor and a ladder that reaches up to a hole in the roof, into and out of which the spirit beings (kachinas) move. Hopi rituals mostly originate in the interiors of these kivas, where long secret practices and teachings are undertaken, occasionally breaking out into surface expressions through public dances and other events. However, the deep Hopi teachings are interior. In other words, everything in Hopi Land is introverted, internal, underground, and womblike. It is perhaps no accident that the Hopi are also matrilineal: women own, men borrow.

Basic to virtually all Hopi rituals is the desire for rain. This manifests itself in many ways. To begin with, moisture (breath) appears and is referred to at every stage. Smoking tobacco (the smoke rises in clouds) is one example; the ancestral dead are called upon because they are "cloud people"; the kachinas (masked figures who inhabit the rituals) are intermediaries between the Hopi and the bringers of rain. Ritual for the Hopi is simultaneously twofold: first, the rituals are designed to bring offerings and prayers to the gods to bring rain, and they must be carried out with pure hearts and spirits; second, the rituals in themselves reproduce the patterns of the cosmic machinery and "kick start" the rain-bringing environment (Loftin 2003). Moreover, there are special spiritual elements to the rituals that need to be considered. First, as mentioned, the kachinas are not just local people masked for the purposes of portraying spirits: they are the spirits who have come to inhabit the masked wearers for the duration of the ritual. Second, the rituals do not just reproduce what the ancestors once did: the rituals re-inhabit the "long ago"—they take place in timelessness.

Interwoven with these rituals are the stories, myths, and practices concerning each clan and village and the natural features of Hopi Land. Different clans—Badger, Bear, Spider—oversee different rituals, and different villages host variations of the rituals. Each clan feels that they are especially close to the character of the animal spirit of their clan (e.g., the Sand Clan feels close to the earth; Loftin 2003). The Hopi stories detail the history of every feature of the landscape, how each mesa and village was settled by which clan and why. The landscape around the mesas is saturated in meaning and story: for example, the San Francisco Peaks in the distance are the homes of the kachinas for six months of the year, and that is where they are called from in order to bring rain; in the Grand Canyon (nearby) is a dome with a hole in it, which is a ritual destination because it was the place where humans emerged into the Fourth World. Every story interpenetrates every other story: you sense that any story could lead to every other story. Just as life begins with corn, so it ends with corn: the Hopi word for a corpse is *qatungwa*, which also means a harvested corn plant. After death—and before becoming a "cloud ancestor"—the dead descend into the underworld and mirror the aboveground rituals, but they are still connected to the living by moisture, breath, rain, and the power of fructifying emergence (time for the Hopi moves along a primary axis of unmanifestation/manifestation; Loftin 2003).

The Gift Economy

There is not much literature on the traditional Hopi economy. The Hopi are currently mostly inside the American economic system, except in their family and ritual activities. Hopi were historically traders, so part of their economic horizon involved barter (Beaglehole 1937). Nevertheless, much of their traditional economic system appears to have fit into the classic description of a gift economy, which we have already discussed—that is, one that is based in reciprocal gift giving, rather than one that is nonreciprocal (Mauss 1925). The reason for this is probably that their environment, while generous according to Hopi notions, is not wildly abundant. This may be compared with the earlier description of the Nayaka in India, who were living in an abundant jungle setting and had developed a gift economy characterized by nonreciprocity. For the Hopi, things are different: their margin for survival necessitates some certainty of return. It is also likely that there is an aspect of the fact that we are dealing with an agricultural people that influences their economic vision: there must be intensive work done—planting, weeding, and hoeing—rather than occasional efforts.

The most obvious example of this economy is the giving of offerings and other valuable commodities to the spirits in the expectation that they will reciprocate by bringing rain. There are leaks in this reciprocity, however. For one thing, there are various times when the kachinas will express the gift equivalent of a flash flood: they will hurl a vast number of goodies at the assembled people without expectation of a return. For another, the lack of reciprocity from the gods (i.e., the rains fail) is universally interpreted as a judgment on the practitioners. Either some individual Hopi (or the whole society) has transgressed in some way—thus causing the rain gods to stay away in punishment—or some celebrant has not been pure; therefore, the ritual machinery did not "take."

Whiteley (2004) has explored the Hopi gift system as it functioned in the nineteenth century; it is clear that that system was, like the rest of Hopi society, ritualized and inseparable from the wider ecology. Because whatever the Hopi made came from local materials and was made up of animals and plants of the region—all of which were alive and interconnected—it was not just that transactions were relational between the two parties to the transaction (as Polanyi [1944] famously argued about premodern economic systems); they were relational among the entire web of the ecosystem out of which the "product" emerged. As an expression of this, it is noticeable that many Hopi gifts were covered in signs (feathers, symbols, colors, sacred sticks) that signaled their role in the spiritual economy of which they were a part. Marx is well known for having made the point that commodities mask the labor that went into them: the Hopi celebrated the labor, the materials, and the spirits that came together to make the object—what one might call a true ecological theory of value (cf. the current interest in tracking the origins of products to assess their ecological footprint).

Conclusion

The Hopi are interesting, not only because they have survived for a thousand years, but because theirs is a very powerful ecological and social—"is as ought; ought as is"—world. What strikes one is that the ethic of such a world view is that it is so all pervasive, infusing every part of their lives. It is not an "add-on"—it is what is meant by a way of life, a "logic of the body" (Bourdieu, cited in Loftin 2003). A Hopi elder phrased it this way:

> Participation in the ceremonies, as we see now with younger kids being initiated and participating, is important. They need to be told in the kivas, in the homes, that the corn is the way the Hopis have chosen; it goes back to our Emergence. As Hopi people, we are fortunate to

have survived this long. It is a privilege to be a part of this complex Hopi community of clans living under this one philosophy of corn, of humility. I think if we can continue to teach that, we'll strengthen the culture as it stands.

(Kuwanwisiwma, quoted in Wall and Masayesva 2004)

10. BALINESE ETHICS/ECONOMICS/ECOLOGY: ETHICS OF THE SUSTAINABLE COMMUNITY

As noted already, Gregory Bateson described Balinese society as a "steady state"; his version of a steady state was one in which social conflicts, or potential conflicts (schismogenesis), were dampened down from the outset by a range of powerful cultural norms.[6] This dovetailed with a general sketch of Balinese individuals and society generated by Bateson (2000a), Margaret Mead (Sullivan 1999), Belo (1970a, 1970b), and others of the first wave of anthropologists in Bali. Essentially, they found that the Balinese themselves were trained from birth to be intensely introverted, and they were taught (expressly or by example) to express themselves in very formal, ritualized, and balanced ways. Their movements and senses of space were characterized as if all Balinese were like dancers or acrobats—always poised—and that in their world, individuals and buildings and villages temples were constantly and powerfully oriented according to the magical cosmological geography of Bali itself (axes of north, south, east, west, mountains, seas, upstream, downstream, etc.).

However, rather than ending up with introverted individuals, it was the community, the village that integrated everyone's existence through village rules (*awig-awig*), commitments to local temples and temple gods, and endless ceremonies saturated in famous artistic practices (gamelan music, dancing, sculpture, flower arranging, etc.). Much of Bali society spent extraordinary amounts of time in these ceremonies and festivals and rituals, mostly designed to propitiate and appease an extraordinary range of gods, goddesses, demons, etc. These spiritual figures were drawn from the very layered—even crazy-quilted—traditions of Bali, which included local traditions, Hindu traditions, other Javanese religious practices, and finally Islam—each of which had at some time arrived on the island. Many of these rituals—and much of daily life—seem to have been based on the need to ward off dangers in multiple forms. The exquisiteness of Bali culture—its poise and artistic sensibility—is woven throughout with this sense of imminent, immanent danger. This danger (sometimes called *seng-hara*, or dissolution, chaos—what we would call "entropy") is to be resisted through collective control and reinforced by the constant collective rituals to the relevant gods.

Briefly, Bali's main economic activity (until tourism arrived) was rice farming, and this seems to have been sustainable for many hundreds of years. Bali's landscape is dominated by volcanoes, some recently active, and its rice fields are prominently on the slopes and along the ravines. There are no large rivers but myriad smaller rivers and streams, many of which flow downslope, interrupted and channeled by a vast array of microdams, weirs, tunnels, etc. Historically, although Bali was ruled by various kings and suffered conquest from a variety of external agents, no large-scale monolithic empire ruled. Rather, power was widely diversified, and decision-making at the village and temple level by local farming committees and organizations was typical. In spite of their poverty, as mentioned, vast amounts of individual and group resources were constantly being spent on "useless" ceremonies.

In the 1980s, research was carried out on the irrigation economy of Bali, which had been (in parts) sustainable for a thousand years. The system was both ecological and spiritual: each element of the irrigation process was overseen by a temple or a shrine to the Rice Goddess/Goddess of the Water. It was discovered that control over the dams, weirs, and diversions—like valves in a complex system of pipes—was mediated by farmers associated with the Temple system itself. Not only were the layers of upstream and downstream irrigation integrated by the timing of festivals that opened and closed seasons, but the temple guardians acting on behalf of the Goddess, through consultations with farmers' associations (subaks), ritual activities, etc., came up year by year with the decisions about who was to get irrigation water, how much, and when. It turned out that this "water temple" system had evolved over the centuries in such a way that it was able to integrate a mass of information and decisions that led to the long-term sustainability of the rice fields.

Lansing (1995; 2006; and in a variety of publications and films) explored the complex adaptive system and how it is managed: a system based on mutually advantageous restraints between upstream and downstream users of water. Roughly speaking, upstream users are tempted to take extra water, but in doing so they are threatened by the possibility of pest outbreaks originating downstream—pest outbreaks are dampened by coordinated flooding of all potential fields at the same time. As a result, the entire system is engaged in constant mutual coordination and mutual restraint. These restraints are underpinned and reinforced by the simultaneous constant offerings and rituals by the farmers at each shrine and temple "valve" dotted at crucial points in the system, all of which also serve to keep the farmers from cheating. Lansing (1995) reported on the work that he and colleagues carried out through modeling exercises associated with the Santa Fe Institute, essentially proving that this system was not only adaptive, but that it could well have been self-organizing in origin. (It is also important to note that the system was very nearly destroyed by the implementers of the Green Revolution in Indonesia, who essentially told everyone to ignore the constraints and plant all the time.)

Discussion and Implications

A review of the Balinese system suggests that it is the most applicable case for considering the interplay of long-term ecological sustainability of a managed system with ethical and religious requirements. Not only has it been well worked out theoretically, but there is a wealth of anthropological data at hand. Lansing (2006) and colleagues are continuing this work, and they are searching for equivalent cases elsewhere in the world where management of an ecosystem over the long term has been interwoven with practical ethics.

11. CONCLUSION

As discussed in the introduction to this book, standard economics was created in the nineteenth century. Part of its magnetic charm was its physico-mathematical abstractedness—a combination of the drive toward a social science equivalent of the natural sciences, the arrival of vast quantities of the new "mass man" that needed to be managed, and the vague hybridization of an "ethics of the greatest good for the greatest number" with an ethics of individual human rights. One could also add into the mix the Marxist analysis of the alienation of labor from its products and the resultant veil thrown over the origins of commodities (Hudson and Hudson

2003) as capitalism expanded. Another contributing element, more and more powerful as time passed, was the use of fossil fuels to eliminate the effort and friction of moving through space—the physical expression and intensification of abstraction.

The case studies in this chapter illuminate a different world—a "deeper" world where economic transactions are embedded in particular spaces, specific relationships, and lived history. Exchanges are rich in understandings; products are tied to their origins and makers; the flow of things (gifts, mutual trades, etc.) echoes and assists the continuation of the universe. Underneath the arcane debates over interest and usury were intuitions— and sometimes more than intuitions—that a disembedded force (money on the move and increase) was dangerous because, among other things, it could get loose. It could act as an acid bath, dissolving deep life connections in ways that could only be initially hinted at in the shifting and encroaching realm of neutral exchange. The agonies and tabus and ritual alternatives to "neutral exchange" were defenses against unleashing such a power, whose ultimate end might be a total, catastrophic dissolution of all categories, touchstones, and limits. These categories, touchstones, and limits can be likened to the semipermeable membrane of a cell, whose integrity depends on being able to determine what is or is not allowed, and which can fail when the defenses are dissolved and overcome with water—or, to alter the metaphor, disembedding takes the brakes off.

By reintroducing the physical realm into economics, ecological economics begins the task of re-embedding economic relationships and processes, to reintroduce "use value" into a world where nature and humanity have disappeared into global webs of exchange. Part of the unfinished journey still to be worked through is what this means for the social sciences more broadly and for the model of the re-embedded human more specifically. These have already been profoundly affected by the disembedding abstracts of the standard economic model and the modernity (and postmodernity) that stems from them (Jameson 1991). We now find ourselves in a situation where human beings are being persuaded on a daily basis to see themselves, and act, as the sort of runaway beings that the model requires in order to make that model work more efficiently and effectively. Ecological economics needs to break that mesmerizing, hegemonic grip, and one contribution of this chapter (it is hoped) is as a contribution toward this goal.

NOTES

1. Besides the references cited herein, this section is also based on Baeck (1994), Gudeman (1986), and Hyde (1983).

2. This section is based on Armstrong, Hellmann, and Short (2002) and Wolf (2003).

3. Besides the references cited herein, this section is also based on Dasgupta (1996), Gandhi (1993), Ghosh (2007), and Iyer (1986).

4. Besides the references cited herein, this section is also based on Choudhury and Malik (1992) and Ghazanfar (2003).

5. Besides the references cited herein, this section is also based on Brandt (1954), Broder (1978), Geertz (1990), Sekaquaptewa and Washburn (2004), Thompson and Joseph (1965), and Waters (1981).

6. Besides the references cited herein, this section is also based on Barth (1993), Geertz (1991), Hobart (1978), and McPhee (1970).

REFERENCES

Afzal-ur-Rahman. 1974. *Economic Doctrines of Islam*. Vol. 4. Lahore, Pakistan: Islamic Publication.

Allen, George. 2008. *Radio Interview by Laura Ingraham*. Accessed September 15, 2008. http://thinkprogress.org/politics/2008/09/15/29200/george-allen-oil/.

Aristotle. 2011. *Aristotle's Nicomachean Ethics*. Translated by Robert C. Bartlett and Susan D. Collins. Chicago: University of Chicago Press.

Aristotle. 1998. *Politics*, ed. C.D.C. Reeve. Indianapolis: Hackett Publishing Co.

Armstrong, Regis J., J. A. Wayne Hellmann, and William J. Short, eds. 2002. *Francis of Assisi: Early Documents*. Vol. 1: *The Saint*. New York: New City Press of the Focolare.

Baeck, Louis. 1994. *The Mediterranean Tradition in Economic Thought*. London: Routledge.

Barth, Fredrik. 1993. *Balinese Worlds*. Chicago: University of Chicago Press.

Bateson, Gregory. 2000a. "Bali: The Value System of a Steady State." In *Steps to an Ecology of Mind*, 107–127. Chicago: University of Chicago Press.

Bateson, Gregory. (1972) 2000b. *Steps to an Ecology of Mind*. Chicago: University of Chicago Press.

Beaglehole, Ernest. 1937. *Notes on Hopi Economic Life*. New Haven, CT: Yale University Press.

Belo, Jane. 1970a. "The Balinese Temper." In *Traditional Balinese Culture*, edited by Jane Belo, 85–110. New York: Columbia University Press.

Belo, Jane, ed. 1970b. *Traditional Balinese Culture*. New York: Columbia University Press.

Berkes, Fikret. 2012. *Sacred Ecology*. 3rd ed. New York: Routledge.

Bird-David, Nurit. 1990. "The Giving Environment: Another Perspective on the Economic System of Gatherer-Hunters." *Current Anthropology* 31 (2): 189–196. doi:10.2307/2743592.

Bourdieu, Pierre. 1997. "The Work of Time.", trans. Richard Nice, In *The Logic of the Gift: Toward an Ethic of Generosity*, edited by Alan D. Schrift, 190–231. New York: Routledge.

Brandt, Richard B. 1954. *Hopi Ethics: A Theoretical Analysis*. Chicago: University of Chicago Press.

Broder, Patricia Janis. 1978. *Hopi Painting: The World of the Hopis*. New York: Dutton.

Brody, Hugh. 2000. *The Other Side of Eden: Hunters, Farmers, and the Shaping of the World*. Vancouver: Douglas & McIntyre.

Caputo, John D., and Michael J. Scanlon, eds. 1999. *God, the Gift, and Postmodernism*. Bloomington: Indiana University Press.

Choudhury, Masudul Alam, and Uzir Abdul Malik. 1992. *The Foundations of Islamic Political Economy*. Basingstoke, UK: Macmillan.

Collins, Steven. 1982. *Selfless Persons: Imagery and Thought in Theravāda Buddhism*. Cambridge, UK: Cambridge University Press.

Courlander, Harold. 1971. *The Fourth World of the Hopis*. Albuquerque: University of New Mexico Press.

Dalai Lama. 2009. "Exclusive Interview: 'I Am a Supporter of Globalization' by Joerg Eigendorf." The Office of His Holiness the Dalai Lama. Accessed September 30, 2014. http://www.dalailama.com/news/post/362-exclusive-interview---i-am-a-supporter-of-globalization.

Dasgupta, Ajit Kumar. 1996. *Gandhi's Economic Thought*. London: Routledge.

Derrida, Jacques. 1992. *Given Time: I. Counterfeit Money*, trans. P. Kamuf. Chicago: University of Chicago Press.

Feyerabend, Paul. 1999. *Conquest of Abundance: A Tale of Abstraction Versus the Richness of Being*. Chicago: University of Chicago Press.

Foltz, Richard. 2006. *Animals in Islamic Tradition and Muslim Cultures*. Oxford: Oneworld.

Friedman, Milton. 1953. "The Methodology of Positive Economics." In *Essays in Positive Economics*, 3–43. Chicago: University of Chicago Press.

Gandhi, Mohandas Karamchand. (1926) 1980. *M. K. Gandhi Interprets the Bhagavad-gita*. Delhi: Orient Paperbacks for the Navjeevan Trust.

Gandhi, Mohandas Karamchand. (1927) 1993. *An Autobiography: The Story of My Experiments with Truth*. Boston: Beacon Press.

Gandhi, Mohandas Karamchand. 2008. *The Essential Writings*. Edited by Judith M. Brown. Oxford: Oxford University Press.

Geertz, Armin W. 1984. "A Reed Pierced the Sky: Hopi Indian Cosmography on Third Mesa, Arizona." *Numen* 31 (2): 216–241. doi: 10.2307/3269955.

Geertz, Armin W. 1990. "Reflections on the Study of Hopi Mythology." In *Religion in Native North America*, edited by Christopher Vecsey, 119–135. Moscow: University of Idaho Press.

Geertz, Hildred, ed. 1991. *State and Society in Bali: Historical, Textual, and Anthropological Approaches*. Leiden, the Netherlands: KITLV Press.

Gellner, Ernest. 1989. *Plough, Sword, and Book: The Structure of Human History*. Chicago: University of Chicago Press.

Ghazanfar, Shaikh M. 2003. *Medieval Islamic Economic Thought: Filling the "Great Gap" in European Economics*. New York: RoutledgeCurzon.

Ghosh, B. N. 2007. *Gandhian Political Economy: Principles, Practice, and Policy*. Aldershot, UK: Ashgate.

Giddens, Anthony. 1990. *The Consequences of Modernity.* Stanford, CA: Stanford University Press.

Golden, Timothy Allen. 2009. *The Relevance of Buddhist Economics: Capitalism, Morality, and the Global Financial Crisis.* Dept. of Economics, Lingnan University. Accessed, December 2014. http://p2pfoundation.net/Literature_Review_of_Buddhist_Economics

Gudeman, Stephen. 1986. *Economics as Culture: Models and Metaphors of Livelihood.* London: Routledge & Kegan Paul.

Guenther, Lisa. 2006. *The Gift of the Other: Levinas and the Politics of Reproduction.* Albany: State University of New York Press.

Harvey, Peter. 1995. *The Selfless Mind: Personality, Consciousness, and Nirvāna in Early Buddhism.* Richmond, Surrey, UK: Curzon Press.

Hobart, Mark. 1978. "The Path of the Soul: The Legitimacy of Nature in Balinese Conceptions of Space." In *Natural Symbols in South East Asia,* edited by G. B. Milner, 5–28. London: School of Oriental and African Studies, University of London.

Hudson, Ian, and Mark Hudson. 2003. "Removing the Veil? Commodity Fetishism, Fair Trade, and the Environment." *Organization & Environment* 16 (4): 413–430. doi: 10.1177/1086026603258926.

Hyde, Lewis. 1983. *The Gift: Imagination and the Erotic Life of Property.* New York: Vintage Books.

Ittoen. n.d. "What Is Ittoen." Accessed September 10, 2014. http://www.ittoen.or.jp/english/.

Iyer, Raghavan, ed. 1986. *The Moral and Political Writings of Mahatma Gandhi.* Oxford: Oxford University Press.

Jameson, Fredric. 1991. *Postmodernism; or, the Cultural Logic of Late Capitalism.* Durham, NC: Duke University Press.

Kaye, Joel. 1998. *Economy and Nature in the Fourteenth Century: Money, Market Exchange, and the Emergence of Scientific Thought.* Cambridge, UK: Cambridge University Press.

Kuran, Timur. 2004. *Islam and Mammon: The Economic Predicaments of Islamism.* Princeton, NJ: Princeton University Press.

Langholm, Odd Inge. 1979. *Price and Value in the Aristotelian Tradition: A Study in Scholastic Economic Sources.* Bergen, Norway: Universitetsforlaget.

Langholm, Odd Inge. 1984. *The Aristotelian Analysis of Usury.* Bergen, Norway: Universitetsforlaget.

Lansing, John Stephen. 1995. *The Balinese.* Fort Worth, TX: Harcourt Brace.

Lansing, John Stephen. 2006. *Perfect Order: Recognizing Complexity in Bali.* Princeton, NJ: Princeton University Press.

Ling, Trevor Oswald. 1976. *The Buddha: Buddhist Civilization in India and Ceylon.* Harmondsworth, Middlesex, UK: Penguin.

Livingston, John A. 2002. *Rogue Primate: An Exploration of Human Domestication.* Toronto: Key Porter Books.

Loftin, John D. 2003. *Religion and Hopi Life.* 2nd ed. Bloomington: Indiana University Press.

Marx, Karl. (1867) 1990. *Capital: A Critique of Political Economy.* Vol. 1. London: Penguin.

Masri, Al-Hafiz B. A. 1989. *Animals in Islam*. Petersfield, UK: Athene Trust.

Mauss, Marcel. 1925. "Essai sur le don." *L'Annee Sociologique*, n.s., 30: 30–186.

McPhee, Colin. 1970. "Dance in Bali." In *Traditional Balinese Culture*, edited by Jane Belo, 290–321. New York: Columbia University Press.

Meikle, Scott. 1995. *Aristotle's Economic Thought*. Oxford: Clarendon Press.

Nishida, Tenko. 1969. *A New Road to Ancient Truth*. Translated by Makoto Chashi and Marie Beuzeville Byles. London: Allen & Unwin.

Nomani, Farhad, and Ali Rahnema. 1994. *Islamic Economic Systems*. London: Zed Books.

O'Kane, Walter Collins. 1950. *Sun in the Sky*. Norman: University of Oklahoma Press.

Ohnuma, Reiko. 2005. "Gift." In *Critical Terms for the Study of Buddhism*, edited by Donald S. Lopez, 103–123. Chicago: University of Chicago Press.

Page, Susanne, and Jake Page. (1970) 2009. *Hopi*. Tucson, Arizona: Rio Nuevo.

Polanyi, Karl. 1944. *The Great Transformation*. Boston: Beacon Press.

Rousseau, Jean-Jacques. 1763. *Emile*.

Ruskin, John. (1860) 2004. "The Roots of Honour, Chapter 1, Unto This Last." In *Selected Writings*, edited by Dinah Birch, 140–153. New York: Oxford University Press.

Sahlins, Marshall. 1972. *Stone Age Economics*. Chicago: Aldine-Atherton.

Schelling, F. W. J. (1809) 2006. *Philosophical Investigations Into the Essence of Human Freedom*. Translated by Jeff Love and Johannes Schmidt. Albany: State University of New York Press.

Schumacher, E. F. 1960. "Non-Violent Economics." *Observer*, August 21.

Schumacher, E. F. 1973. *Small Is Beautiful: Economics as If People Mattered*. New York: Harper & Row.

Schumacher, E. F. 1977. *A Guide for the Perplexed*. New York: Harper & Row.

Schumacher, E. F. 1979. *Good Work*. New York: Harper & Row.

Schumpeter, Joseph A. 1950. *Capitalism, Socialism, and Democracy*. 3rd ed. New York: Harper & Row.

Schumpeter, Joseph A. 1934. *The Theory of Economic Development: An Inquiry into Profits, Capital, Credit, Interest, and the Business Cycle*. Translated by Redvers Opie. Cambridge, MA: Harvard University Press.

Sekaquaptewa, Emory, and Dorothy Washburn. 2004. "They Go Along Singing: Reconstructing the Hopi Past from Ritual Metaphors in Song and Image." *American Antiquity* 69 (3): 457–486. doi:10.2307/4128402.

Shari'ati, Ali. 1971. *Fatima Fatima Ast*. Tehran: Husainiyah Irshad.

Sullivan, Gerald. 1999. *Margaret Mead, Gregory Bateson, and Highland Bali: Fieldwork Photographs of Bayung Gedé, 1936–1939*. Chicago: University of Chicago Press.

Talayesva, Don C. 1942. *Sun Chief: The Autobiography of a Hopi Indian*. Edited by Leo W. Simmons. New Haven, CT: Yale University Press.

Thompson, Laura, and Alice Joseph. 1965. *The Hopi Way*. New York: Russell & Russell.

Timmerman, Peter. 2010. "Boundary Matters: Buddhism and the Genetic Prospect." *Worldviews: Global Religions, Culture, and Ecology* 14 (1): 68–82. doi:10.1163/156853510X498069.

Tripp, Charles. 2006. *Islam and the Moral Economy: The Challenge of Capitalism*. Cambridge: Cambridge University Press.

Tucker, Mary Evelyn, and Duncan Ryuken Williams, eds. 1997. *Buddhism and Ecology: The Interconnection of Dharma and Deeds*. Cambridge, MA: Harvard University Press.

Turnbull, Colin M. 1983. *The Mbuti Pygmies: Change and Adaptation*. New York: Holt, Rinehart, and Winston.

Wall, Dennis, and Virgil Masayesva. 2004. "People of the Corn: Teachings in Hopi Traditional Agriculture, Spirituality, and Sustainability." *American Indian Quarterly* 28 (3/4): 435–453. doi: 10.2307/4138926.

Waters, Frank. 1963. *Book of the Hopi*. New York: Viking Press.

Waters, Frank. 1981. *Pumpkin Seed Point: Being Within the Hopi*. Athens: Ohio University Press.

Weber, Thomas. 1999. "Gandhi, Deep Ecology, Peace Research and Buddhist Economics." *Journal of Peace Research* 36 (3): 349–361. doi: 10.2307/424698.

Whiteley, Peter M. 2004. "Bartering Pahos with the President." *Ethnohistory* 51 (2): 359–414.

Wittgenstein, Ludwig. (1953) 2009. *Philosophical Investigations*. Edited by P. M. S. Hacker and Joachim Schulte. Translated by G. E. M. Anscombe, P. M. S. Hacker, and Joachim Schulte. 4th ed. Chichester, West Sussex, UK: Wiley-Blackwell.

Wolf, Kenneth Baxter. 2003. *The Poverty of Riches: St. Francis of Assisi Reconsidered*. Oxford: Oxford University Press.

Wood, Diana. 2002. *Medieval Economic Thought*. Cambridge, UK: Cambridge University Press.

Woodburn, James. 1982. "Egalitarian Societies." *Man* 17 (3): 431–451. doi:10.2307/2801707.

Wrigley, E. A. 2010. *Energy and the English Industrial Revolution*. Cambridge, UK: Cambridge University Press.

Zaehner, R. C. 1969. *The Bhagavad-Gītā: With a Commentary Based on the Original Sources*. Oxford: Clarendon Press.

We might summarize our present human situation by the simple statement: In the 20th century, the glory of the human has become the desolation of the Earth. And now, the desolation of the Earth is becoming the destiny of the human. From here on, the primary judgment of all human institutions, professions, and programs and activities will be determined by the extent to which they inhibit, ignore or foster a mutually enhancing human–Earth relationship.

—Thomas Berry

CHAPTER TWO

Ethics for Economics in the Anthropocene

PETER G. BROWN

In this chapter, I offer an attempt to start over by regrounding our moral and metaphysical beliefs in a scientific understanding of the world—and more particularly in how we understand the person and the nature of our knowledge. Ethical systems typically have at least five features, which may be weighted and function very differently: a foundation or justification, premises, structured principles or rules, virtues, and a guiding metaphor or ethos.

1. FOUNDATIONS

As with most Western ethical traditions, I ground my argument in an affirmation of human life; however, I place the understanding of what life is in a broader and deeper context. Beyond this, I want to both affirm and deeply question the emancipation project—the drive, initially articulated in Europe in the seventeenth and eighteenth centuries, that we can practically and legitimately free ourselves from nature, use "lesser" people as we wish for our own purposes, and alter the nature of the human self to fit the desires of the powerful (Horkheimer and Adorno 2002). This concept found fertile soil in antecedents in Western culture, such as the idea of a chosen people from the Hebrew Bible, the idea that humans are different in kind from all other life-forms found in both Judeo-Christian and Greek sources, and the suggestions in Platonic texts such as *The Republic* and Aristotle's *Ethics* and *Politics* that the human self should be designed to serve the overall needs of the society as perceived by the wise and well educated.

I suggest that we should regard the emancipation project as a source of great insight and liberation, as well as the justification for an enormous, unjust, and perhaps fatal hubris. This hubris has legitimated the enslavement and extirpation of the many of the world's peoples, decimated natural living and nonliving systems (which took billions of years to evolve), and ultimately enslaved ourselves to a false conception of who we are.

Fortunately, contemporary science offers a fresh starting point for rethinking our relationship with life and the world. In the last two centuries, but particularly since World War II, we have made monumental advances in understanding the universe and our place in it. At the same time, deep mysteries remain, such as the nature of dark matter, which composes a substantial portion of the universe; how things at the quantum level relate to macro phenomena, as described by the laws of special and general relativity; and the like. Nevertheless, at this time in history, we are positioned to ask these very simple questions: what is life, and what makes it possible? In the twentieth century, more light was shed on these questions, beginning with Erwin Schrödinger's seminal essay published in 1945, entitled "What is Life?" Schrödinger placed the questions within the domain of physics—more particularly, within the current understanding of thermodynamics. He gave humanity the stepping stones for solving one of the great scientific puzzles of the twentieth century: how do far-from-equilibrium systems, like living organisms, begin and then maintain themselves in an entropic universe? Schrödinger (1945), Prigogine (1968), Schneider and Kay (1995), and others were able to both answer this question and connect the answer to the origins and evolution of the universe, as they are now widely, but not universally accepted in the scientific literature.

If one is to respect life and its flourishing, one must respect what makes life possible. The universe as a whole, the planet, and the other life-forms with which humans have coevolved should be respected. Unsurprisingly, this mindset may lead one to the conclusions of many of the world's religions (although the deep dualism of these—particularly the Judaic, Christian, and Muslim traditions—have also led us astray). Today's human world, and the economic forces that drive it, are made possible by physical forces that also dwarf us—and which we see, to our peril, as something different than us. The universe has a beauty and majesty that we can only glimpse. The human mind and spirit are integral parts of this vast system; we can experience this oneness if we can escape the dictatorship of the ego—a common goal of religious teachings and meditations. Scientific research has confirmed that we live in a world of continuous creation, where a young universe is constantly developing new properties and possibilities. The universe is, in a sense, learning as we are. Failing to respect these realities is foolish: it diminishes each of us and undercuts the flourishing of life.

Three questions are essential to answer if we are to construct a sane and safe future for ourselves and the other life-forms with which we share heritage and destiny:

1. What is the nature of the person?
2. What do we know about what we know?
3. What should we do and not do?

The answers to these questions are intertwined. As they emerge, they will constitute the tissue of a new understanding of our relationship with life and the world.

This chapter is a preliminary attempt to set the moral foundations of this emerging understanding by exposing some of the premises on which ecological economics must rest. I then trace how this understanding may develop the other four elements of the ethical underpinnings of the discipline: premises, principles or rules, virtues, and an overall ethos.

1.1. What Is the Nature of the Person?

The concept of the person has to be one of the cornerstones of ecological economics. The "rational person," who coolly seeks to maximize his or her own interests and assumes everyone else does the same, is a cornerstone of neoclassical economics. This mythological figure has been repeatedly challenged over the years, most recently by behavioral economics and psychology, although there has been only limited success in changing mainstream thinking. This conception of the person is a partial mixture of rationality (as conceived during the Enlightenment) and the hedonism of thinkers such as Bentham and Mill. In its neoclassical version, this idea contains an individualistic notion of "the good," where notions of compassion and empathy, as well as community and connections, have largely been stripped away. In its place is a self-interested consumer who is engaged in choosing among alternative goods that satisfy personal preferences, no matter how fleeting or unconsidered, under conditions of scarcity.

1.1.1. The embedded permeable person

The continued findings from evolutionary biology, cognitive science, quantum physics, and systems theory will assist us in answering the questions of who humans are and to what we can aspire. Quantum physics, for example, provides a different conception of the human self than that found in neoclassical economics. Because events at the quantum level cannot be directly perceived by the human senses, we are not normally aware that all physical reality emerges through the interaction of fields and quanta. However, from the perspective of our most advanced scientific knowledge, this is the grounds for our existence in physical reality. As Robert Nadeau put it, the part we call "self" emerges from and is embedded in a seamless web of activity that is the entire cosmos; any sense we have of being separate or disconnected from this ground of being is an illusion that is not in accord with the actual character of physical reality (Nadeau 2013).

Systems theory tells us there are no individuals as the concept is normally understood; human beings live in complex, interlocking environments with other life-forms, which overlap with one another (Wheeler 2006). We live in a world swarming with hitchhikers and symbionts, such as bacteria and viruses. Most of this activity is well below the level of consciousness. These systems influence and, at times, dictate our behavior. For example, our immune system identifies and re-identifies what is us and what is not us at every moment. From the perspective of contemporary systems science, the human self is highly sensitive to initial conditions, subject to multiple feedback loops, and given to wide variation in subjective/behavioral outcomes. For example, a huge variety of factors—the weather, a sudden collision with another life-form, an indisposition—can change our course of action. We are relational and permeable with respect to energy and matter. We live in a world of shared semiotic meanings. Conscious reasoning is not the primary motivator of our actions—and a great deal of what we think is our knowledge is tacit and creaturely. The self is emergent in, and entangled by, the brain, body, environment, culture, and cosmos.[1] If we understood from the beginning that the human self is fully embedded, our policies with respect to toxins, for example, might be very different. We would not necessarily regard the world as something to be exploited, but rather as part of who we are. In our culture, how we think of the self largely blocks the understanding of our embeddedness in the biophysical world.

1.1.2. How the market manufactures the person

Once we recognize the inherently embedded character of the human self, it should come as no surprise that this self is also shaped by the institutions and cultural assumptions that surround us. As Stephen Marglin put it:

> Markets organize the production of goods and services, but at the same time markets produce people: they shape our values, beliefs and ways of understanding in line with what makes for success in the market. Markets thus exist in a kind of symbiosis with the discipline of economics, shaping people to fit the assumptions of the discipline even as economists shape the world in the textbook image of the self-regulating market. A new economy will need a new economics, which goes beyond the calculating, self-interested, individual to take account of community, compassion, and cosmos.[2]

To escape the lethal grip of the "market civilization," we need a fresh conception of the person that reflects both our ontological and cultural embeddedness.

As just one element in this reconceptualization, we can consider the emerging understanding of the brain and its role in consciousness and behavior. The brain is a complex and adaptive system, which makes it malleable. Many of the behavioral characteristics of human beings, whatever the source of their reinforcement, are enabled through the establishment of neurological pathways that become ingrained. The more the pathways are used in the present, the more they will be used in the future—they are the biological basis of habits. The more these pathways are associated with the pleasure centers of the brain, the stronger the incentives are to increase their use—something that has been long understood, at least on some level, by advertisers. What humans and other animals do (e.g., how much time people spend on the computer) actually changes how their brains are constructed. Understanding how the brain works can help us to grasp how the emancipation project has been massively refined in our time to construct nations of consumers while the concept of citizenship fades into distant memory (Curtis 2002).

1.2. What Do We Know About What We Know?

According to contemporary science, the idea of a certain and predictable world is, at best, an approximation of reality. This view is a legacy of the scientific revolution and the European Enlightenment. However, this view, which holds that the world is made up of quantifiable and stable parts, has been modified by nineteenth- and twentieth-century science that emphasizes relations and systems. Indeed, systems theory is now an established foundation of a scientific understanding of the universe (Kauffman 1995).

1.2.1. The importance of uncertainty

The systems that make up the universe have multiple, interactive feedback loops, as well as both fragile and robust initial conditions. The universe itself is a complex adaptive system that is "a creative advance into novelty," to quote Whitehead (1978). Hence, equilibria or static, predictable states— the centerpiece of the neoclassical model—are rare, perhaps even delusional. The overall reality, as Heraclitus wrote, is change. Surprises should not surprise us. In a world of complex systems, attempting to maximize a single variable, such as the gross domestic product, is sure to bring chaos and instability in its wake by causing perturbations in other changing and evolving systems. For example, international commitments to economic growth are destabilizing the climate system, even though the climate system is typically not part of mainstream economic discourse or, worse still, of any concern to macroeconomic policy.

1.2.2. Knowledge is approximate and provisional

The human ability for abstract reasoning is one of our great adaptive capacities, but it also has many shortcomings. Consider the example of a map of Quebec. A map should tell us as much as possible about the land and water in the province. The more the map approaches the size and complexity of what it is trying to depict, the more accurate the depiction is. Thus, an ideal map of Quebec would be the size of Quebec itself, but then it would not be very useful. Therefore, much is subtracted from maps, and they are typically reduced in size. In the same manner, our abstractions leave out most details, even though much of what we need to know is lost in achieving the benefits of abstraction. Our sensory systems naturally edit out the blooming, buzzing confusion of the world: we operate through gestalts. The more those gestalts are simplifications, the more hazardous they may become.

1.2.3. Is intelligence lethal?

One thing we need to consider is the question of the adaptive advantage of intelligence. Noam Chomsky summarized Ernst Mayr's arguments on this question as follows:

> [What] he basically argued is that intelligence is a kind of lethal mutation. And he had a good argument. He pointed out that if you take a look at biological success, which is essentially measured by how many of us are there, the organisms that do quite well are those that mutate very quickly, like bacteria, or those that are stuck in a fixed ecological niche, like beetles. They do fine. And they may survive the environmental crisis. But as you go up the scale of what we call intelligence, they are less and less successful. By the time you get to mammals, there are very few of them as compared with, say, insects. By the time you get to humans, the origin of humans may be 100,000 years ago, there is a very small group. We are kind of misled now because there are a lot of humans around, but that's a matter of a few thousand years, which is meaningless from an evolutionary point of view. (Chomsky 2011)

From this perspective, what is remarkable about our intelligence is how it is both highly adaptive while at the same time quite stable. Humans are inventive creatures: new products, discoveries, and scientific and technical insights pervade our age. But concurrently, we fail to examine the underlying assumptions of the emancipation project. We are insightful within it but not about it. These inquiries are blocked in part because they touch on religion, which is contentious terrain—terrain that sparked the devastating religious wars in Europe during the sixteenth and seventeenth centuries. The current internecine wars and armed struggles have many of the same sources, with religious and ethnic groups pitted tragically against each other. As a result of these wars, people tried to disentangle religion from politics during the seventeenth and eighteenth centuries, which was a key to the subsequent success of liberal societies.[3] The same strategy is still followed today by some in the desire for the establishment of "secular societies." However, this disentanglement was not and is not without cost; it allowed the development of superficial narratives that were ungrounded in an empirically substantiated sense of place in the universe. In North America, the task of taming (liquidating?) the continent and improving our material circumstances took precedence. It seemed more desirable and safer to concentrate on technique (i.e., the means) rather than to reflect on the ends (Lowi 1969). Our culture is profoundly and tragically allergic to any self-conscious metaphysics; as a result, we have embraced a decadent materialism that now is undercutting life's prospects.

The Western philosophical project has not been self-correcting. This is not to say that in the late nineteenth and twentieth centuries there were not philosophers who tried to exhibit the limits and dangers of the emancipation project. These philosophers included Bergson (1911), Schweitzer (1987) and Whitehead (1978), but they were marginalized. What took center stage was the less threatening, narrow debate between the Kantians (who advocated rights and strong related duties) and the utilitarians (who argued that maximizing human happiness is the sole standard of conduct), which dominated ethics as well as much of the social and political philosophy in the Anglophone world. This debate was conducted in the main, without anyone noticing that it took place *within* the emancipation project. The challenge in these early days of the Anthropocene is to prove Mayr and Chomsky wrong—to show that intelligence can be self-correcting and highly adaptive. Can some kind of collective response be formulated to respond to the emerging abyss of the Anthropocene? Hopefully, a global narrative informed by current science can emerge to ground and shape a collective response. If we do not succeed, life's prospects will be dim indeed.

1.2.4. What should we do and not do?

Contemporary sciences call into question one of the key ideas of economic and political liberalism: that each person is free to act as he or she wishes so long as that action does not harm other people. Two important sources of this idea are John Locke's *A Letter Concerning Toleration* ([1685] 1983) and John Stuart Mill's *On Liberty* ([1859] 2011). Locke held that our religious beliefs are internal matters and hence should be beyond the legitimate reach of the state, whose principal tasks are external—to secure "life, liberty and property." Mill held that the state has no right to interfere in what he called "purely self-regarding acts"—although interpreting this phrase has proved contentious, even for Mill. Yet, despite the pedigree of these two philosophers, the assumptions their ideas contain have become problematic as foundations for economic and political liberalism alike.

Locke's ideas about what one thinks is private have been transformed into the idea that one can live however one wants. Also, in the tradition of Veblen, Clive Hamilton (2010) and others have noted that consumptive goods became key markers of social status in the twentieth century. When we connect these foundational principles of political liberalism to the basic laws of chemistry and earth systems science, there are two distressing implications. First, in normal chemical reactions, matter is neither created nor destroyed. This means that the carbon released when fuel is burned in, say, a Toronto traffic jam, directly affects the interests of people and the composition of ecosystems around the world. Second, the process of burning fuel inevitably creates waste heat (most of which is radiated into space), causing a net decline in the stocks of useable energy on the earth.

We must see that how we live is potentially harmful to others. There are no actions that only affect us alone. The conceptual and moral underpinnings of economic and political liberalism were flawed from the beginning. We have no choice but to recognize that, as Thoreau articulated, "our whole life is startlingly moral. There is never an instant's truce between virtue and vice" (2004:210). It is critical that we interrogate the ideas of liberty and freedom (Fischer 2004; see also chapter 10 in this volume). What do they mean once we consider them in the context of a scientific understanding of the world? What are their relationships to justice? In a world of limits, liberty may only be legitimately exercised if one is using only one's fair share of low-entropy sources and sinks. Properly understood, true liberty lives in a modest room within the mansion of justice. Hence, "justice" for us must be understood, as it was for Aristotle, both as a particular virtue and as one overarching concept that holds the rest of morality in balance.

2. PREMISES

Henceforth, one can see that ecological economics rests on at least three rather simple and interconnected premises. I call them premises because, within the scientific paradigm, they are self-evident.

2.1. Membership: Escaping the Undertow of the Emancipation Project

The Western tradition has become globally hegemonic. It has also come to embrace a form of "exemptionalism": the idea that human beings are special, are in some miraculous way not a part of nature, and are therefore not subject to its sanctions, controls, and limitations. This has led to absurd ideas, such as thinking we can control "pests" with compounds that will affect them but not us. From a scientific perspective, humans are fully embedded in, and creatures of, the world. Humans are related to all earthly life; like all life on this planet, humans have coevolved with the earth itself. Recognition that we share heritage and destiny with all other people and all other life on this planet, as well as the dependence of life on physical and chemical evolution, must lead us to expand the moral community. The attitude of domination of the world and its peoples must be replaced with respect and reciprocity toward all that is. Humans are members of, not masters over, life's commonwealth (Leopold 1949). It is the flourishing of all people and life-forms that matters.

Of course, we are members of human communities of enormous variety, complexity, and size, and we have duties and privileges in these communities. "Membership," as I have defined it, recognizes this. However, membership also aims to redirect our attention to the natural world that makes the human world possible, as well as to treat it with respect and reverence for its own sake. It seeks to avoid, for instance, the idea that ethics is primarily confined to humans and to human relationships with God, as found in documents such as the Ten Commandments. The internalization of these ideas may have played a large role in the "success" of Western culture in reaching toward global hegemony. However, as Hans Jonas (1984) pointed out, these restricted ethical codes offer limited traction for responding to long-term macroscale crises, such as global climate change—good for a world with a few people in a large place but not sufficient compass for the Anthropocene.

Contemporary science has completely overturned the ontological dualism that has dysfunctionally underpinned Western culture for more than two thousand years (Singer 1985). The new (or rediscovered) "relational" ontologies do not need to labor to justify concern for the commonwealth of life of which we are a part (Bateson 2002; Kohn 2013). Also discarded is the "great chain of being"—the idea that there is a hierarchy of value, with a creator god at the top and rocks at the bottom. This schema is backward. Mind permeates the natural world and is the product of nearly fourteen billion years of evolution—not, to the best of our knowledge, its source. The teeming cities of India; the basement of an abandoned barn with spiders, snakes, and rotting timbers; the water trickling into a pool where brook trout lurk; the Milky Way; and the vast reaches of the universe—these are all our community.

The alleged superiority of the West is a conceit that has legitimated slaughter, slavery, empire, the appropriation and liquidation of precious sources of low entropy, and the filling of absorptive sinks locally and globally. The most recent manifestation of this global pathology is the world economic system, dominated by elites who are the primary beneficiaries, underpinned by assassination, terror, and covert military operations (Henry 2003). All people in all cultures have equal moral claims to flourishing, which are constrained and enhanced by the claims of other species for their place in the sun. We are not the chosen species or the chosen people. This, if you like, is the new emancipation.

2.2. House Holding

The idea of natural resources needs to be radically revised. When humans see themselves as members of the natural community, the idea that the earth and all of its life exist solely for our utility becomes absurd (Brown 2004). The world is not a collection of sources for satisfying human desires and a place to legitimately dispose of the waste stream inevitably created by those satisfactions. Rather, it should be considered a commonwealth where all species interact with each other and the planet's biophysical systems in a manner that facilitates the thriving of life. Ultimately, this thriving ought to be allowed and enabled to continue on its metaphysical journey into novelty.

The idea of the earth as a collection of resources and waste receptacles must give way to that of the earth as life's household (*oikos*). The house need not have an owner; however, it does represent a world and locus, familiar to many traditional communities, of reciprocity and respect. Over time, everything is a resource to everything else. In this conception, private property can legitimately exist only if it enriches the flourishing of life's commonwealth. This understanding, if adopted, would be an all-species version of John's Locke's justification of private property in the *Second Treatise of Government*—that private property is justified as a means to enhance what he took to be the divine mandate to preserve all mankind (Locke [1690] 1980).

2.3. Entropic Thrift

Low-entropy stocks and flows, as well as the sinks for high-entropy waste, must be used judiciously and with respect. Like all other systems that are far from equilibrium, life depends on low entropy—a fundamental good. Low entropy is the preservation of the earth's capacity to support flourishing human and natural communities; it makes all life possible. Broadly defined, energy is a fundamental good that underlies all other goods. It enables autocatalytic living organisms that are far from equilibrium to exist and thrive. This repositions the eminent twentieth-century philosopher John Rawls's (1971) concept of primary goods, such as income, wealth, and opportunity, to a secondary status because they all depend on energy. In an ecological political economy, wasting resources that make life possible is a fundamental moral wrong (Odum 2007). The earth's limited capacity to construct and maintain systems that are far from equilibrium implies that there are limits to the legitimate human appropriation of energy and sinks. Energy must be conserved for future human generations and for the flourishing of life itself.

Considerable interest in the restoration of the commons has been spurred by the work of Peter Barnes (2001), Joshua Farley (2010), and many others. One way to think of successful common property systems, such as those described in Ostrom's *Governing the Commons* (1990), is that they manifest a resting point or temporary halt—perhaps lasting as long as millennia on the entropic highway. They keep their stocks and flows of low-entropy sources in balance, and they avoid filling their sinks. They are masters at moderation.

Perhaps the clearest example of entropic thrift is a culture that lives within the current solar flow and does not fill its sinks any faster than natural systems can handle the waste. However, the idea of entropic thrift offers no definite answer to the question: how far into the future is of moral concern? The idea of a discounted present value—the centerpiece of much neoclassical analysis of duties to the future (Nordhaus 2008; Stern 2007)—is irrelevant in answering this question when the viability of the earth's life support systems is at stake.

The ideas of *membership* and *house-holding* set the reference points for defining our responsibility. How thrifty should we be with energy sources and sinks? The mandate should be as follows: be thrifty enough to persist and flourish among the other members of life's commonwealth. The Iroquois idea of being mindful unto the seventh generation sets the right tone. We are members of historical human and natural communities that have persisted for millennia, enabling us and defining who we are. We, the living, are custodians of a trust with no definite beginning and no discernible end.

There may be technological escapes from living within the solar flow—for example, ways of extracting and using fossil fuels that, over time, responsibly access the huge stores of hydrocarbons that remain in the earth's crust. There also may be ways of using these sources that allow for their burning without contributing to overfilling the sinks. These empirical questions will only be answered in the future. Ethics requires that we behave on the basis of reliable knowledge, not some fanciful dream of what might happen as promulgated by technological optimists. The current fossil fuel orgy clearly violates our minimal obligations to the flourishing of life's commonwealth.

3. PRINCIPLES

Ecological economics has an implicit structure of concern (Daly 1996). These concerns are matters of the scale or size of the economy relative to the earth's capacities, the fair distribution of these capacities, and efficiency in allocation, which provide the foci for the consideration of ethics within the paradigm.

As the principal insight of ecological economics, the size or scale of the economy relative to the earth's biophysical systems must be explicitly addressed in formulating and implementing economic policy. This has at least two dimensions: (1) maintaining the low-entropy sources of life and (2) not filling sinks with more high-entropy waste than they can process. The first point, as emphasized by Nicholas Georgescu-Roegen, Herman Daly, and others in this tradition, is a fundamental concern; it entails the prudent use and restoration of low-entropy stocks and flows. The second point requires an acknowledgment, understanding, and respect for the fact that nature has only limited capacity to process the high-entropy waste stream produced by society; the atmosphere and water sources can be overloaded with carbon dioxide, phosphorous, nitrogen, and the like. A failure to operate the economy within the "resilience limits" of the earth's sources and sinks can bring about massive—even catastrophic—instability that is incompatible with life's flourishing.

Although it is useful analytically to separate issues of scale, in practice this is quite misleading (Malghan 2010). Transgressions of the earth's resiliency limits are heavily dependent on purchasing power (or lack of it). The 500 million or so well-to-do individuals are the main current contributors to climate change. These people are in continuous violation of the golden rule. The billions of radically poor citizens contribute very little to climate change, although poverty often stimulates local ecological degradation. In some countries, the wealthy have simply bought the political process and use their control to prevent desperately needed climate-related legislation.

A core concern within ecological economics is the idea of intergenerational fairness. One generation has no right to deplete the sources or fill the sinks required for the flourishing of future generations. The position in time is morally irrelevant. We have a duty to pass along a world that is at least as good as the one we found. In the contemporary scientific narrative, the position in space is also morally irrelevant. Everyone alive today is a descendant of a small group of people; all of us share common DNA. We have the capacity for complex symbolic thinking and consequent participation in cultural narratives. In our time, we share a common, emerging, global narrative—and, of course, a collective (and perhaps tragic) destiny. The cultures of the world and their physical circumstances have selected for different talents, skin color, customs, and the like. Insofar as our differences are the result of biological and cultural evolution, there is no evidence that one group is more deserving than another.

Within the evolutionary tradition, matters of fair distribution are best informed by the idea of flourishing. One could start with Amartya Sen's (1999) work on capabilities and functioning, which is very aware of the dimensions of the Western emancipation project that legitimated empire on the presumed basis of European superiority. However, in two other closely connected respects, Sen remains entangled in it: (1) the human-focused individualism of the neoclassical model, and (2) an instrumental-ist, arms-length view of the natural world. Hence, Sen nods in the right direction, but the reconstruction agenda of ecological economics requires a fundamental rethinking of fair distribution—the foundations of which lie beyond the purview of the emancipation project.

With respect to climate change, Henry Shue (1993) has suggested the following four questions of justice:

1. Who pays the costs of avoiding future global warming?
2. Who pays the costs of making the adjustments to global warming that are not, or cannot be, prevented?
3. What background conditions of wealth and power would make bargaining over the first two questions fair?
4. What rights are there to future emissions within some cap designed to stabilize concentrations to avert runaway climate change (if such a course is still open to us)?

The reconstruction agenda of ecological economics suggests that the analysis of questions of fair distribution must be rethought in a broader context. Words like *who*, *pay*, and *cost* must be thought of in a different context. Let's begin with *who*: Within the evolutionary framework in which ecological economics finds its home, there is no reason why only human flourishing should matter. Rather, the flourishing of all of life must be a principal moral concern of economic and other policies. Regardless of the distributive framework, the challenge tabled by ecological economics is that we must have an account of the fair shares of the earth's life support capacity for all members of life's commonwealth.

Ideas such as *cost* and *pay* must also be thought of in a broad context that includes, but goes well beyond, money. Our understanding must be grounded in the use of the earth's life support systems. A logical place to begin is with the use of low-entropy sources and sinks—in the present, as well as in the past and the future. There can be no question that people in industrialized nations owe a huge climate debt to those in the South. In part, this debt can be paid in technology transfer and money, but in both cases the postulate of entropic thrift must be front and center. If money transferred to Kenya to combat drought is derived from pumping and burning fossil fuels, then we have defeated the purpose. The use of money for any purpose, including the discharging of debt, must be done in a manner that is constrained by the limits of the earth's life support systems, with an obligation to protect and enhance a flourishing Earth.

Efficiency is a core concept in ecological economics, but it is nested. Efficiency in this discipline is more complex than, and quite diverse from, the neoclassical conception. Once scale and distribution concerns have been met, a highly modified version of the neoclassical conception of efficiency comes into play. It is essential to see that efficiency is a derivative concept—typically depending on assumptions that are not explicit; however, in all cases, it is something that is done in support of an end or objective. To be true to its objectives, ecological economics must reject a foundational idea of the neoclassical model—that efficiency is the maximization of individual preferences. Instead, efficiency is enacted by fostering and supporting people with the virtues set out in section 4, as well as by maintaining, restoring, and enhancing the well-being of other species and their enabling conditions. It is essential that we replace the current goal of constructing and motivating consumers to maximize consumption. Instead, the construction of ecological citizens must be a core goal of economic policy.

4. VIRTUES

The postulates and principles of ecological economics require the cultivation of a new body of virtues. This section presents some examples.

4.1. Courage

The ethic argued for herein is out of step with contemporary culture. Those who endorse and follow it will be thought of as outcasts, killjoys, or outside agitators. Courage will be required. As the Iroquois put it: "The thickness of your skin shall be seven spans—which is to say that you shall be proof against anger, offensive actions and criticism. . . . Self interest shall be cast into oblivion."[4]

Like the crafters of the civil rights movements, we can expect to be out of step. However, the voices calling for a new human–Earth relationship are moving from scattered voices into a chorus—small but growing. A thorough regrounding of the human project and prospect is essential. We cannot rise to the challenge set out by ecological economics by simply extending vocabulary from the worldview we are trying to overturn, such as in the concepts of natural capital and ecosystem services or ill-defined concepts of sustainability. A complete rethink is required of the language, structures, practices, and guiding principles that inform our current system; an urgent action agenda requires mobilization, direct action, and the like.

4.2. Epistemological Humility

Uncertainty and unpredictability should ground epistemological humility. The scientific fact that all human knowledge is partial and provisional has profound implications for action. It should lead us to treat the urge to manage complex systems with enormous caution while at the same time recognizing that, at the present level of overshoot of ecological capacity, some sort of orderly pullback is essential. In *Water Ethics*, Jeremy Schmidt and I have called for a compassionate retreat—a concerted effort to reduce the human impact on the earth's life and its life support systems (Brown and Schmidt 2010).

4.3. Atonement

A quest for atonement is necessary for what we have wrought in the domination of the natural world and our fellow humans. Ecological economics argues that the human enterprise has grown too large. This is expressed by a number of measures, some of which are discussed in part 2 of this volume. These measures all carry the same basic message: there are limits to the earth's life support systems, we are approaching these limits, and these limits are likely already past their ability for self-renewal. We need to be prepared to respond to the collapse of whole systems, such as what is currently underway with the acidification of the oceans, the ongoing disintegration of the Antarctic ice sheets, and the destabilization of the climate system. Life's flourishing is in decline, and the rate of deterioration is rapidly increasing. Once we recognize that humans, like any other native life-form, are in a reciprocal relationship with the earth, the duty to help restore the massive damage to the earth's living system caused by our species comes into clear and central focus (van Hattum and Liu 2012). Atonement for our lack of respect and responsibility in the past must inform every action of the children of a new and regrounded enlightenment.

We must not fall back into the trap of overmanaging and forcing complex human or natural systems—much of our current trouble is a result of this attitude. Rather, we must enable the reconstruction of nature and societies and stand aside (often in awe) as they flourish afresh. The reconstruction agenda of ecological economics stands both in reference to, and apart from, the very idea that we ought to have an agenda in the governance of complex systems. We cannot take our first ethical step upon presumed ideas about what we ought to do. Rather, it must emerge from an empirical understanding of our place in a broader community of life.

Atonement, in this case, is more like being a midwife than a surgeon. Atonement can be achieved through activities such as regreening our planet by enabling the flourishing of damaged or destroyed ecosystems. In the future, this will require radical reductions in human consumption in the so-called developed countries, increases in consumption in many places, and reductions in fertility virtually everywhere. The education of women and access to reproductive health services are key to advancing toward a humane, but urgently needed, lowering of the human population. Of course, not all cultures have the same debts to come to terms with, not even remotely. The legacy of unjust carbon emissions and imperialism of the North is immense.

4.4. Fair Share

We must treat all living beings justly and leave them a fair share so they may flourish. Marine protected areas, Ramsar sites, and the like are only very small steps in this direction. In part, the human enterprise should be structured to live interstitially with life's commonwealth. Pockets of this are already arising. For instance, the Chicago Wilderness alliance works to preserve and integrate natural waterways and other ecosystems with the urban/suburban landscape of Chicago (Chicago Wilderness 2014). Although these steps are in the right direction, they are too modest. If life on Earth is to flourish, it must be given the space to do so. The present projections for the global growth in consumption are 2 or 3 percent per year, which mean that by 2100 the economy would be, respectively, seven or eighteen times greater than it is today (Garver 2009). Similarly, the projected increase of the human population to nine billion (or more) by the middle of this century is incompatible with membership, house-holding, and entropic thrift. In the next two to three centuries, our descendants will be living in a world undergoing a substantial—and perhaps very rapid—sea level rise, along with changes in patterns of food production. It serves us and the rest of life's commonwealth to reduce the human footprint on Earth with all deliberate speed.

4.5. Respect

We must respect all that is. As most or all of the world's religions recognize, we should understand ourselves as citizens of the cosmos, as finite actors in an infinite narrative. We are custodians of a tiny fragment of this story. When we gratuitously alter the earth, such as to fuel the North American automobile and truck fleets (and their attendant land-use patterns), we show disrespect for the sources of our being (Santayana 1905). On the other hand, when we tread lightly on the earth, with modest family sizes, low carbon footprints, and the like, we show respect.

The state of the virtues is an economic indicator. In our culture, where the economy is the central and ubiquitous institution, consumption is a major force in the construction of self-identity and expression. It is thus helping to produce people whose characteristics are at cross purposes with what is needed for life to flourish in the Anthropocene. One of the principal tests of an economy must be the kind of citizens it produces; in this context, a major goal must be the cultivation and maintenance of the aforementioned virtues (Madiraju and Brown 2014). Getting these virtues front and center will require a major rethink of macroeconomics, as outlined in chapter 8 of this volume. Also very worthy of attention are a number of movements trying to escape the hegemony of growth above everything else, such as the degrowth movement, transition towns, and *buen vivir* (born in the Andean regions of South America).

5. ETHOS

The Western idea of progress, which arguably took firm root within the emancipation project approximately 300 years ago, informed our expectations that the future will be better in some way than the present. It formed the overarching idea of our culture (subject to many interpretations, to be sure) and grounded a sense of optimism. Progress, which today tends to be understood as increased consumption by a massive human population, is now in the process of devouring its own possibility. The sustainability discourse that has become prominent since the Brundtland report (World Commission on Environment and Development 1987) has revealed its impotence. It was not even a timid step aside from the hegemony of the emancipation project. It was an accommodation with the forces and worldviews that are perpetrating the crises in which we are now entrapped. It can be thought of as a failed quest for another ethos, whose comforting rapprochement with the status quo wasted at least two crucial decades.

Ecological economics offers an escape from the emancipation project— an invitation to end to the tyranny of the market over humanity and nature alike, as well as the celebration of our citizenship in a universe that is ever evolving into novelty. The attraction of ecological economics is that it allows us to re-envision the future of life's commonwealth by calling on us to undertake a thorough re-examination of our relationship with life, the world, and the universe. We are alive in the adolescence of the universe— full of energy and possibilities. By accepting the invitation of ecological economics, we can discover our place with humility and respect. *This* is the meaning of emancipation in our troubled age.

NOTES

I am indebted to Julie Anne Ames, Margaret Brown, Holly Dressel, John Fullerton, Geoff Garver, Janice Harvey, Bruce Jennings, Suzanne Moore, Robert Nadeau, Alex Poisson, Jeremy Schmidt, Julie Schor, Gus Speth, Dan Thompson, and Laura Westra for comments on various drafts of this paper; to my colleagues Mark Goldberg and Tom Naylor at McGill; and to the students in the classes we have taught together.

    1. This description of the self is taken, with modifications, from Wendy Wheeler's *The Whole Creature* (2006), and draws heavily on conversations and correspondence with Robert Nadeau.

    2. I am indebted to Stephen Marglin (2010) for this idea.

3. The most notable philosophical defense of these ideas is found in John Locke's *Second Treatise on Government* ([1690] 1980) and *A Letter Concerning Toleration* ([1685] 1983).

4. The Constitution of the Iroquois Nations, Article 28. http://www.indigenous-people.net/iroqcon.htm. Accessed 25 February 2015. Prepared by Gerald Murphy (The Cleveland Free-Net – aa300). Distributed by the Cybercasting Services Division of the National Public Telecomputing Network (NPTN).

REFERENCES

Assadourian, Erik. 2010. "The Rise and Fall of Consumer Cultures." In *State of the World 2010: Transforming Cultures; From Consumerism to Sustainability*, edited by Linda Starke and Lisa Mastny, 3–20. Washington, DC: Worldwatch Institute.

Barnes, Peter. 2001. *Who Owns the Sky? Our Common Assets and the Future of Capitalism*. Washington, DC: Island Press.

Bateson, Gregory. 2002. *Mind and Nature: A Necessary Unity*. Cresskill, NJ: Hampton Press.

Bergson, Henri. 1911. *Creative Evolution*. Translated by Arthur Mitchell. New York: H. Holt and Company.

Brown, Peter G. 2004. "Are There Any Natural Resources?" *Politics and the Life Sciences* 23 (1):12–21. doi: 10.2307/4236728.

Brown, Peter G., and Jeremy J. Schmidt. 2010. "An Ethic of Compassionate Retreat." In *Water Ethics: Foundational Readings for Students and Professionals*, edited by Peter G. Brown and Jeremy J. Schmidt, 265–286. Washington, DC: Island Press.

Chicago Wilderness. 2014. "Chicago Wilderness." Accessed September 17, 2014. http://www.chicagowilderness.org.

Chomsky, Noam. 2011. "Human Intelligence and the Environment." *International Socialist Review* 76. Accessed December 30, 2014. http://www.isreview.org/issues/76/feat-chomsky.shtml.

Curtis, Adam (director). 2002. *The Century of the Self* (documentary series). London: RDF Television. DVD, 235 min.

Daly, Herman E. 1996. *Beyond Growth: The Economics of Sustainable Development*. Boston: Beacon Press.

Farley, Joshua, 2010. "Conservation Through the Economics Lens." *Environmental Management* 45, 26–38.

Fischer, David Hackett. 2004. *Liberty and Freedom: A Visual History of America's Founding Ideas*. New York: Oxford University Press.

Garver, Geoffrey. 2009. "The Ecor: An International Exchange Unit for Fair Allocation of Ecological Capacity." Paper presented at Conference on the Human Dimensions of Global Environmental Change, Amsterdam. http://www.earthsystemgovernance.org/ac2009/papers/AC2009-0158.pdf.

Hamilton, Clive. 2010. "Consumerism, self-creation and the prospects for a new ecological consciousness." *Journal of Cleaner Production* 18 (6): 571–575.

Henry, James. 2003. *The Blood Bankers: Tales from the Global Underground Economy*. New York: Four Walls Eight Windows.

Horkheimer, Max, and Theodor W. Adorno. 2002. *Dialectic of Enlightenment: Philosophical Fragments.* Edited by Gunzelin Schmid Noerr. Translated by Edmund Jephcott. Stanford: Stanford University Press.

Jonas, Hans. 1984. *The Imperative of Responsibility: In Search of an Ethics for the Technological Age.* Chicago: University of Chicago Press.

Kauffman, Stuart A. 1995. *At Home in the Universe: The Search for Laws of Self-Organization and Complexity.* Oxford: Oxford University Press.

Kohn, Eduardo. 2013. *How Forests Think Toward an Anthropology Beyond the Human.* Berkeley: University of California Press.

Leopold, Aldo. 1949. *A Sand County Almanac and Sketches Here and There.* New York: Oxford University Press.

Locke, John. (1685) 1983. *A Letter Concerning Toleration.* Edited by James H. Tully. Indianapolis, IN: Hackett.

Locke, John. (1690) 1980. *Second Treatise of Government.* Edited by C. B. McPherson. Indianapolis, IN: Hackett.

Lowi, Theodore. 1969. *The End of Liberalism: Ideology, Policy, and the Crisis of Public Authority.* New York: Norton.

Madiraju, Kartik Sameer, and Peter G. Brown. 2014. "Civil Society in the Anthropocene: A Paradigm for Localised Ecological Citizenship." In *Power, Justice and Citizenship: The Relationships of Power,* edited by Darian McBain, 135–148. Oxfordshire: Interdisciplinary Press.

Malghan, Deepak. 2010. "On the Relationship between Scale, Allocation, and Distribution." *Ecological Economics* 69 (11): 2261–2270.

Marghlin, Stephen. 2010. "Policy Statement: Premises for a New Economy: An Agenda for Rio+20." *Conference on the Challenges of Sustainability.* Stephen Marglin, co-convener. United Nations Division for Sustainable Development, New York, May 8–10. PDF

Mill, John Stuart. (1859) 2011. *On Liberty.* Accessed December 30, 2014. http://www.gutenberg.org/ebooks/34901.

Nadeau, Robert. 2013. *Rebirth of the Sacred: Science, Religion and the New Environmental Ethos.* Oxford: Oxford University Press.

Nordhaus, William D. 2008. *A Question of Balance: Weighing the Options on Global Warming Policies.* New Haven, CT: Yale University Press.

Odum, Howard T. 2007. *Environment, Power, and Society for the Twenty-First Century: The Hierarchy of Energy.* New York: Columbia University Press.

Ostrom, Elinor. 1990. *Governing the Commons: The Evolution of Institutions for Collective Action.* Cambridge: Cambridge University Press.

Prigogine, Ilya. 1968. *Introduction to Thermodynamics of Irreversible Processes.* New York: Interscience Publishers.

Rawls, John. 1971. *A Theory of Justice.* Cambridge, MA: Harvard University Press.

Santayana, George. 1905. *Reason in Religion (Vol. 3): Life of Reason, or, the Phases of Human Progress.* New York: Charles Scribner's Sons.

Schneider, Eric D., and James J. Kay. 1995. "Order from Disorder: The Thermodynamics of Complexity in Biology." In *What Is Life: The Next Fifty Years. Reflections on the Future of Biology,* edited by Michael P. Murphy and Luke A. J. O'Neill, 161–172. Cambridge: Cambridge University Press.

Schrödinger, Erwin. 1945. *What Is Life? The Physical Aspect of the Living Cell*. Cambridge, UK: The University Press.

Schweitzer, Albert. 1987. *The Philosophy of Civilization*. Translated by C. T. Campion. Amherst, NY: Prometheus Books.

Sen, Amartya. 1999. *Development as Freedom*. New York: Knopf.

Shue, Henry. 1993. "Subsistence Emissions and Luxury Emissions," *Law & Policy*, 15(1): 39–59.

Singer, Peter, ed. 1985. *In Defence of Animals*. New York: Blackwell.

Stern, Nicholas. 2007. *The Economics of Climate Change: The Stern Review*. Cambridge: Cambridge University Press.

Thoreau, Henry. 2004. *Walden: A Fully Annotated Edition*. Edited by Jeffrey S. Cramer. New Haven, CT: Yale University Press.

Van Hattum, Rob, and John D. Liu (directors). 2012. *Green Gold* (documentary film). Hilversum, The Netherlands: VPRO. Online video, 49 min. http://tegenlicht.vpro.nl/afleveringen/2011–2012/Groen-Goud.html.

Wheeler, Wendy. 2006. *The Whole Creature: Complexity, Biosemiotics and the Evolution of Culture*. London: Lawrence & Wishart.

Whitehead, Alfred North. 1978. *Process and Reality*. Edited by David Ray Griffin and Donald Wynne Sherburne. New York: The Free Press.

World Commission on Environment and Development. 1987. *Our Common Future*. Oxford: Oxford University Press.

CHAPTER THREE

Justice Claims Underpinning Ecological Economics

RICHARD JANDA AND RICHARD LEHUN

1. INTRODUCTION

The project of constructing an economy that operates within the planet's ecological boundaries is a matter of justice. Justice has to do with rendering what is due, and thus with a proper accounting for individual and collective conduct.[1] Ethics has to do with good character and manners, and thus with proper behavior. Justice holds ethics to account. It is both a subset of ethics and the architecture of ethics. Thus, whereas one can discuss the ethical foundations of economics in general and of ecological economics in particular in the sense of seeking to identify forms of good behavior that would characterize the economy, the related justice inquiry sets the stage for identifying forms of governance, modes of accountability, and norms to discern fair shares of resources.

The existing economy is out of joint. Past and present generations have exploited resources and caused environmental and social externalities on a scale and at a pace that imperils the future prospects of life—human and nonhuman alike. To state this is to make a justice claim against the existing economy. It is also to contemplate a more just economy. Thus, any effort to build ecological economics requires an examination of the following questions: What justice is the existing economy purporting to render? What are the justice claims being made against the existing economy? And, what justice would be rendered by the ecological economy that would substitute for it?

1.1. The Illusory Search for a Unifying Ethic

These three questions immediately pose a difficult conceptual challenge for us. To approach them and to begin to make sense of them, a shift in expectations as to the kinds of answers we would hope and expect to find is also at stake. That is, we typically hope to find a basis for persuasion, so as to gain consent and adherence for the kind of answer being offered. Thus, in formulating answers to these questions, we are often seeking intuitions or evidence of the kinds of arguments that we believe are most likely to produce collective assent. If only the golden arguments could be found to which all would spontaneously adhere, we tell ourselves, we could be enabled simply to produce, in an act of collective will, a change from the existing economy to an ecological economy that would function within planetary boundaries. Thus, for example, we imagine that clear arguments can be made showing that the existing economy is focused too narrowly on the pursuit of consumption and growth, that this is unjust because it fails to take account of claims of nature against us, and that therefore adherence to a land ethic in which we view ourselves as stewards of nature could be made socially compelling. Sometimes, less ambitiously, we hope at least to discover the compelling arguments that some reasonable people could follow to guide their own choices when confronting a dystopic world. In this vein, we might hope to find guidance as to whether it is ethical to consume meat, own a car, live in the suburbs, or travel by plane.

Unfortunately, we will not get far looking for ideas about justice and ethics that could flow into public debate and produce collective, conscious, and reasoned assent. The very forum we would seek to persuade on the basis of individual choice and adherence is now utterly colonized by a conception of choice that elevates consumption and property to a right; thus, it resists inherently any appeal to make a choice that would circumscribe or scale back choices to consume or appropriate. Furthermore, the kinds of arguments we would employ in liberal ethics—be they consequential-ist (looking to the outcomes of our acts) or deontological (looking to the responsibilities we have)—are viewed as simply existing on a menu of possible ethics to be selected by individuals. For example, it is standard in liberal ethics discourse to situate arguments within typologies such as act consequentialism, prioritarianism, Lockean libertarianism, rule consequentialism, capabilities, and pure deontology so as to discuss how different arguments can be made from these various standpoints within liberalism (e.g., see Schlosberg 2012). If one seeks then to produce an even broader moral overlapping consensus, to borrow a term from Rawls (1993), that could encompass not only versions of arguments from liberalism but also from outside liberalism (religious positions, Marxist positions grounded in radical social critique, deep ecologist positions grounded on a moral priority to nature, and so on), all that one ends up generating is yet another contribution to the marketplace of ideas, from which individuals will select their preferred position. The proposed solution—an argument that would generate consensus—becomes a further standpoint to take account of and to absorb in producing a consensus and therefore a further impediment to achieving it.

If what we faced was an unencumbered, boundless field of social endeavor in which the test of ideas could ultimately be whether they were freely chosen and retained, we could simply wait for the inherent appeal of arguments—their strength or weakness—to win the day. Even if we acknowledged, more realistically, that the inherent rationality of argument will not determine what in fact gains social consensus, we might nevertheless persist in setting out and defending the rational social ideal we envisage, and then seek to counter politically whatever forces we believe exist to impede the triumph of the right and the true.

1.2. The Question Put in This Chapter

These approaches are utterly inadequate to overcome the collective action problem we now face with respect to the biosphere, and arguably they have long since proven inadequate to confront the pathologies of collective action concerning the equitable distribution of social goods. What we face, in short, is a need for global coordination that will not be solved by seeking to find and develop a single set of ethical commitments to which everyone will adhere. We must pose our question in the following way: given deep ethical divergences, variable but persistent commitments to the protection of individual preferences to consume and appropriate, and the impossibility of gaining acceptance of a unified set of norms within the proximate time frame needed to produce collection action, what kind of legitimate justice foundations could be given to an ecological economy that would emerge from the existing economy?

Two preliminary objections to this question should be considered. A first, hard-line objection would challenge the question as irrelevant because it fails to envisage the possibility of overcoming ethical diversity through nondemocratic means. A second, soft-line objection would claim that the question is irrelevant because conventional liberal democratic justice theories are already flexible enough to absorb the vast array of existing social views; also, in any event, whatever is proposed as an answer to the question of ecological justice would have to survive democratic scrutiny. Before setting out the plan for this chapter, therefore, these two threshold objections will be canvased briefly in turn.

1.3. The Hard-Line Objection

According to the hard-line objection, if the production of democratic consensus around a common ecological framework is impossible, then undemocratic methods must ultimately be adopted to pursue it. The point is not to cope with the plurality of ethical positions but to overcome them. Hence, we should expect authoritarian solutions to our ecological crisis, perhaps (to use a current example) originating in China (Beeson 2010).[2]

In fact, there is no necessary connection between assuming, on the one hand, that it is impossible to produce a consensus upon a singular normative framework, and pursuing undemocratic methods as a result. It is only if we affirm that a singular ethic is required, despite the impossibility of achieving it, that we would seek recourse to authoritarian means. Furthermore, a deeper critique of authoritarianism reveals that it is incapable of producing a singular ethic and especially incapable—despite the myth of noblesse oblige—of producing coordinated stewardship of planetary systems. The corruption to which authoritarianism is inherently subject and the incentives that it faces to gain and concentrate wealth are far from providing any guarantee that its imposed ethic will respect planetary boundaries. On the other hand, because authoritarian regimes compose part of the global context of ethical pluralism, the justice frameworks and accountability standards needed to arrange a network that assures the provision and safeguard of environmental public goods will have to be operable also in such contexts.

1.4. The Soft-Line Objection

According to the soft-line objection, authoritarian, undemocratic means will not produce a legitimate ecological economy; therefore, we are left inevitably to rely upon democracy to produce whatever framework is proposed (Burnell 2009).[3] In the end, despite protests to the contrary, the approach amounts to a version of the conventional effort to formulate an argument that will be subject to liberal democratic deliberation.

Stated in this way, the soft-line objection also proceeds on the assumption that we are aiming to produce a singular normative framework. The very thing democracy cannot achieve is purported to require democracy. If the objection is redirected to affirm simply that democracy is about producing legitimate outcomes from social pluralism, it fails to address a critical feature of our collective action context: there is no single democratic forum, and indeed no monopoly of democratic forms—let alone a common understanding of democracy—within which collective decisions about global collective action can be achieved.

Even if one were to grant that global democratic institutions could produce a global collective action framework (although this is dubious given the capture to which existing democratic institutions are prone), the soft-line objection would nevertheless dissolve. First, no such global institutions exist. In fact, we find ourselves within a governance context in which a network of multiple regimes must be steered and coordinated—not only those of nation-states but also those of corporations and other nonstate actors. Second, even if the international state system were viewed as our closest available substitute for global democratic institutions, it has so far proven incapable of producing the requisite level of coordination. This is not to say that the state system should be eliminated or bypassed—that too would be a naïve and insufficiently multidimensional analysis. Rather, it is to say that what we require is a way to model the coordination of multiple justice frameworks—and in particular, the deficiencies and inadequacies that will be produced in seeking to render multiple justice frameworks that are interlinked and in some measure interoperable—rather than a single theory of justice to which we would all hope to adhere.

1.5. The Need for a Metatheoretical Approach

The production of such a justice model is what is meant by a metatheoretical approach to justice. If our only hope to produce the requisite collective action proximately is to deploy a global social network for the coordination of behavior, and if by definition such a social network will have to operate across multiple ethical "platforms," then a metatheoretical model for the operation of such a network is needed.

A metaphor to represent this is "swarm intelligence": the collective behavior of self-organized systems—democratic or otherwise—that interact through common signaling functions. Swarms can behave like locusts to produce environmental devastation, or they can behave like bees and help to maintain and reproduce their ecosystems. The contribution of a metatheoretical justice model is to identify the deficiencies of existing signaling across multiple justice platforms, producing our locust-like behavior. The term "model" rather than "theory" is used because the goal of the analysis is not to offer another theory seeking adherence. Rather, it is to produce an aggregate, macroscopic assessment of the justice deficits currently being produced by the interacting signals of behavior in the existing economy. Metatheory is to theory as signaling is to behavior. Metatheory concerns the orientation of justice frameworks as they interact. The justice deficits of the price mechanism, which is the principal signaling we currently rely upon, thus becomes a starting point for the model. Identifying those deficits relates ultimately to identifying a more adequate signaling function, reflecting a gift relationship with nature and coordinating fiduciary roles.

1.6. Outline of the Argument

The metatheoretical standpoint inscribed in ecological economics is the starting point for this chapter.[4] Having identified what the metatheoretical approach is, the chapter turns to an application of the metatheoretical model so as to assess the justice deficits of the existing economy. These considerations point us toward what shall be called a second Copernican revolution in the modeling of justice, a modeling that requires us to discover how to introduce a signaling function into the existing economy that will allow us to take up the role of stewards or fiduciaries of the entire oikos, not just gatherers and exploiters of available (and disappearing) low-entropy resources. The chapter then seeks to identify a fiduciary methodology consistent with the metatheoretical approach that could overcome the justice deficits of the existing economy, against the backdrop of what the existing economy leaves unaccounted. The chapter concludes by raising a justice claim against the fiduciary methodology we seek to develop: the transition from the existing economy to an ecological economy will itself present the dramatic challenge of shifting our own relationship to subjectivity and freedom.

2. A METATHEORETICAL APPROACH TO JUSTICE

To establish justice foundations for an ecological economy, we need a theory of theories of justice, rather than just another theory. This is because we need to align and interrelate our disparate norms of just behavior—those that now actually guide us—with what we can gather from science about our place in, disruption of, and steering capacity toward planetary systems and boundaries. The challenge of ecological economics is not one that can be borne simply by those who adhere to one particular model of justice. It must be borne by all. Any individual theory or model of justice will seek to organize and rank justice claims so as to produce the maximum social aggregate of justice. In the wake of accumulating justice deficits after two world wars and the Holocaust, post-Enlightenment (postmodern) conceptions of justice put into doubt whether any singular, totalizing theory of justice is possible (Horkheimer and Adorno 1982). Furthermore, we are confronted with the challenge of producing global collective action with respect to planetary systems out of overlapping and conflicting social affirmations of justice.

At the outset, a metatheoretical approach is to be distinguished from normative pluralism. Normative pluralism consists of identifying all the partial sets of justice conceptions, seeing them as operating in relation to each other, acknowledging that individuals negotiate their way among them, and adhering sometimes to multiple conceptions within the multiple normative orders in which they find themselves (Kleinhans and Macdonald 1997). Although this response can be a prolegomenon to setting justice foundations to ecological economics, it is in itself inadequate to the task. We are faced with overcoming the collective action problem posed by the aggregate impacts of our individual choices on the planet. Our plural conceptions of justice are not spontaneously aligning toward the reduction and elimination of those impacts. On the contrary, we lack a common set of signals for behavior about the value to be placed in collective goods. The pluralism of justice conceptions operating in conflict is in fact impeding the emergence of such signals.

Take the example of climate change and how the global economy should be oriented toward it. For those in developing countries, access to and deployment of the resources, technologies, and modes of production that developed countries have exploited is a matter of justice (Sen 2009). The gaping divide in standards of living between parts of the world, with attendant incapacity to provide for basic needs, produces overwhelming justice claims. Yet as China's recent experience demonstrates, the path to development is marked by growth in greenhouse gas (GHG) emissions and environmental devastation. Developing countries claim that the burden of reducing these impacts should be most heavily borne by those who have most exploited resources over time: a climate debt is owed. Leading developed countries claim that those living in the present cannot be held to account for what was done by others in the past: no climate debt is owed (see Posner and Weisbach 2010). These are conflicting justice conceptions. Their coexistence in a pluralism of conceptions is a problem to be solved rather than a contribution, if you will, to the richness of social life.

A metatheoretical approach goes further than pluralism and seeks to identify the strengths and inadequacies of particular justice conceptions— what they reveal and what they conceal; what they enable and what they disable. In particular, a metatheoretical approach would hold to account the deployment of any justice conception in arriving at social outcomes by asking what injustices will thereby be done in addition to, or as the result of, implementing that particular form of justice.

A metatheoretical approach can be modeled according to three dimensions of justice: (1) the metaphysical (that imbuing justice from without— originally the sacred), (2) the subjective (that ascribed to justice by our own lights), and (3) the conventional (that given to justice by society). A fourth dimension running through the other three—that of emancipatory justice—relates to enabling our freedom from any fixed justice constraint established in the other dimensions. Emancipatory claims emerge from the subjective dimension of justice. However, insofar as they are a negation of existing inadequate justice, they cannot simply be mapped within the three foregoing dimensions. They therefore add a fourth dimension running through and in tension with the other three (figure 3.1). Whereas metaphysical, subjective, and conventional claims map out "horizontally" the possible constitution of justice claims, the emancipatory dimension maps out "vertically" how we stand in relation to those constituted claims— either in making them so as to overcome justice burdens weighing upon us (the positive emancipatory claim) or in seeking to bear the weight of those burdens by submitting to them (the negative emancipatory claim).

For example, we could take up a relationship of stewardship to the effects of our choices upon planetary boundaries through either a positive or negative emancipatory claim. The positive emancipatory claim would seek to free ourselves and future generations from the justice burden being imposed by our existing choices and to enable a social transformation to accomplish this. The negative emancipatory claim would seek to bring ourselves and future generations under the aegis of what is required of us by

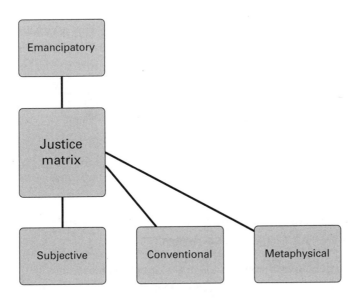

Figure 3.1. Four dimensions of justice.

our relationship to what is other, namely the conditions of life, and to conform to those requirements.

Emancipatory claims have become central to contemporary justice models. They are notably embedded in both the position of developing countries and of leading developed countries just described. Both seek to maintain the greatest possible scope for overcoming limits upon our choices by insisting on the priority of economic development. Amartya Sen (1999) captured this notion in the phrase "development as freedom."

We refer to the interplay and tension among all four dimensions through which specific justice claims are engendered as a "justice matrix." A matrix shares with justice (the right) the quality of providing a structure, support, or architecture within which a claim can reside. Its purpose is to be able to map all possible sources of justice deficiency that can be articulated—as a claim—upon our collective efforts to render justice. Because the dimensions of justice in this matrix do not take their orientation from any substantive theory of justice but rather seek to locate how justice claims can arise, the test that must be put to them is whether they can provide a complete mapping. Because the dimensions include self-constituted, socially constituted, and supervening nonhuman claims—as well as the justice claims that arise out of the effort to break free from the constraints of any of the foregoing—they are designed axiomatically to map all possible claims.

A metatheoretical analysis cannot be used simply to restore a grand balance to justice by reasserting justice claims that have been cast aside and asserting a notional equality to all justice claims. That would be to shift from the effort to map and identify all possible claims to the effort to rank and order them—in short, constructing a new theory of justice. However, a metatheoretical analysis does reveal that the justice claims weighed by the existing economy align most with the subjective dimension of the justice matrix—that is, the marketplace or battleground of contested views. The next section turns to an immanent critique of how that arose.

3. THE JUSTICE CONSIDERATIONS UNACCOUNTED FOR IN THE EXISTING ECONOMY

As documented in chapter 2, it is not inherent to an economy that it functions according to transaction and market exchange. Yet the existing economy has taken on a self-reproducing and autonomous quality by purporting merely to channel the inherent characteristics of economic actors. It is in this sense that, within the justice matrix, it aligns most with subjective claims. Because economic actors are understood inherently to pursue self-interest and even make gifts to gain favor, the only kind of justice that can be rendered is one that the actors themselves would each affirm for themselves.[5] No other justice claim is held to be realistic or socially reproducible. This is the world of transaction and contract. The economy thus comes to contain the aggregate of justice simply by definition because it aggregates all of the individual affirmations of outcomes sought in transactions and achieved in contracts. It achieves the justice of exchange.

The justice of exchange is perfected in two ideas: freedom and efficiency. All are free to pursue their own interests, and gains in efficiency are achieved as long as the self-interest of one is advanced without that of another being prejudiced. An efficient economy—one that succeeds in aggregating and aligning all individual self-interests (or at least notionally compensates for prejudice) is just because no other form of justice is possible, in the sense that no other form of justice coheres with real, existing subjectivity. Hegel's dictum—that the rational is the actual and the actual is the rational—is transmuted into the following: justice is the actual outcome produced by self-interested rational actors. Indeed, a sign of the question-begging quality of the justice that is affirmed for the existing economy qua aggregator of individual self-interest is that justice drops out of the equation. The idea of justice really adds nothing to the simpler formula: self-interested rational actors produce the actual outcome of the economy. Its justice is efficiency. It is thus an important feature of the existing economy as an autonomous social field that it succeeds in occluding its own justice affirmations. The dictum becomes "justice is what justice is," which is another way of saying that justice considerations are irrelevant and give way to brute facts. An increase in the aggregate of justice becomes simply having more of what the economy already is: growth of the economy, which is called a gain in welfare. The single metric for justice is efficiency and the increase of its aggregate is growth.

It has been commonplace since Aristotle, however, that the justice of exchange is separable from the justice of distribution. The binary corrective justice achieved between parties to a transaction will not incrementally and spontaneously aggregate into the geometric proportions of distributive justice. Corrective justice can coexist with distributive injustice but will not necessarily do so. Thus, an economy that narrows its justice consider-ations to the outcomes produced by the binary affirmations of individual actors in transactions will not have any clear footing in distributive justice. Indeed, in this respect, the economic sphere has come to be separable from the social sphere. Political investment is made in the social sphere so as to produce a modicum of distributive justice modulating the outcomes of the economy. However, the economic sphere has priority over the social sphere in the sense that the efficient outcomes of the economy are taken as given and are only to be adjusted by social regulation to produce greater equity, understood as fairness.

The traditional Aristotelian categories of justice—corrective justice, distributive justice, and equity—can be characterized within the dimen-sions of the justice matrix. Each of the Aristotelian categories participates in each of the dimensions, although equity tends to concentrate in the metaphysical, corrective justice in the subjective, and distributive justice in the conventional. Aristotle conceived of the two domains of corrective and distributive justice as themselves leaving unfulfilled the entire conceptual space of justice: thus, the notion of equity (*epikeia*) had to be added to the other two as a general adjustment for the inadequacy of the justice that would be produced, in aggregate, through the pursuit of norms of correc-tive and distributive justice. However, the justice rendered by the existing economy serves to narrow the constellation of possible justice consider-ations by reducing its justice to efficiency (a version of corrective justice, especially given the Pareto principle) and reserving the rest of justice to "equity"—that is, giving priority to the former over the latter. In effect, equity becomes collapsed with distributive justice. Any supervening, meta-physical conception is emptied out, largely because it becomes identified with mythology and unreason (see Horkheimer and Adorno 1982). Any notion that justice is owed to what has been given to us (in the "given" that is the earth) is relegated to private faith. Aristotle's three-dimensional justice becomes a flattened two-dimensional justice. In short, the justice conception operating within the existing economy narrows and occludes the constellation of possible justice outcomes.

Viewed within the justice matrix, the contemporary market economy relies upon a conception of justice that reduces "equity" to the production of conventions of distribution (equity) and subjective (efficiency) dimen-sions. A metatheoretical approach, by situating justice claims in relation to a matrix of possible claims, helps to give an account of what is left out of any specific set of claims: that is, any strong relationship to supervening metaphysical claims.[6] In particular, it helps to reveal that the metaphysical claims against which subjective claims have emerged in tension and from which they have sought emancipation are occluded—flattened—by our current configuration of the justice matrix. This insight had already been suggested in chapter 1, which showed how alternative conceptions of the economy tend to give precedence to justice claims arising outside of ourselves—from God, nature, or our place in the whole.

In a flattened configuration of the justice matrix, the metaphysical justice claims not made out within the subjective and conventional dimensions remain present but repressed. They are aggregated within other sets of claims and in some measure orient them, but are not themselves signaled. For example, those who would seek to repair or restore the earth out of a sense that past and future generations have claims upon us will tend to articulate these claims as best they can within the language of distribution or efficiency. Thus, sustainable development claims are often made out to involve fair distribution of access to the resources needed for development, and the proposed tools to accomplish this are often markets for the purchase and sale of property rights in the environmental damage being done. It is very difficult, within justice conceptions associated with what can best be called modernity, to articulate claims made directly on behalf of the earth.

Against this backdrop, one task for ecological economics becomes to elaborate what an economy operating on a more fully differentiated justice conception would look like. The term "ecological economics" is promising for this task, but it often fails to go in that direction, particularly when it restricts itself, like "environmental economics" to finding ways to internalize externalities generated by the economy back within the economy. The attempt to internalize externalities flows logically from the attempt to resolve self-defeating justice claims made for the existing economy (i.e., immanent critique): if the existing economy is meant to produce efficiency but does not do so because it undersupplies the public goods (ecosystem services) necessary for the functioning of the economy, there is a "market failure" internal to the economy. Such a result is a kind of paradox, of course, because it amounts to saying that the entire market is failing to be a market.

If a more fully differentiated justice conception is not sought, a solution may be to make the market behave like a market by putting a price on "ecosystem services," thereby preserving the efficiency logic of the existing economy but correcting its market failure. Price remains the one signal that is relied upon to coordinate the multitude of preferences and ethical investments in the environment or any other social good because it is taken to be a universal and interoperable measure of choices and value—however incommensurable. Internalizing externalities does not in itself raise a fundamental justice critique; it simply holds the economy to its own standard and does not challenge the monopoly of the price function on the signaling of activity in the economy.

The beginning point in immanent critique is nevertheless promising for deepening the justice considerations at stake in the economy. The failure of the economy to perform as an economy on its own terms (market failures), despite the fact that it purports to be a self-reproducing aggregation of efficient (just) outcomes, raises the possibility that there is a set of justice considerations unaccounted for in the existing economy and signals of justice claims that are muted or nonexistent. This prompts a reconsideration of the term "ecological economics" as going beyond market correction, and this is the standpoint we take in this book.

4. OVERSTEPPING PLANETARY BOUNDARIES: TOWARD A SECOND COPERNICAN REVOLUTION IN JUSTICE THEORY

The contemporary configuration of the justice matrix, with the priority it gives to subjective claims and its flattening out of metaphysical claims, can be traced to the Enlightenment (see Horkheimer and Adorno 1982). The first "Copernican Revolution" of justice theory, developed by Kant, placed man at the center and measure of all things by identifying the human subject as determining the grounds of knowledge, and human agency as determining the contours of justice (Kant 2003). Unlike the Copernican Revolution in science, which reoriented us toward our place in the cosmos, the revolution in justice theory oriented us away from our place in the cosmos. In this regard, the Enlightenment project became one of overcoming any artificial and archaic limits upon knowledge: only those limits to knowledge that were inherent to the human subject itself were to be accepted, and even those were contestable. Human subjectivity as the ground of knowledge, and human agency as the measure of justice, were combined into an equation articulated by Spinoza and rearticulated repeatedly since: knowledge is power. The mythologized mysteries of nature were to be revealed as scientific laws, and the subservience to fate and authority was to be overcome. Free inquiry and free choice were the watchwords of the Enlightenment. Emancipation was its theme. The free market was its progeny.

The goal of justice theory became to throw off the shackles of repressed self-determination and to model the enabling of emancipation. There are two main contemporary models of that sort. The dominant liberal model is about enabling the freedom to exploit, produce from, appropriate, and consume resources. Justice is achieved as long as whatever interferes with that freedom is held to account using the counterweight of discipline, power, and authority, notably as deployed by the state in protecting rights. The other, discredited socialist countermodel is about emancipating us from exploitation, private control of production, disproportionate appropriation, and overconsumption by some so that the means of production can be controlled and shared by all. Justice is achieved when discipline, power, and authority are brought to bear, for redistributive purposes, on the past outcomes of the liberal model. Both models seek to harness human mastery over nature so as to maximize the sphere of enabled agency.

One can state two truisms in this regard. The dominant liberal model invests most justice resources in the emancipation of individual agency. The countermodel invests most justice resources in the emancipation of collective agency. They are dialectically intertwined. The dominant model postulates a justice dividend of enhanced and enabled collective agency if individual agency is emancipated. The countermodel postulates an enhanced and enabled individual agency if collective agency is emancipated. Both are premised on the inadequacy of the countermodel, and neither is designed to confront the inadequacy of emancipation itself. Neither attempts to model the disabling that is produced through what is emancipated. Both simply pass that disabling on as an externality for succeeding generations.

Now, we confront the fact that our centuries-long emancipatory quest has succeeded in producing a form of human mastery over nature. Indeed, Žižek has argued that we have eradicated nature as a category because at least the biosphere can no longer be viewed as a causal sphere distinct from human agency (Žižek 2008:433–443). However, this mastery is impotent to address its own self-destructive capacity. Our emancipated science allows us to determine that we have overstepped the planetary boundaries requisite for the survival and flourishing of life, but nevertheless our emancipated agency surges onward, increasing the very exploitation and consumption that produces this outcome.

What is revealed about the justice foundation of our current economy that it has been constructed so as to overstep planetary boundaries? Nothing less than that a second Copernican Revolution in our modeling of justice is required. A second Copernican Revolution in justice theory would again reorient us toward the relationship between our subjectivity and justice, but now with our subjectivity as the source of the injustice to be overcome. Like Copernicus' own revolution, it would decenter us—and place us back into a cosmic narrative from which the Enlightenment wrenched us. What justice theory would now model is how to rectify injustices done in the pursuit of emancipated economic agency. It would recognize that the celebration of freedom by the economic agent, when it operates at the present scale, enslaves the other both now and in the future by altering the biogeochemical processes of life. In particular, the scale of time over which the externalized burden of our emancipation accumulates is longer than a human lifetime. For the human agent making choices now, the injustice of that emancipated agency appears ghost-like or spectral (Derrida 1994). It is a shadow of the past and a legacy to haunt the future, but it is not recognized to be a living presence now. Yet it is precisely if that injustice is allowed to remain merely spectral that it can eventually erupt into the present as an overwhelming and insurmountable burden. Is it not remarkable to realize that fossil fuels are the freed and burnt ghosts of dead creatures from long ago?

The second Copernican Revolution in modeling justice is arguably already under way as the narratives of cosmic and biological evolution profoundly change our understanding of what it means to be human (see Chaisson 2006; Wilson 2004). We are now seeking to recapture ancient wisdoms that understood human embeddedness. In this volume, Timmerman recovers justice conceptions that were delegitimated as emancipation gained preeminence. Yet it would be paradoxical and ultimately unsuccessful to deny the Enlightenment legacy in an effort to overcome the legacy of the Enlightenment. As we identify the justice deficit of the contemporary economy, we are not only continuing to live with the legacy of the

Enlightenment that will be left for generations, but we are also applying the knowledge drawn from what the Enlightenment made possible. That legacy is not one of a particular continent or culture, but it is now borne and shared by all of life.

Thus, the second Copernican Revolution is not so much a substitution for the first as it is a repositioning of it. If the metaphor may be pursued, whereas Copernicus shifted perspective as to what lay at the center of our own solar system and empowered us to see better where we stand in relation to the universe around us, subsequent learning, building on Copernicus, has repositioned the significance and centrality of our own solar system in relation to the constellations (billions of galaxies) around us. A second Copernican Revolution in modeling justice also would reposition the centrality of our own pursuit of freedom in relation to a wider constellation of justice considerations (the justice matrix), which bear upon both our stewardship of the conditions for the possibility of life itself and the incontrovertible contemporary reality that each of our free choices impinges upon every other being.

These last affirmations deserve additional context. When the Enlightenment project took hold, the environment in which our freedom was to be exercised seemed inconceivably vast and capable of absorbing indefinitely any human intervention. Furthermore, it appeared plausible to imagine that there was a sphere of individual autonomy in which one's own choices could be held discrete from effects on others—so much so that the formula "my freedom ends where yours begins" did not seem to beg a question, namely whether there was any point at which my freedom can be disassociated from yours. Common examples to illustrate that the formula made sense included that certain economic preferences of mine—to purchase or consume a widget—had no bearing on your freedom and thus give rise to a protected sphere of right.

It is true that, at least by the time of Malthus in the eighteenth century, human agency generating environmental catastrophe had crept into the ethos of Enlightenment thought. Thus, the capacity of the environment to absorb human action began to take on perceived but, it seemed, distant limits. Similarly, the externalities generated for others by individual choices came more and more sharply into view as "market failures." These were nevertheless treated as exceptions to the general principle that discrete choices could be assembled into free transactions that contained the meeting of the minds—the idea of an efficient contract. However, Malthusian warnings and incremental efforts to regulate externalities have ceded now to a starker and more radical post-Enlightenment ethos. Human action not only affects the environment, it now has responsibility for it. Externalities are not exceptions to free transactions; they arise from all transactions. This was not always so. It is the result of the Faustian bargain we have struck, perhaps unwittingly, to gain mastery of nature.

There is thus a deep justice deficit in the operation of the existing economy. What is being held to account as the singular justice metric is the capacity for free choice. How the exercise of free choice disables future generations and encroaches upon a safe operating space for life is not held to account. One prerequisite for shifting from the existing economy to an ecological economy would be to provide an accounting for the injustice we do in exercising free choice. This would require, in particular, that we be signaled in our transactions, not only as to the resources we will have to provide so as to allow our choice (price), but also as to the resources we will have to forgo so as to respect planetary boundaries and as to the resources we will have to invest to restore what we are in the midst of damaging. Each transaction would have to carry with it not only a price signal but also signal as to its environmental and social cost.

Consistent with the metatheoretical approach described previously, we should resist the temptation to erect for ecological economics a new singular theory of justice to replace the myriad efforts to establish a theory of justice underpinning a market economy functioning within liberal democratic institutions. Indeed, if respect for planetary boundaries were to be consecrated as the new grounding principle to replace freedom or emancipation, this too could and indeed would work injustice. There is a particular form of blindness that can come with rendering justice purely out of fear of destruction. Everything else can fade from view. Anything might be done in the name of holding off apocalypse. We can escape from this paradox, not by escaping the laws of the universe—from which there is no escape—but by enabling a repositioning of human agency, to use Bruce Jennings' term, as relational and embedded. This cannot be achieved simply by having us all adopt such a position—which we will not do—but rather by enabling and reinforcing signals of behavior that will move us in this direction. A justice model for ecological economics would seek to hold the economy to account for producing outcomes consistent with planetary boundaries, but it would also hold it to account for having to address other legitimate claims, including those arising from human agency and inter-subjective needs and relations.

Furthermore, a justice model for ecological economics will be centrally focused upon holding the economy to account in its transition to an eco-logical economy. This means accounting for the burden of accumulated debt from the existing transgression of planetary boundaries, such as the enormous carbon debt owed by the industrialized countries for filling the earth's carbon sinks to overflowing. The just way to manage and discharge that debt as well as to address the accumulation of new debt becomes a cen-tral governance concern for ecological economics. It also means account-ing for the way our modes of governance will address the resistance of existing economic actors to any transition. It means finally accounting for the inevitable deficit that will arise—and is already arising—from the gap between what we imagine an ecological economy might be and what it actually will produce.

The transition to an ecological economy will require new kinds of eco-nomic actors, heightened signaling capacity to steer multiple indicators of just economic outcomes, and a highly differentiated model of fair shares of planetary resources. The forms of stewardship this entails suggests that a fiduciary concept focused on these requisite forms of economic capacity will lie at the center of the justice model. This fiduciary concept will have to bear the burden of accounting for and rendering justice to the incommen-surable claims made upon the economy. Not only can the avoidance of one planetary boundary encroach upon another, but the avoidance of planetary boundaries will encroach upon other deeply held justice claims, such as the claim that all have a right to gain their highest possible remuneration.

5. HOW IS THE SHIFT FROM MARKET ECONOMICS TO ECOLOGICAL ECONOMICS A SHIFT FROM THEORY TO METATHEORY?

A capitalist market economy that seeks growth is tied to the liberal justice model that focuses on the emancipation of individual agency. It leaves out of account our stewardship of the environment and the economy's systemic creation of externalities. The socialist justice model, although it reverses the locus of emancipation, is nevertheless tied to the same process of mastery of nature. The contemporary confluence of capitalism and socialism in a global economy—not only in China but in Brazil, India, and Russia—signals that the older ideological locus of justice antinomies has lost its force (Hardt and Negri 2000). With the fall of communism, the justice matrix shifted its axis away from the antinomies of individual and collective emancipation toward the antinomies of positive (self-affirming) and negative (other-regarding) emancipation. Capitalism globalized, but it was not the end of history as Fukuyama (2011) would have it. The putative triumph of individual (bourgeois) emancipation over collective emancipation revealed with even greater stringency the deficit of other-regarding, self-denying, negative emancipatory claims confronting the legacy of self-affirming bourgeois freedom.

Ecological economics is something other than capitalism or socialism. It is an attempt to bring to bear what we know and do not know of the household we manage—now our whole environment (oikos-logos)—on the norms governing our free choices (oikos-nomos). We seek to steer those free choices according to the contours of scientific knowledge and uncertainty concerning the effects we produce on life and its prospects. The effort to place economy within ecology is metatheoretical, not simply in the sense that it is transdisciplinary (linking social and natural sciences). In a more radical sense, it requires theorizing how bodies of thought can be made to interact and interoperate.[7]

An illustration may help to explain the foregoing. Suppose that an oil company executive explains that the problem of crossing planetary boundaries for atmospheric CO_2 and other GHGs, as important and legitimate as it is to address, must simply give way to the economic reality. Given increasing energy demand and the slow pace at which renewables are coming on stream, we will have increasing global fossil fuel consumption over the next fifty years. In response, a climate scientist responds that, as important and legitimate as it is to consider how to address increasing energy demand, we cannot allow increasing fossil fuel consumption over the next fifty years if we are to stabilize GHGs at acceptable levels. As it stands, these two discourses are superimposed upon each other, and the market discourse trumps by default. Ecological economics seeks to bring these two discourses together and to establish principles of justice. According to these principles, the scientific discourse could gain capacity not only to trump the market discourse, but also to take account of what is lost when one discourse trumps the other.

How does this debate look from a metatheoretical perspective? The justice claims made by the executive are situated principally within the subjective dimension (efficiency imperative), with some reliance on the conventional (freedom of contract) and implicit reference to the emancipatory (provide for human energy needs). The justice claims made by the climate scientist are principally in the metaphysical dimension (life is sacred), with almost no anchoring in the conventional (given limited climate norms and a gap in global governance) and a countervailing positive emancipatory claim (free future generations from the unjust burden of the legacy we are bestowing upon them) as well as a negative emancipatory claim (we must confine ourselves to planetary boundaries). Within the existing justice matrix, the oil company executive wins, largely because the only signals of value that are made and received have to do with price and its relation to supply and demand. On the conventional economy's mapping of efficiency and equity, the sanctity of life and what we owe to future generations (principally the metaphysical) remains tributary and ancillary to the subjective and the conventional; it is not even signaled in transactions.

The question becomes whether and how the contemporary justice matrix could be reconfigured to give a certain priority to the metaphysical over the subjective where planetary boundaries are at stake. At minimum, a metatheoretical approach reveals that if countervailing subjective justice claims were overcome, justice would have to be rendered to those whose claims were thus sacrificed. More ambitiously, a metatheoretical approach raises the question as to whether there is inadequate signaling within the existing economy for justice claims that are currently in deficit.

6. HOW THE FIDUCIARY PRINCIPLE OPERATES WITHIN A METATHEORETICAL CONCEPTION OF JUSTICE

To reconfigure the justice matrix, we must transform how we imagine our relationships with others and the world. This will require a new relationship to time and to the use and exchange of resources that eclipses the current formal use of law. Legal norms themselves will have to shift in real time in feedback to indicators warning of approaching planetary boundaries. All actors will have to be enabled to respond collectively to these signals. Fiduciary relationships can be used to model the transformation required.

A fiduciary relationship does not involve a self-reproducing norm that cuts off justice claims. Rather, it demands of all participants, even the courts, to conceive of the relationship, both in its process of articulation and its outcomes, as determined at the peril of the moment: exactly as the justice burdens we create are encountered. Indeed, that moment arises when authority is exercised in real time rather than in relation to a fixed norm, because to stand in that moment is to acknowledge the inadequacy of any outcome. For the fiduciary standard of care to be upheld, all of the fiduciary's real-time resources must effectively be made available to address all claims. Despite the fact the fiduciary will rely upon bodies of specialized knowledge and metrics to measure relevant impacts, there is no way of fully automating either the fiduciary process or outcome. The fiduciary and the beneficiary are involved in a relationship—the reproduction of which subjects them both to running risks that can negate the legitimacy of the entire interaction. This is true of doctor and patient, for example, but it has also become true of the relationship between all of us and future generations.

The fiduciary relationship necessarily binds the resources of all participants, in contrast to delimited legal norms, outside the scope of immediately identifiable needs. The fiduciary relationship represents disparate justice claims or moments, and it is the means by which an otherness as relationship is maintained in requisite tension. The fiduciary's obligation is to an outcome that exceeds the limits of fulfilling a prescriptive catalogue of duties; the prescriptive dimension, despite at times being questioned, is upheld regularly in the courts' review of fiduciary inadequacies.

Thus, legal recognition of the fiduciary form comes closest to a methodology that would preserve the differentiation of incommensurable claims at the expense of formal, delimited law. It is a kind of productive paradox within the law, but one that can dissolve readily if the law cedes to the temptation to delimit narrow fiduciary norms.

To return a second time to the example of the oil executive and the scientist, to place the resolution of the conflict between discourses into a fiduciary setting would require differentiating and rendering justice to both sets of claims. This might mean, for example, charging the executive (or allied social interests) with the task of investing resources to enable a decrease in consumption and charging the scientist (or allied social interests) with the task of meeting energy needs more quickly through sustainable means at the same time as helping to identify unsustainable consumption that is to be excluded.

An articulation of the fiduciary principle is central to the justice founda-tions of ecological economics because it allows market outcomes to be situ-ated against what we know concerning the implications for the prospects of life of each transaction. Because we do now know that every transaction does indeed produce (often) negative and (sometimes) positive social and environmental externalities, each transaction also comes bundled, so to speak, with a fiduciary burden. If this fiduciary burden is to be met, each of us must receive through other fiduciaries a signal as to the scope of the burden we are creating by our choices, such that each of our choices could themselves become fiduciary in character.

7. HOW CAN THE CRISES OF THE EXISTING ECONOMY ENABLE THE TRANSITION TO AN ECOLOGICAL ECONOMY?

If we must enable all transactions in an ecological economy to carry a fiduciary burden respecting the prospects of life, the difficult question of transitional justice arises: how can all of the investments made in the exist-ing market economy be shifted to an ecological economy within a genera-tion and with justice? Habermas's treatment of legitimation crises focuses on capitalism's almost infinite malleability to withstand contradictions (Habermas 1974). This stands in contrast to earlier Marxist accounts of capitalism that suggested that the crises of capitalism would produce the conditions to overcome it. History now suggests that capitalism is resilient and can shift its forms and functions. The working hypothesis of this sec-tion is that ecological economics would thus seek to penetrate and imbue capitalism rather than to overcome it.

Typically, the problem of transitional justice has been addressed retro-spectively: given a great past injustice, such as apartheid, and a new social order, how is justice to be done (e.g., truth and reconciliation committees, lustration, trials, institutional reform)?[8] However, the problem of transi-tional justice faced by ecological economics is prospective: given that the existing market economy is not ecologically sustainable, how is the injus-tice now being done to the future to be addressed?

There are two premises underlying this version of the problem of transi-tional justice. First, it is assumed that we can have a clear understanding of what an ecological economy would be—namely, one sustainably operating within planetary boundaries. It is not presumed, however, that we have a full understanding of the kinds of social and institutional arrangements that would allow an ecological economy to come into being and persist. Second, given our current scientific knowledge of the existing and widening transgression of planetary boundaries, it is assumed that the transition to an ecological economy is not in the nature of an incremental reform of existing economic practice. That is, if it could be assumed that the existing market economy is operating in a manner that is close to achieving an ecological economy and only requires regulatory adjustments, or that the transition to an ecological economy could take place gradually over a long period of time, our question would not focus on proximate transitional justice for the economy as a whole. Rather, the question would focus on incremental adjustments to regulatory regimes and managing implementation.

To illustrate the second premise, if it were assumed by contrast that the existing market economy could be brought sustainably within planetary boundaries by creating a carbon market and analogous markets for all ecosystem services—by putting a price on all environmental externalities—then the narrower transitional justice question would become as follows: how can the design of such markets be achieved, and what justice claims have to be resolved as those markets are brought into being? Such justice claims would include addressing the problem of common but differentiated responsibilities for the accumulated externalities (i.e., must there be compensation from those who have yet to produce externalities to those who have been and are producing them?), sharing the burden of sunk costs and stranded assets in economic processes being phased out (i.e., will that social burden lie entirely where it falls?), and identifying fair timelines for implementation (i.e., what is the reasonable planning horizon within which existing economic actors can adjust to new costs?). Such questions have to do with transitional justice in the face of incremental change. However, this is not our circumstance.

There is a point at which problems of transitional justice in the face of incremental change accumulate so as to become problems of transitional justice in the face of systemic change. Thus, for example, the Montreal Protocol on Substances that Deplete the Ozone Layer (Montreal Protocol) gave rise to and sought to solve the transitional justice problems just identified for a discrete set of sectors of the economy. The phase-out of chlorofluorocarbons (CFCs) and hydrochlorofluorocarbons (HCFCs) affects the manufacture of refrigerants, solvents, and a narrow range of other processes. The adjustments required by the Montreal Protocol, while dramatic for those sectors, have been readily absorbed by the market economy as a whole. The Montreal Protocol does not therefore in itself give rise to systemic change. By contrast, the United Nations Framework Convention on Climate Change (UNFCCC) gives rise to and seeks to solve transitional justice problems for the economy as a whole, given that GHG emissions arise from the production and consumption of energy and that the entire economy relies upon energy infrastructure. Even this observation would not yet entail that the UNFCCC gives rise to the problem of transitional justice in the face of systemic change. It could simply mean, for example, that a carbon or GHG tax would have to be applied so as produce an incremental shift in price signals, thereby allowing the existing market economy to align its use of resources with avoidance of the costs of climate change.

However, the UNFCCC does present certain hallmarks of the problem of transitional justice in the face of systemic change and in turn provides an indication that the even more substantial shift to an ecological economy—which could not be achieved by successful implementation of the UNFCCC alone—would indeed represent systemic change. Whereas the Montreal Protocol had to overcome strategic behavior and collective action on the part of those who were extracting rents from GHGs and HCFCs (e.g., there was a period of industry investment in ozone depletion denial and organized effort to discredit the science underpinning the Convention), the application of cost-benefit analysis taking account of the precautionary principle was straightforward: the potential harm to many overcame the vested interests of a few. By contrast, the significant range of social investments—rational and irrational—in existing modes of consumption (and indeed, in shared hopes for the prospects of future consumption) are widely perceived to be threatened by the UNFCCC process. Furthermore, the collective willingness to put off to future generations the accumulating and manifest damage to the biosphere suggests that we are in the midst of constructing a proximate future crisis on the implicit calculation that we should all seek to withdraw rents from the existing economy while we can. Finally, it is telling that there is increasingly common recourse to arguments that we need not address the looming catastrophe now because we will find and deploy cheaper technological solutions in the future to reverse climate change. Betting on the future to redeem the past suggests that we do not believe that present shifts in the economy would be incremental in the relevant sense. That is, we do not accept the anticipated costs of mitigating climate change now because we assume that this would impair our ability to benefit from the existing economy. In short, we are signaling the present incapacity of the existing market economy to absorb and avoid the costs of climate change and therefore are anticipating or courting systemic, albeit dystopian, change.

Systemic change from a market economy to an ecological economy will involve continuity with the existing market economy. This hypothesis may seem paradoxical, but the opposite hypothesis—discontinuity with the existing market economy—is on close inspection an impossibility. This is to say that the institutional and normative arrangements underpinning the market economy, as well as (and indeed most notably) the shared and generalized participation in consumption through the market economy, will not vanish at some discrete point of transition to an ecological economy. Even if the existing market economy were to be marked by energy or food crises, for example, it would still operate within those crises and establish transactions profiting from them. To date, the market economy has been capable of internalizing environmental externalities only if economic growth has not been at stake. It has proven incapable of a wholesale shift away from allowing externalities to be generated, because those externalities are intimately connected to economic growth.

Systemic change will nevertheless be accompanied by (and to a significant degree, given impetus by) a crisis of the existing market economy. Precisely because incremental change is unmanageable, the incapacity of the existing market economy to bring itself within planetary boundaries will manifest itself to the point of socially undermining the existing economy before systemic change is undertaken. That is, only at the point where there is general social disinvestment in the outcomes produced by the market economy will an ecological economy arise. The paradox to be confronted here is that consumers will cling to the market as citizens divest from it—but the consumers and citizens are one and the same.

Stewardship—or the exercise of fiduciary duty within the context of systemic change—involves seeking to fulfill a burden that can never be completely fulfilled. That is, because we cannot and should not seek to produce a discontinuity between the existing market economy and the ecological economy, the task will be to do what justice can be done within and out of a crisis of the market economy. We cannot simply posit the end state of an ecological economy and demand that it be substituted for the market economy.

It is the transformation—the in-between state—rather than the end state that becomes the focus of fiduciary duty and resources. The duty is owed to those who still rely on their consumer subjectivity and its legacy of claims. It is also owed to those seeking to overcome it and to those in the future who will not be able to rely upon it. Ascribing fiduciary resources to legacy justice claims is pure waste from the vantage point of producing an ecological economy, yet it is unavoidable as a matter of justice. Ascribing fiduciary resources to those now seeking to overcome consumer subjectivity is inherently beset with failure. Ascribing fiduciary resources to the future means enabling the fulfillment of unknown needs.

Although incremental change toward an ecological economy is here presumed to be beyond reach, given the need for systemic change, stewardship now involves in part identifying which sorts of incremental change might facilitate or at least be most consonant with systemic change. Thus, for example, whereas costing ecosystem services will not produce the requisite transition to an ecological economy because it will still place the transgression of planetary boundaries within a cost-benefit framework rather than setting a limit on transactions themselves, it would nevertheless bring within legal norms and institutional arrangements of the market economy enhanced capacity to steer within planetary boundaries. Another example is to build up and expand the fiduciary duties of existing economic actors themselves to consider, disclose, and account for their contribution of the transgression of planetary boundaries. Fiduciary stewardship also involves analyzing which existing economic processes and institutions are most prone to early system breakdown (e.g., perhaps the agricultural or energy sectors) and thus possible targets for more ambitious reform.[9]

Fiduciary stewardship also involves anticipating legal norms and institutional arrangements that could emerge through a process of systemic change and identifying their nascent forms in the existing market economy. The task would then become to help engender or reinforce those nascent forms. For example, we can safely assume that an ecological economy will require rigorous measurement of whether and to what degree planetary boundaries are being transgressed with feedback mechanisms to ensure that activities that collectively transgress are scaled back. Even if sustainable development indicators do not yet perform this function, they should be prepared and institutionalized with a view to enabling them to do so in the far more sophisticated way that will be required in the future.

Finally, and most problematically, fiduciary stewardship also involves beginning to work now on the implications of transgression of planetary boundaries—not simply from the standpoint of fending that off but also from the standpoint of modeling all of the justice claims that will arise as that unfolds. In this regard, the Cancun and Durban Summits speak volumes because the lion's share of announced new funding are targeted to "adaptation" (although whether promises will be fulfilled is another matter). The investments being made in geoengineering are another sign.

Mitigation of climate change is already being cast institutionally as beyond reach. Our current fiduciary task is taken up, therefore, from the perspective that we will continue to consume up to and past the point where the planet has been made dramatically unliveable. The starkest problem of transitional justice is to take up the burden of planning for that outcome, both because its violent consequences cannot be pretended away and because a successful transition to an ecological economy might only be achieved with such consequences in view.

NOTES

1. A classic account of the basic forms of justice can be found in chapter 5 of Aristotle's *Nicomachean Ethics*. See Aristotle (2011).

2. See also Shearman and Smith (2007), where the authors defend the thesis that democratic dysfunction over environmental issues could well lead toward authoritarianism in the absence of a major overhaul of democracy. See also Leo Hickman's 2010 interview with James Lovelock, "Fudging the Data is a Sin against Science" (*Guardian*, March 29, 2010), where Lovelock is quoted as stating: "We need a more authoritative world. We've become a sort of cheeky, egalitarian world where everyone can have their say. It's all very well, but there are certain circumstances—a war is a typical example— where you can't do that. You've got to have a few people with authority who you trust who are running it. And they should be very accountable too, of course. But it can't happen in a modern democracy. This is one of the problems. What's the alternative to democracy? There isn't one. But even the best democracies agree that when a major war approaches, democracy must be put on hold for the time being. I have a feeling that climate change may be an issue as severe as a war. It may be necessary to put democracy on hold for a while" (Hickman 2010).

3. See also Held and Hervey (2009) and Stehr (2013). Stehr wrote: "Climate researchers have evidently been impressed by Diamond's deterministic social theory. However, they have drawn the wrong conclusion, namely that only authoritarian political states guided by scientists make effective and correct decisions on the climate issue. History

teaches us that the opposite is the case. Therefore, today's China cannot serve as a model. Climate policy must be compatible with democracy; otherwise the threat to civilization will be much more than just changes to our physical environment. In short, the alternative to the abolition of democratic governance as the effective response to the societal threats that likely come with climate change is more democracy and the worldwide empowerment and enhancement of knowledgeability of individuals, groups and movements that work on environmental issues" (citations omitted).

4. See Žižek (2006). Any economy purports to render justice. It is the collective management of resources for the *oikos*—our common household or home; the sphere we inhabit. The outcomes produced in the *oikos* with those resources are affirmed as legitimate as long as its foundational principles of allocation, production, development, and governance are just. Indeed, we should observe that the terms ecology and economy suggest two standpoints on the same sphere of *oikos*. One is the *logos* (science, knowledge) of what we inhabit; the other is the *nomos* (custom, law) of what we inhabit. Both standpoints give rise to their own theories (*theorein*: to look attentively on what appears). Thus the double standpoint of logos and *nomos* (ecological economics, which could also be economic ecology) suggests a metatheory—a theory of theories.

5. See Adam Smith's *The Theory of Moral Sentiments* (1812), at paragraphs 9–13, where Smith makes clear that fellow-feeling is but the projection of our own sentiments upon the circumstances of the other should we bear them ourselves, notably against the backdrop of fear of death. This empties out the notion that we could act deeply or essentially for another.

6. On the relationship between justice and being held to account, see Butler (2005).

7. See in particular Latour (2013), who undertakes an ambitious modeling of how bodies of thought are networked but also remain opaque to each other.

8. See, for example, the work of the International Center for Transitional Justice (ictj.org).

9. Note, for example, that Canada's National Round Table on the Environment and the Economy (NRTEE), which had arguably sought to identify such opportunities, shifted toward identifying how to prosper from climate change: see National Round Table on the Environment and the Economy (2011), *Climate Prosperity: The Economic Risks and Opportunities of Climate Change for Canada*. As it was being abolished, NRTEE did publish a final series of four reports in 2012 signaling the need for significant change in the economy in response to the climate crisis.

REFERENCES

Aristotle. 2011. *Aristotle's Nicomachean Ethics*. Translated by Robert C. Bartlett and Susan D. Collins. Chicago: University of Chicago Press.

Beeson, Mark. 2010. "The Coming of Environmental Authoritarianism." *Environmental Politics* 19 (2): 276–294. doi:10.1080/09644010903576918.

Burnell, Peter. 2009. "Should Democratization and Climate Justice Go Hand in Hand?" *Böll Thema* 2: 17–18.

Butler, Judith. 2005. *Giving an Account of Oneself*. New York: Fordham University Press.

Chaisson, Eric. 2006. *Epic of Evolution: Seven Ages of the Cosmos*. New York: Columbia University Press.

Derrida, Jacques. 1994. *Specters of Marx: The State of the Debt, the Work of Mourning, and the New International*. New York: Routledge.

Fukuyama, Francis. 2011. *The Origins of Political Order: From Prehuman Times to the French Revolution*. New York: Farrar, Straus and Giroux.

Habermas, Jürgen. 1974. *Legitimation Crisis*. Boston: Beacon Press.

Hardt, Michael , and Antonio Negri. 2000. *Empire*. Cambridge, MA: Harvard University Press.

Held, David, and Angus Fane Hervey. 2009. *Democracy, Climate Change and Global Governance: Democratic Agency and the Policy Menu Ahead*. London: Policy Network. http://www.policy-network.net/publications_detail.aspx?ID=3406.

Hickman, Leo. 2010. "James Lovelock: 'Fudging Data Is a Sin against Science.'" *The Guardian*, March 29. http://www.theguardian.com/environment/2010/mar/29/james-lovelock.

Horkheimer, M., and T. W. Adorno. 1982. *Dialectic of Enlightenment*. Translated by John Cumming. New York: Continuum.

Kant, Immanuel. 2003. *Critique of Pure Reason*. Translated by Norman Kemp Smith. Revised 2nd ed. Basingstoke, UK: Palgrave Macmillan.

Kleinhans, Martha-Marie, and Roderick A. Macdonald. 1997. "What Is a Critical Legal Pluralism?" *Canadian Journal of Law and Society* 12: 25–46.

Latour, Bruno. 2013. *An Inquiry into Modes of Existence: An Anthropology of the Moderns*. Cambridge, MA: Harvard University Press.

National Round Table on the Environment and the Economy. 2011. *Climate Prosperity: The Economic Risks and Opportunities of Climate Change for Canada*. Accessed January 8, 2015. http://collectionscanada.gc.ca/webarchives2/20130322143042/http://nrtee-trnee.ca/climate/climate-prosperity.

Posner, Eric A., and David A. Weisbach. 2010. *Climate Change Justice*. Princeton, NJ: Princeton University Press.

Rawls, John. 1993. *Political Liberalism*. New York: Columbia University Press.

Schlosberg, David. 2012. "Climate Justice and Capabilities: A Framework for Adaptation Policy." *Ethics & International Affairs* 26 (4): 445–461. doi: doi:10.1017/S0892679412000615.

Sen, Amartya. 1999. *Development as Freedom*. New York: Knopf.

Sen, Amartya. 2009. *The Idea of Justice*. Cambridge, MA: Harvard University Press.

Shearman, David J. C., and Joseph Wayne Smith. 2007. *The Climate Change Challenge and the Failure of Democracy*. Westport, CT: Praeger Publishers.

Smith, Adam. 1812. *The Theory of Moral Sentiments*. 11th ed. London.

Stehr, Nico. 2013. "An Inconvenient Democracy: Knowledge and Climate Change." *Society* 50 (1): 55–60. doi: 10.1007/s12115-012-9610-4.

Wilson, Edward O. 2004. *On Human Nature*. Revised ed. Cambridge, MA: Harvard University Press.

Žižek, Slavoj. 2006. *The Parallax View*. Cambridge, MA: MIT Press.

Žižek, Slavoj. 2008. *In Defense of Lost Causes*. London: Verso.

PART II

MEASUREMENTS: UNDERSTANDING AND MAPPING WHERE WE ARE

INTRODUCTION AND CHAPTER SUMMARIES

A prerequisite step in our shift to an ecological economy—one that recognizes and undertakes to preserve the environmental and social foundation of the human economy—is a rethinking of concepts and goals for human beings and actions in the biophysical environment. We create an awareness of the condition of our biophysical environment and our relationship to it by developing and maintaining models in our mind. In our current modern world, this description of the environment is accomplished through scientific modeling, although alternative ways of describing reality exist in other societies, past and present, as Peter Timmerman noted in chapter 1. It has also been one of the many claims of standard economics that price represents an ideal measure or "metric"—anything that cannot be measured according to this metric is (temporarily) in outer darkness. Part of the alternative proposed by ecological economics is to present other metrics that incorporate aspects of the world that are underincorporated or unincorporated in current economic thinking and practice. Shifts in measurement accompany or prompt shifts in worldviews.

The current, widely shared scientific model of the environment—indeed, of reality—is continually updated through sensory feedback from the environment. The immense scale of our environmental concern and our scientific method of fact-based reasoning mean that a good part of this feedback is obtained through quantification. Quantification began its ascent in Western society during the late Middle Ages and the Renaissance, which led to the advancement of science and technology thereafter (Crosby 1997; Frängsmyr, Heilbron, and Rider 1990). Nowadays, the quantitative toolkit ranges from simple counting to sophisticated instrument detection and statistical techniques, yielding data that is at times copious and abstruse.

The challenge of handling quantitative data is to transform them into useful information, then use this information to advance knowledge and inform human action. In other words, how do we interpret and use the data at hand in a meaningful way? The steps entailed in quantitative data analysis—from problem framing to study design, actual data collection and analysis, and finally to interpretation and synthesis with existing knowledge—require sound judgment on the part of the researcher. This latitude in quantitative data analysis for human judgment turns the skills and techniques involved into, as it were, an art.

Measuring and assessing the condition of the biosphere—a complex system—further complicates the matter of quantitative data analysis. It involves issues such as identification of key parameters for monitoring, managing multiple temporal and spatial scales, interpretation of measurements, and the construction and use of composite indicators. The use of measurement systems to monitor and rein in human impact on the biosphere requires the identification of thresholds, an act that is not without its ethical implications. Indeed, the "planetary boundaries" created by Rockström et al. (2009) is an example of the type of metrics that could be created for framing and directing our response to our environmental predicament.

The abstract modeling of the environment and quantification of environmental data should not distract us from the fact that the human individual and economy are embedded in the biophysical environment. As we rely on scientific methods, with their reductionist tendencies, to describe and understand the environment, we need to remain cognizant of the fact that human existence is inextricably linked to the environment—from our need for air, water, and food. Because human society is a subsystem of the environment, the condition of the former is tied to that of the latter, so ultimately we are concerned with the joint condition of the two.

A possible way of imagining this connection is through the concept of health. The health concept is commonplace in the context of humans and most people would have an understanding of it, at least from experiencing their own health or lack thereof. Beginning around 1990, the use of the health concept has been increasingly applied to ecological systems, especially in reference to their conceptualization as "ecosystems," which gained a foothold in the scientific community at the same time. The use of health to describe the functioning of complex systems like ecosystems is appealing from an intuitive perspective, but it is problematic from a strictly scientific perspective. Health refers to a (perceived) state of being that is based on interpretation of information from an array of quantitative and qualitative indicators. However, due to variations in context and human judgment, no exact and universal definition of health exists. In other words, health is a connotative rather than a definitive concept, which calls for a circumspect and judicious approach. This means that we need to be cautious when using ecosystem health as a scientific concept and be cognizant and open to the multitude of possibilities to which ecosystems can evolve.

SUMMARY OF CHAPTERS

"Measurement of Essential Indicators in Ecological Economics," by Mark S. Goldberg and Geoffrey Garver

The authors propose a methodological framework for measurements to support the development of indicators relevant to ecological economics, and they discuss the issues in developing and interpreting composite indicators. For indicators in ecological economics to accurately gauge progress and assess impacts, it is essential to understand what processes or factors (often referred to as "drivers") underlie or are associated with the values they take. To fully appreciate these processes and interrelationships, sound scientific principles must be followed so that the measurement process leads to accurate values; that is, the measurements must be both valid (measure what they purport to measure) and reliable (the variability in the values obtained are sufficiently small to provide a meaningful interpretation). These principles apply to measuring both the key processes and their drivers. The authors discuss the elements that constitute their proposed methodological framework for measurements and indicators: the context of ethics, justice, and governance; the scope of parameters under consideration; the temporal and spatial scale; the commensurability of constituent measurements in forming a composite indicator; the nature and purpose of the measurement itself; and the issue of uncertainty and interactions of complex systems from which measurements are taken. They discuss the inherent problems in using indicators that are mixtures of disparate variables and recommend that indicators be used with great care and in limited circumstances.

"Boundaries and Indicators: Conceptualizing and Measuring Progress Toward an Economy of Right Relationship Constrained by Global Ecological Limits," by Geoffrey Garver and Mark S. Goldberg

In their chapter, Garver and Goldberg discuss the complex considerations behind designing a new set of indicators for governance in a new, ecological economy—in particular, the issues of scale, distribution, and efficiency of the human economy. Using the concept of safe operating space and planetary boundaries developed by Rockström et al. (2009) and the premise of the right relationship between humans and the biosphere developed by Brown and Garver (2009) as the contextual framework, Garver and Goldberg focus on the issue of governance and argue that the planetary boundaries call for a governance regime using a more refined set of indicators. They propose ten features that these indicators for economic and ecological governance should possess. Three aspects of planetary boundaries—atmospheric concentration of greenhouse gases, nitrogen loading, and biodiversity—are discussed in detail to illustrate the process of developing indicators. Garver and Goldberg conclude with a call for an iterative approach to developing and using indicators to keep up with an evolving understanding of the planetary system and a dynamic governance context.

"Revisiting the Metaphor of Human Health for Assessing Ecological Systems and Its Application to Ecological Economics," by Mark S. Goldberg, Geoffrey Garver, and Nancy E. Mayo

In this chapter, the authors discuss the notion of health as applied to humans and to ecosystems, and they explain how in both domains the state of health cannot be adequately defined or assessed using scientific terms and measures. They show that the notion of human health is elusive and the various definitions that have been attempted have serious shortcomings. Importantly, the myriad attributes and domains that make up the concept of human health cannot be measured uniquely in any individual, and there is no consensus as to how to uniquely define human health. Health goes beyond the internal signs reflected by physiological and pathological parameters measured by physicians, and even goes beyond the exteriorized signs of disability; and can even include the concept of "well-being," and how a person "feels" about their health. Moreover, health is an evolving process, and individuals change in different ways through time. In short, human health in its entirety cannot be measured in a specific individual.

The authors thus conclude that defining ecosystem health by appealing to the analogy of human health is incorrect—and certainly incorrect when considering only physicians as diagnosticians and healers. Ecologists acting as physicians to diagnose and correct pathology are of course correct and essential. Ecologists have developed myriad indices to measure various attributes of ecosystems. In parallel with humans, it is unlikely that a finite set of indicators can be developed or measured to be able to claim that an ecosystem is "healthy." More importantly, assessments of ecosystem function and states do not require a clear definition of ecosystem health. In particular, complex indices that combine elementary ones to measure ecosystem health cannot measure all of the dimensions in complex ecosystems, and the use of complex indicators must be benchmarked; claiming that an ecosystem is healthy based on these types of indices can be fraught with error. The authors conclude that using these indicators in ecological economics—especially in terms of monitoring the effects of human activities on ecosystems and species at the local, regional, and global levels—requires a judicious choice of objectives as to which indicators are to be measured for the purposes of remediation and for making statements of policy.

"Following in Aldo Leopold's Footsteps: Humans-in-Ecosystem and Implications for Ecosystem Health," by Qi Feng Lin and James W. Fyles

In this chapter, Lin and Fyles retrace the thinking of Aldo Leopold on conceiving humans as plain members and citizens of the biotic community, perceiving the intrinsic character of the land as "land health," and considering how these two ideas led to his famous "land ethic." Leopold's land ethic urges people to expand their relationship with land beyond economics to include "integrity, stability, and beauty." The authors then apply his thinking to the context of ecosystem and ecosystem health by considering humans as part of ecosystems, and the resulting implications for ecosystem health. With this thinking, the authors return to Leopold by studying one of his essays in *A Sand County Almanac* (1949). In "A Mighty Fortress," Leopold mentioned how his woodlot, having been visited by various tree diseases, became a rich habitat for wildlife. This essay underscores the multiplicity of perspectives in an ecosystem and its complex nature, which in turn challenges humans to learn the richness and meaning of the concept of health. Put another way, and echoing the view of the previous chapter, health cannot be adequately portrayed by using only medical science. The authors conclude with a call for supplementing the scientific, rational mode of perceiving reality and human action with thinking from the arts and the humanities.

REFERENCES

Brown, Peter G., and Geoffrey Garver. 2009. *Right Relationship: Building a Whole Earth Economy*. San Francisco: Berrett-Koehler.

Crosby, Alfred W. 1997. *The Measure of Reality: Quantification and Western Society, 1250–1600*. Cambridge, UK: Cambridge University Press.

Frängsmyr, Tore, J. L. Heilbron, and Robin E. Rider, eds. 1990. *The Quantifying Spirit in the 18th Century*. Berkeley: University of California Press.

Leopold, Aldo. 1949. *A Sand County Almanac and Sketches Here and There*. New York: Oxford University Press.

Rockström, Johan, Will Steffen, Kevin Noone, Asa Persson, F. Stuart Chapin, III, Eric F. Lambin, Timothy M. Lenton, Marten Scheffer, Carl Folke, Hans Joachim Schellnhuber, Bjorn Nykvist, Cynthia A. de Wit, Terry Hughes, Sander van der Leeuw, Henning Rodhe, Sverker Sorlin, Peter K. Snyder, Robert Costanza, Uno Svedin, Malin Falkenmark, Louise Karlberg, Robert W. Corell, Victoria J. Fabry, James Hansen, Brian Walker, Diana Liverman, Katherine Richardson, Paul Crutzen, and Jonathan A. Foley. 2009. "Planetary Boundaries: Exploring the Safe Operating Space for Humanity." *Ecology and Society* 14 (2): 32.

CHAPTER FOUR

Measurement of Essential Indicators in Ecological Economics

MARK S. GOLDBERG AND GEOFFREY GARVER

1. INTRODUCTION

Ecological economics is concerned with incorporating into economic systems ecological and other constraints that neoclassical economics ignores. In this chapter, we discuss essential methodological considerations in making measurements to derive indicators that can be used to monitor the course of economic activity and the state of the environment so as to determine whether these constraints are being met.

For indicators to be useful, it is first important to define clearly why one wishes to measure them (in scientific inquiries, this would be referred to as the "objectives" of a research study or program) and then to ensure that the measurements are indeed measuring what they purport to measure (i.e., validity) and are consistent (i.e., reliability) (Koepsell and Weiss 2004). Contextual considerations affect the framing of these objectives and usually go to the heart of the questions being asked. Understanding what processes underlie indicators and what other factors or processes are associated with them (often referred to as "drivers") are also important.

For example, Ehrlich and Holdren (1972) suggested that the size of the human population, its affluence, and technology are essential drivers of environmental impacts, although many others will be important in specific circumstances. An example comes from air pollution and its effects on ecosystems (Lovett et al. 2009), where it is clear that combustion-related pollutants are due primarily to human activities. The interrelationships between different processes are often complex, requiring systems analysis or other techniques to help unravel the intricacies.

Challenges in measurement are compounded by the scale dependency of many ecological processes, varying from the local to the regional to the global. Many local processes scale up so as to have important aggregate effects at the global level. For example, local anthropogenic emissions of greenhouse gases, which vary geographically, in the aggregate are leading to changes in climate and other sequelae that are important at regional and global scales; in turn, those regional or global changes lead to impacts on life systems that vary across the globe (Intergovernmental Panel on Climate Change 2007). Measurement of the impact on ecosystems and species is also complicated by their ability to adapt. As well, there are important challenges in comparing metrics from ecosystems with different characteristics (commensurability). Heterogeneity of ecosystems ("a forest is not just a forest") makes for diversity, but it may make difficult the interpretation of measurements and affect their generalizability to other systems, especially if the natural history of the ecosystem is not well understood or if there are no benchmarks for comparison. The heterogeneity of ecosystems and their interactions complicates the development of appropriate indicators that are important, valid, and reproducible.

A proposed methodological framework for measurements to support the development of indicators relevant to ecological economics is presented in the following sections. The framework has five elements: (1) contextual considerations, (2) considerations of scale and dimension, (3) considerations of scope, (4) considerations of commensurability, and (5) considerations of the interacting systems so that the paths leading to the indicator can be identified and targeted. We also discuss methodological aspects regarding what constitutes accurate measurements, as well as the myriad problems in interpreting indicators that comprise other primary ones (i.e., complex indicators).

2. CONSIDERATIONS OF CONTEXT

Contextual considerations concern the relationship of indicators to desired outcomes, as defined in terms of the objectives of ecological economics, including those derived from criteria based on considerations of ethics and justice. Questions regarding the application of indicators in a governance context are important; indicators must be developed not only in view of ethical and justice criteria, but also in view of practical issues regarding their application in governance and the means by which they are adopted and communicated.

One school of thought suggests that the state of ecosystems can be defined in terms of "ecosystem health" (Costanza 1992; Costanza and Mageau 1999; Jørgensen, Xu, and Costanza 2010; Rapport 1992; Suter 1993). Unfortunately, we show in chapter 6 that this concept, which has been developed in analogy with human health, is in fact vague and in practical terms of limited usefulness because of the difficulty of measuring all key domains. The essential consideration is that health, while an important concept, cannot be measured uniquely: it comprises multidimensional indicators that cannot be easily defined or measured. Indeed, Jorgensen indicated that "it is clear today that it is not possible to find one indicator or even a few indicators that can be used generally, or as some naively thought when ecosystem health assessment . . . was introduced" (Jørgensen 2010b). Thus, although it has been suggested that "ecosystem health [is] a comprehensive, multiscale, dynamic, hierarchical measure of system resilience, organization, and vigor" (Costanza and Mageau 1999), we argue that these constructs of resilience, organization, and vigor cannot be defined simply nor measured easily so as to portray the full range of function. Affirming these three specific attributes, or indeed any general combination, does not imply health.

To assess the contextual basis of the human–Earth relationship, biological and thermodynamic indicators (e.g., those in the eight categories in Jørgensen et al. [2010, 12–14]), more macroscale indicators like human appropriation of net primary production and ecosystem indicators,[1] and social and economic indicators are certainly relevant but they do not define the entire space of function. In some sense, this should be obvious, although complex: the planet functions on many different spatial and temporal scales, and the processes that define these functions are myriad and highly interrelated, with many feedback loops built in. Moreover, the heterogeneity of ecosystems as well as "natural" secular changes makes it difficult to find norms by which to compare.

In light of these challenges, objectives derived from principles of ecological economics may provide context so as to allow the range of complexity with which indicators must contend to be reasonably managed. For example, the planetary boundaries concept (Rockström et al. 2009a, 2009b) discussed here and in the next chapter, provides a framework for deriving indicators based on the objective of maintaining the human economy so as to avoid changing the state of the planet to one that is not suitable for survival of many species, including humans.

3. CONSIDERATIONS OF SCOPE

Considerations of scope concern the range of parameters for which indicators are sought. For example, if the context is defined by an ethic of right relationship (Brown and Garver 2009), then it may be relevant to examine a broad range of human and social parameters in addition to environmental ones.

Issues of scope are thus related to issues of context. Presumably, the ethical and justice-based context for indicators will entail examining ecosystems and biogeophysical systems—not just in terms of biological, physical, and chemical parameters, but also parameters regarding the relationship of humans to each other, as well as their relationship to ecosystems and to the planet.

4. CONSIDERATIONS OF SCALE

Considerations of scale refer to the spatial and temporal scale of indicators. Scale is an essential element for consideration in developing indicators: some problems are global, such as climate change, yet both the changes that will occur and the local drivers of change will vary geographically and temporally. Other problems are local but may be present in several places in a region. For example, acid rain is a regional problem but its effects on lakes may be local; blue-green algae blooms are local to lakes and other waters but may be widespread in certain watersheds. Thus, it is essential to consider what indicators are appropriate and useful at different scales and how indicators at different scales interrelate. Indicators of the overall environmental condition of the planet may be useful in some contexts, but they may not necessarily be used to estimate the condition of a specific ecosystem or those in a region.

Scale also refers to time: some processes change slowly in time or have delayed feedbacks, whereas others may be more rapid (Rockström et al. 2009a, 2009b). Processes that usually work on geological time scales that are now showing changes over a few centuries, or even decades, are important to identify. For example, the increase in atmospheric carbon dioxide, which usually occurs over millennia, has been changing dramatically over a few decades. Evolution is clearly a slow process, but losses in biodiversity are occurring at much faster rates.

At the local ecosystem scale, the eight classes of biological and thermodynamic indicators in Jørgensen et al. (2010a) may be particularly relevant (see note 1) but nevertheless are incomplete. At this scale, one can attempt to characterize ecosystems by homeostasis, absence of disease, diversity or complexity, stability or resilience, vigor or scope for growth, balance between system components, and ecological integrity ("the interconnection between the components of the system"; Jørgensen 2010a). Specific indicators would need to be identified for each of these axes, and selection of indicators may vary with the context. Measures of function ("the overall activities of the system") could be added, as well as other specific indicators that could be dependent on the type of ecosystem. It is not correct to think, however, that these indicators would comprise the universe of function.

At the global scale, relevant indicators of planetary function are those identified recently as useful for establishing "planetary boundaries for estimating a safe operating space for humanity with respect to the functioning of the Earth system" (Rockström et al. 2009a, 2009b). We discuss these in depth in the following chapter. The boundaries relate to climate change, ocean acidification, stratospheric ozone, global nutrient cycles, atmospheric aerosol loading, freshwater use, land use change, biodiversity loss, and chemical pollution. The global systems clearly interact in complex ways with systems at smaller scales, with complex feedback loops connecting them.

Another example comes from the indicator of human appropriation of net primary production, which can be relevant at local, regional, and global scales, and its correlation with species-energy curves, which relate the survival of species to the total production of available energy in a region. For example, by examining the effect of human appropriation of net primary production on the total energy available, Wright (1990) derived estimates of the percentage of species expected to be extinct or endangered; the estimates were found to be generally consistent with observations.

5. CONSIDERATIONS OF COMMENSURABILITY

Considerations of commensurability concern whether different measures can be reduced to a common one. The incommensurability of the value of thriving coral reefs with the value of fifty shares of ExxonMobil stock is at the heart of the debate over valuation of "ecosystem services" in terms of monetary value, which carries so much baggage of the standard neoclassical economic framework in which the value of money is defined. Indeed, the much-cited paper by Costanza et al. (1997) in which monetary values were assigned to ecological systems, while an attempt to gain ecological economics some weight in current policy debates, is clearly incompatible with the ethic and underlying rationale of ecological economics. Another example derives from the estimate that "wild" insects are worth $57 billion, and the conclusion that this value is "an amount that justifies greater investment in the conservation of these services" (Losey and Vaughan 2006). The absurdity of this specific assessment and conclusion—or for that matter any other monetized valuation of ecosystems or species—becomes clear when one considers the problems associated with the extinction of all or selected insects and the implicit assumption that insects, like all other goods, are replaceable at the cost of $57 billion or any other amount. The planet will be a dramatically different place should all of the bees become extinct (some research suggests that certain pesticides, neonicotinoids, may be the cause of bee deterioration; Henry et al. 2012; Whitehorn et al. 2012), and no amount of money could replace them and their essential functions. The challenge, then, is to avoid this problem of incommensurability in considering indicators of right relationship, and to develop ways to work toward desired objectives using decision-making techniques that involve consideration of indicators that are incommensurable with each other (Martinez-Alier et al. 2010).

Issues of commensurability raise questions of whether and how different ecosystems can be assessed in the aggregate, as well as how societal issues can be so assessed, and whether scorecard indicators that contain composites of incommensurable indicators provide useful information. Related to these issues are issues of controlling variables or indicators: Are some indicators more important than others in drawing conclusions about the functioning of an ecosystem or the planet? Can these indicators be used to develop sound policies that have a low probability of backfiring or simply failing? Also related to the question of commensurability is the nonlinear nature of the evolution of ecosystems, which complicates not only the reducibility of indicators to common metrics but also the use of at least some indicators prospectively. Lastly, this category of considerations includes issues of benchmarking and reference points, which are used extensively in assessing components of human health and in, for example, ecosystem restoration; however, they may be problematic in other areas.

Issues of incommensurability can be illustrated when asking whether composite indicators may be more useful than sets of indicators for the purposes of understanding functional status and for setting actions that may help improve function or mitigate environmental problems. Composite or aggregate indicators are those that express a complex aggregation of parameters, either in one unit of measure as does money or the ecological footprint, which makes use of derived functions to compute hectares of productive land; as a unitless index, such as the Sustainable Society Index; or as an indicator with uninterpretable dimensions (any quantities that have different units and are combined, either multiplicatively, additively, or in more complex ways will be difficult if not impossible to interpret). Examples of composite indicators include the following:

- Monetary value
- Gross domestic product, gross national product, and gross world product (expressed in monetary units, such as dollars)
- Ecological footprint (http://www.footprintnetwork.org/en/index.php /GFN; expressed in real or fictional hectares or acres of productive land, which includes the near-shore sea)
- Genuine Progress Indicator (http://www.rprogress.org/sustainability _indicators/genuine_progress_indicator.htm; expressed in monetary units, such as dollars)
- Various indicators used in material flow accounting (expressed in mass units [of biomass, minerals, etc.], per capita mass units, or percentages of the whole)

- Sustainable Society Index (http://www.eoearth.org/article/Sustainable _Society_Index#2.1_____Definition_of_sustainability)
- Index of Sustainable Economic Welfare (http://www.neweconomics .org/gen/newways_about.aspx)
- Dashboard of Sustainability (http://www.iisd.org/cgsdi/dashboard.asp)
- Environmental Performance Index (http://epi.yale.edu)

Many of these composite indicators take individual items (e.g., air quality, population growth, other composite indicators), multiply the value of each item by an assumption-laden weighting factor, and then add the products together. A generic indexing algorithm has been developed by Jollands, Lermit, and Patterson (2003). We have serious problems with this and any other methodology that combines disparate quantities measured in different units in essentially an arbitrary fashion. Our concerns are described below.

5.1. Limitations of Composite Indicators

While composite indicators are being used widely, most if not all suffer from serious issues of interpretation, including the following:

1. *Mixing disparate quantities*: The interpretation of a composite index in biological, physical, statistical, or other specific terms is often not possible because "apples and oranges" are mixed together. What does it mean to have a 7 for air quality added to a 5 for gender equality? That is to say, there is no clear construct that the index is purporting to measure. Moreover, without some benchmark or set of reference values, there is no way of interpreting composite indices except in a very general way, such as in concluding that a rising index is "good" and a falling index is "bad" (or vice versa); certainly, no policy decisions can be made using these indicators. Comparisons in terms of changes in time or comparisons between regions are always possible, but the meaning of the comparisons is obscure because of the difficulty in appreciating which components of the complex indicator are changing; this is usually impossible without detailed information about the individual components, and thus begs the question entirely as to the rationale for combining indices.

2. *Combining information*: Adding or multiplying components together, or using other more complicated mathematical functions (either attaching weights or not), is laden with assumptions that are not usually fully

acknowledged, nor are their implications appreciated clearly. A key question is: what construct is actually being measured and how is it to be interpreted? In addition, uncertainties in the values of the index (e.g., statistical sampling variability, errors in measurement) may be overlooked or not identified. Other technical questions relate to the choice of the function to aggregate the different components and how to determine whether one mathematical function is superior to others.

3. *Ambiguity in interpreting changes*: Changes in a complex index, either temporally or spatially, cannot easily be attributed to the change of a specific component. For example, in the Sustainable Society Index, each of the twenty-two indicators is rescored on a scale from 0 to 10, then added together using assumption-laden weights to come up with a set of five summary values for each category ("personal development," "health environment," "well-balanced society," "sustainable use of resources," "sustainable world"). The scores for the categories are added together to obtain a final cumulative value. Such an index is impossible to interpret. For example, if one component increases to the same extent that another decreases, then there would be no change recorded. If there is change, it is not possible to determine where the change has occurred by simply inspecting the aggregate.

Here is a simple example: If two independent variables are summed or multiplied together, it is not possible to distinguish the meaning of any particular value of the sum or product. For example, how does one interpret an index that is computed as $x_1 + x_2$? If $x_1 = 4$ and $x_2 = 5$ or if $x_1 = 5$ and $x_2 = 4$, then the same result (9) is obtained, even though there may be large differences in what 9 means in the two cases.

4. *Use in policy*: Specific problems and solutions cannot be addressed using composite indicators, unless they are developed in such a way as to measure an important concept that has face or construct validity. For example, let us assume that "water quality" could be defined by adding together the concentration of individual pollutants (e.g., volatile organic substances, *Escherichia coli*, metals, turbidity). If the reason for measuring water quality is to prevent health problems in people, then interpretation of such an index would be obscure because some of the elements relate to long-term health effects (metals, volatiles), whereas others relate to short-term effects (*E. coli*, turbidity). There is no clear interpretation of adding concentrations of such disparate, and possibly correlated, quantities

because it is impossible to benchmark the index—namely, determining what is a "safe" threshold.

On the other hand, various composite indices have been developed in public health policy, but these are used mostly for the purposes of communication with the general population to reduce their risk of contracting specific diseases. For example, the Canadian air quality index (Stieb et al. 2005; http://www.hc-sc.gc.ca/ewh-semt/air/out-ext/air_quality-eng.php; http://www.ec.gc.ca/cas-aqhi/default.asp?Lang=En) and the ultraviolet light index (http://www.epa.gov/sunwise/doc/what_is_uvindex.html) are used to inform people on a daily basis to reduce their burden of developing acute health effects from poor air quality and to reduce the long-term risk of developing skin cancer and cataracts from excessive ultraviolet light, respectively. These indices are not generally used to develop environmental policies. For example, the air quality index is derived from the actual concentrations of ozone, fine particles, and other pollutants; the associated health effects are used, with various political and economic constraints, to develop regulatory "acceptable" limits and actions. (The index has been used to estimate burden of disease [Stieb et al. 2005], but otherwise its main use is for the purposes of education and prevention.)

Some authors have suggested that composite indicators may be useful in a policy framework:

> The costs of an aggregate indicator are that, if one is not careful and informed, one can be ignorant of where the numbers came from, how they were aggregated, the uncertainties, weights, and assumptions involved, etc. It's not that one "loses" the more detailed information— usually it is possible to look at the details of how any aggregate indicator has been constructed—but rather that decision-makers are too busy to deal with these details. The beauty of the aggregate indicator is that it does that job for them. Even given this advantage of aggregate indicators, no single one can possibly answer all questions and multiple indicators will always be needed (Opschoor 2000), as will intelligent and informed use of the ones we have.
> (Costanza 2000)

This assessment understates the problem of composite indicators, not least because it dismisses too easily the problem of busy decision makers misusing them. A composite indicator, unless benchmarked and under-stood in terms of the construct it is measuring, cannot be used appropri-ately in decision-making processes because it is unclear what is actually being measured. How can appropriate decisions be made on this basis? If, to have value in the decision-making process, composite indicators must be broken down into their individual components, then why not use the individual components to begin with? The fact that some politicians and bureaucrats cannot deal with sets of indicators does not justify the use (or, more likely, misuse) of composites.

Part of the problem that policymakers have in interpreting sets of indi-cators is that they lose sight of the reasons and clear objectives as to why they were collected. Thus, the policymaker is confronted with a morass of numbers that become uninterpretable and are out of context. Specifying in advance what are the problems and what needs to be measured maintains the context of the indicators.

6. CONSIDERATIONS OF THE MEASUREMENT PROCESS

The previous discussion on composite indicators leads naturally to a dis-cussion of issues related to measurement and interpretation. As we indi-cated, one needs to be very clear about the reasons why something is being measured; in developing scientific projects, we refer to these as "operational objectives." There is no reason to measure something just because it can be measured or is available, such as through administrative databases. An important adage is: "Not everything that counts can be counted, and not everything that can be counted counts" (sign hanging in Einstein's office at Princeton University).

The objectives lead naturally to the specific constructs that one wishes to measure; this is often dictated by purpose but may be limited by knowledge and technology. An example of a well-defined construct is temperature: it defines the average kinetic energy of a system, is proportional to its heat content, and is bounded at the low end (0 Kelvin). It has face validity (see box 4.1 for a detailed description of the concepts of validity and reliabil-ity): that is, experts agree on what it measures. It also has content validity and construct validity: we understand from physical theory what is being measured. Depending on the actual instrument used, the variability in measurements of a system with exactly the same temperature can be very small (referred to as reliability, variability, or uncertainty). After repeated measurements by one or more unbiased instruments, the average will converge to the "true" temperature. Thus, the measurement of temperature is both valid and reliable (the two concepts together are often referred to as accuracy). Because temperature is valid and reliable, one can discern small trends through time, and it can be used in many instances as a sentinel indicator. Valid and reliable indicators are essential, not just in science but for the program of ecological economics.

Of course, the extent of the validity and reliability of a measurement will depend greatly on the inherent accuracy of the measurement instrument (figure 4.1). A judicious choice of instruments is needed for specific instances, taking into account how accurate the measures need to be. (In quantum experiments, one would never use a mercury bulb thermometer; to measure ambient temperatures, one would never use a device designed to measure at an accuracy of one in a billion.)

The case of global warming is instructive. To show that global warming is occurring, one needs to be certain that temporal trends cannot be explained

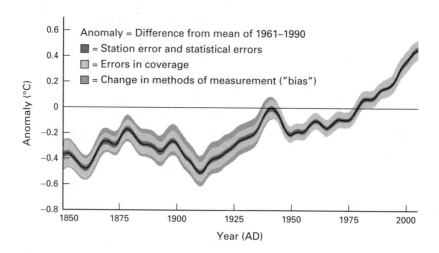

Figure 4.1. HadCRUT3 global temperature anomaly time-series (C) at smoothed annual resolutions. The solid black line is the best estimate value, the dark gray inner band gives the 95 percent uncertainty range caused by station, sampling, and measurement errors; the light gray band adds the 95 percent error range due to limited coverage; and the dark gray outer band adds the 95 percent error range due to bias errors. *Source*: Reproduced from Brohan et al. (2006).

by errors in measurement; that is, changes over time of the average measurements of ambient temperature across the planet are in fact larger than the errors of measurement (Brohan et al. 2006). Annual global mean temperature is assessed through monitors placed throughout the globe, so that the temperature recorded is representative of the temperature in a small area. The uncertainty in a "gridbox mean" caused by estimating the mean from a small number of point values is referred to as the sampling error. Because of costs and feasibility, not all parts of the globe are covered (i.e., coverage errors). There will also be uncertainties in the measurements at each station (i.e., station errors). Lastly, uncertainties in large-scale temperatures will arise because of changes in methods of measurement (i.e., bias error).

Mean global temperatures are built up as the average of the monitor-specific monthly or annual average. To appreciate whether the apparent secular, monotonic increase in temperature is increasing over and above error, an analysis of the global record can be undertaken. This analysis should account for the above-mentioned errors in measurement and coverage, such that the goal is to determine whether the increase in global mean temperature is greater than the total errors associated with the measurements. Figure 4.1 shows the results of an analysis combining the various errors, where the observed trends were greater than measurement error; therefore, the conclusion that the planet has been warming since 1850 was confirmed (Brohan et al. 2006). (A formal statistical analyses of these data leads to the same conclusions as perusing this figure.)

Of course, this particular example says nothing about regional changes in temperature. In fact, we know that areas of the north are warming much faster than near the equator (figure 4.2) (Intergovernmental Panel on Climate Change 2007). Thus, global mean temperature is a restricted indicator of planetary change that obscures strong spatial heterogeneity. As a physical indicator, global mean temperature is nevertheless important because it reflects the extra energy added to the atmosphere from increased climate forcing, which is due mostly to increased anthropogenically-induced greenhouse gas emissions.

This example of global warming shows clearly the importance of analyzing measurements by time to determine trends; it also shows the intimate link between measurement and statistics. The paleoclimatic record suggests that the current concentration of greenhouse gases is far greater than what existed over the last 600,000 years, and there is indeed concern that

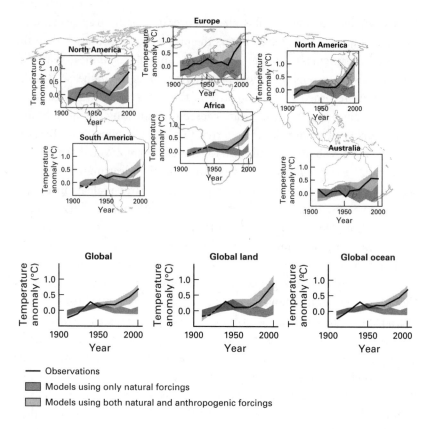

Figure 4.2. Comparison of observed continental- and global-scale changes in surface temperature with results simulated by climate models using either natural or both natural and anthropogenic forcings. Decadal averages of observations are shown for the period 1906–2005 (black line) plotted against the centre of the decade and relative to the corresponding average for the 1901–1950. Lines are dashed where spatial coverage is less than 50 percent. Dark gray bands show the 5 percent to 95 percent range for nineteen simulations from five climate models using only the natural forcings due to solar activity and volcanoes. Light gray bands show the 5 percent to 95 percent range for fifty-eight simulations from fourteen climate models using both natural and anthropogenic forcings. *Source*: Reproduced from Intergovernmental Panel on Climate Change (2007).

continued increases in greenhouse gas emissions will lead to global warming that could melt the polar icecaps. Such a nonlinear, irreversible change (a "tipping point") could lead to a major catastrophe, by causing crops to fail and flooding of much of the earth's land masses. One could then think that, in fact, there is a "normative" standard that can be identified by

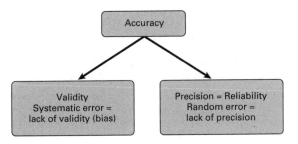

Figure 4.3. The paradigm of accuracy.

perusing the paleoclimatic record; some suggest that it is close to 350 ppm of carbon dioxide (Rockström et al. 2009a, 2009b), and we have gone well past that (approximately 400 ppm as of December 2014; http://co2now .org). With a trend of annual increases of almost 2 ppm, the world is heading for concentrations on the order of 450–500 ppm before long.

7. MODELS AND PREDICTION

Scientific measurements are often used to develop models that can make predictions where there are no data, such as projecting into the future or to other situations. All models have a range of validity, and no statistical models or physical models are perfectly accurate. The question is whether the models are useful. Understanding their limitations is obviously critical if they are to be used in in ecological economics and in policy. All models need to be validated.

The difference between models and measurements should be made clear: measurements are a real-time representation of some construct (e.g., concentrations of a pollutant in a watershed), whereas models can be developed to predict concentrations where measurements were not made (e.g., a spatial prediction model) or, with additional information, to project into time what these concentrations may be. To interpret these models correctly, all of the assumptions built into these models need to be made clear. The value of these models comes only when further measurements are made that show that the model is correct (or not). Sometimes, the models provide generally correct projections, but their estimates may not be

100 percent accurate (this is the usual case). In such an instance, it is likely that the model incorporates the correct science but may be lacking details, such that its quantitative accuracy is not as good as one would hope.

It is important to distinguish between the different types of models that can be developed. Although it is impossible to discuss all of the types of models that are used, we can present a few examples that may provide the reader with some notion of how these are used. Returning to our example of global temperature, one can develop statistical models to project into time what the expected global temperature will be in a certain year in the future (e.g., 2100). Such a model would make use of all of the available historical data (inputs) and would attempt to find the "best-fitting" function to describe these data (output). One of the simplest functions is a straight line, but in fact this does not represent the best line because the rate of change of temperature increased dramatically around 1970–1980 (the so-called hockey-stick graph). However, once one has found the correct function (if one assumes that the trend remains the same through time), then one can project the value of the function through time to arrive at a prediction for 2100 (this projection would also incorporate the associated statistical variability). Clearly, this is a crude model because of the assumption that the conditions that led to changes in temperature will remain the same in the future.

Climate modelers have developed far more sophisticated models, called global circulation models. These models make use of the known physics, including various feedback mechanisms, to integrate possible changes to concentrations of greenhouse gases through different sets of scenarios of anthropogenic emissions. Not every model will arrive at the same prediction because each model has different assumptions built in; also, because of computational complexity, some models may try to make the physics a bit simpler than in others. The models are being continuously updated to include new physics and new information and are tested on past data.

Comparing predictions to observations tests the extent of the usefulness of models. The general circulation models have been used to back-project temperatures in the twentieth century; that is, by using estimates of emissions during the twentieth century (input), the models were run to see if they predicted the set of already observed global temperatures (comparison of predicted to observed values). Indeed, the average across the models was shown to follow observed temperatures. This validation gives confidence that the models projecting into the future will provide plausible predictions; it provides some range of variability, but does not guarantee that the future will correspond with these predictions (Intergovernmental Panel on Climate Change, 2007).

Not all indicators in ecological economics need be used for prediction. However, they will clearly be needed for assessing secular and spatial trends, and perhaps for regulating economic activity. Therefore, knowing the attributes of the indicators, in terms of validity and reliability, is essential.

8. CONSIDERATIONS OF COMPLEX INTERACTING SYSTEMS

As part of defining any indicator, one must be cautious in its interpretation because it may derive from complex processes within and between interacting systems. There are always limitations to measurement and also to predictability; complex systems are subject to unpredictable behavior and thus unpredictable outcomes. This is not just a limitation of measurement, it is an inherent behavior of complex systems. This view contrasts with one stating that "there is an objective truth out there; if we only had enough time and money, we could get enough measurements and build good-enough models that we could model the future with precision." Complex systems theory includes the idea of "irreducible uncertainty." Thus, much care must be given to defining indicators, the models that they are based on, and their interpretation. The precautionary principle, discussed in the next chapter, provides one way in setting policy to deal with uncertainty, especially as it relates to decisions of policy.

Many of the indicators that we consider in the next chapter are global in nature. Nevertheless, the underlying processes act on local and regional scales, and there are many complex feedback loops. We will discuss the concept of setting constraints on the human economy in the form of boundary values that have been proposed to allow the planet to function safely for humanity. To be able to make these boundaries into normative values that should not be crossed, it is essential to be able to have a top-down approach that is able to set limits at small geographic scales.

Two examples should suffice. Again, using climate change, we know that anthropomorphic emissions of carbon occur locally and that international agreements are based on national quotas. Much work has been done to show the link between various sectors and emissions, as well as between atmospheric concentrations of greenhouse gases (globally more or less uniform) and emissions levels. Thus, global targets for the atmospheric concentration of greenhouse gases can be met by meeting national targets for emissions.

An example of the phosphorus cycle also may be useful. This is a key indicator (see chapter 5), and phosphorus is used as an essential nutrient in agriculture. Overuse and runoff has led to massive build-ups in various water systems, including in northwest European countries (Ulén et al. 2007), the Gulf of Mexico from runoff from the Mississippi and other river systems that flow to the Gulf of Mexico, and also in lakes where there is runoff from farms. This has led to eutrophication, apoxia (oxygen deficiency), and growth of algae blooms, such as the toxic blue-green algae (Ulén et al. 2007). Phosphorus is a useful indicator in ecological economics because its build-up in local waterways can be compared to levels that avoid those problems using other methods of farming, and those limits can be scaled up to establish a global boundary for phosphorus.

For ecological economics to be successful in its goal of recognizing the embeddedness of human activities in the planet, setting global constraints on phosphorous is essential to reducing the damage on ecosystems. Setting and enforcing constraints can only be done effectively by assessing regional patterns (for the Gulf of Mexico, this entails thousands of square kilometers), which requires accurate measurements. It also requires a full understanding of the pathways by which phosphorus enters ecosystems, which requires complex models, either based on empirical data (statistical models) or mechanistic models. Validation of the models is a sine qua non for their use, but once in place they can be used to set and enforce regional and local constraints. Having set constraints is, of course, insufficient; one needs alternative modes of agriculture so that reductions can be implemented, as well as methods to buffer runoff. Some efforts to reduce population are needed, as this is a key driver. Educational programs and possibly incentives to implement these new agricultural policies would also be required.

9. CONCLUSIONS

We have discussed the essential properties that indicators need to have for their appropriate use in ecological economics and in making policy decisions. In particular, we showed that clear objectives are required so that one understands exactly what should be measured and why. As well, we discussed various issues associated with the measurement process, but we also discussed the elements that can be used to help define what indicators should be measured by considering context, scale and dimension, scope, and commensurability. As part of this process, one needs to consider interrelationships between variables and attempt to recognize the complexity that may affect the interpretation of the measurements. We indicated that sound scientific principles must be followed so that the measurement process is accurate (valid and reliable). The extent of the inherent variability of the measurements will be dictated by the circumstances, but certainly appreciating the errors will facilitate understanding. We have cautioned strongly about using complex indicators because they will rarely measure a meaningful construct; thus, the interpretation of their values will typically be obscure. These principles apply to both the key processes and their drivers.

BOX 4.1. ISSUES RELATED TO MEASUREMENT

Use of any indicator is dictated solely by the objectives. Without clear objectives, interpretation is difficult and development of specific measures logically impossible. A prima facie criterion for any indicator is that it is valid and reliable (Koepsell and Weiss 2004). This means generally that the indicator measures what it purports to measure and does with a known level of uncertainty. Thus, not only must one determine what and how to measure a quantity, but one must also assess level of error associated with the measurements.

There are many aspects to validity; we present here a few of the main concepts. Face validity of a measure refers to whether it is obvious that it will measure what it is supposed to measure, such as measuring physical quantities. Experts can agree that a certain index measures what it purports to measure. This type of validity often applies to physical and certain biological measures, but often, even in physics, measures are developed that have a certain interpretation but are derived from other measures. Entropy, an essential concept in thermodynamics, is a derived quantity, but theory has developed to make it an essential quantity.

One can say that a measure such as entropy corresponds to the concept of construct validity. It refers to the degree to which inferences can be made from an index with respect to the theoretical construct it is supposed to measure. In another example, back pain has many etiologies and manifests itself differently in people. Is there a way of measuring the extent to which back pain affects one's daily life? One index that has been in use for many years is the Roland-Morris Low Back Pain and Disability Questionnaire (Roland and Morris 1983a, 1983b). It comprises twenty-four questions, as follows:

(continued on next page)

(Box 4.1 continued)

THE ROLAND–MORRIS LOW BACK PAIN AND DISABILITY QUESTIONNAIRE

I stay at home most of the time because of my back.
I change position frequently to try to get my back comfortable.
I walk more slowly than usual because of my back.
Because of my back, I am not doing any jobs that I usually do around the house.
Because of my back, I use a handrail to get upstairs.
Because of my back, I lie down to rest more often.
Because of my back, I have to hold on to something to get out of an easy chair.
Because of my back, I try to get other people to do things for me.
I get dressed more slowly than usual because of my back.
I only stand up for short periods of time because of my back.
Because of my back, I try not to bend or kneel down.
I find it difficult to get out of a chair because of my back.
My back is painful almost all of the time.
I find it difficult to turn over in bed because of my back.
My appetite is not very good because of my back.
I have trouble putting on my sock (or stockings) because of the pain in my back.
I can only walk short distances because of my back pain.
I sleep less well because of my back.
Because of my back pain, I get dressed with the help of someone else.
I sit down for most of the day because of my back.
I avoid heavy jobs around the house because of my back.
Because of back pain, I am more irritable and bad tempered with people than usual.
Because of my back, I go upstairs more slowly than usual.
I stay in bed most of the time because of my back.

Each item refers to the construct of functioning in the "presence of back pain." The degree of back pain could also be measured by asking questions that rate the levels of pain. The Roland-Morris questionnaire is often used giving a response of "yes" the value of 1; a "no" response is given the value of zero. Adding-up the scores across the twenty-four questions, giving equal weight to each question, leads to an index that is bounded on the low end by zero (no problems) to twenty-four (problems with all functional assessments). This index is thus a simple count (no units) of problems that an individual is having at the time of the assessment. Of course, one can always assess the individual indicators by themselves. This composite index is of course not useful for decisions about diagnoses or treatment, but it is very important for the individuals suffering from back pain and is also very useful in assessing treatments and differences between groups of individuals. Indeed, a comparison of the mean of this complex indicator between two populations would lead to a statistical inference as to whether one population has more functional problems with back pain than another. Such an assessment would not state specifically what the functional problems are; only assessing each item would provide that information. Assessing a population in time would allow the answer to the question regarding longitudinal physical functioning because of back pain. This index has construct validity, but it is not the only index that can be used to measure back pain or functioning. When to use this or any other index is driven by the questions that one wants to answer.

There are a number of formal definitions of construct validity. For example, the US Environmental Protection Agency (EPA) defines it as "the extent to which a measurement method accurately represents a construct and produces an observation distinct from that produced by a measure of another construct" (www.epa.gov/evaluate/glossary/c-esd.htm). The previous examples meet this definition.

The Roland-Morris questionnaire contains a number of related items that are combined into a composite index. Each item corresponds to a slightly different aspect of functioning with back pain, but the main concept is to measure functioning with back pain. This is referred to as content validity; the EPA defines it as "the ability of the items in a measuring instrument or test to adequately measure or represent the content of the property that the investigator wishes to measure" (www.epa.gov/evaluate/glossary/c-esd.htm).

There are other aspects related to validity (e.g., criterion validity; sis.nlm.nih.gov/enviro/iupacglossary/glossaryc.html), but it is not necessary to discuss these any further here.

Of considerable interest, however, is the issue of reliability (uncertainty or variability), which is related to measurement error but also to statistical variability. There is an inherent error associated with any measurement process, and one of the key ideas of measurement theory is to develop instruments that have a minimum amount of measurement error. In measuring temperature, we discussed in the main text the concepts of "station error" and "bias error," which described inherent errors from the measurement process across the temperature monitors across the globe and through time. Indeed, if these errors were too large, then any real increase in global temperature would not be identifiable. Reliability can thus be defined as the inherent amount of variability in the measurement process.

Other forms of reliability exist. If one has different individuals assessing a trait, then each individual's assessment of a specific case may vary in time because of subtle changes in the process. This is referred to as an interrater agreement. When there are many raters, even if the same criteria are being used, there will also be variability between raters, and this is referred to as inter-rater variability.

In statistics, the concept of reliability is often referred to as statistical variability or statistical error. Statistical error refers to sampling variability: the results of a study on a specific population will not be the same if a different population was selected randomly. If one had enough resources, one could repeat a study many times and the average of the results across the studies would reflect the best estimate; the amount of variability in these estimates is the sampling error. There is a clear link between this concept and measurement error; if the errors in measurement are not known, then they will be subsumed within statistical variability. However, if the measurement error is known, then it can be partitioned from the total variability. Essentially, statistical variability refers to variation that cannot be readily explained; indeed, one of the goals of statistics is to explain this variability. (The identification of drivers, say in the IPAT formulation [Ehrlich & Holdren 1972], could be viewed in this light if data are available.)

Often the concepts of validity and reliability are viewed together, and this is often referred to as accuracy (see figure 4.3). On the other hand, the term "accuracy" is often used to represent validity.

(continued on next page)

(Box 4.1 continued)

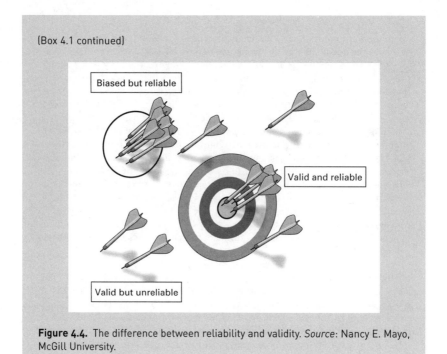

Figure 4.4. The difference between reliability and validity. *Source*: Nancy E. Mayo, McGill University.

NOTE

1. Jørgensen et al. (2010, 12–14) identified eight levels of indicators for the assessment of "ecosystem health," based on the following: (1) the presence or absence of specific indicator species; (2) the ratio between classes of organisms; (3) the presence, absence, or concentrations of specific chemical compounds, such as phosphorous, in regard to trophic level or toxic chemicals in regard to toxicity; (4) the concentration of trophic levels, such as the concentration of phytoplankton as an indicator of eutrophication of lakes; (5) process rates, such as the rate of primary production in relation to eutrophication or the rate of forest growth in relation to forest health; (6) composite indicators, including E.P. Odum's attributes and various indices, such as total biomass or the ratio of production to biomass; (7) holistic indicators, such as biodiversity, resilience, buffer capacity, resistance, and turnover rate of nitrogen; and (8) "superholistic" thermodynamic indicators, such as exergy (the energy in the ecosystem available to do work), energy, and entropy.

REFERENCES

Brohan, P., J. J. Kennedy, I. Harris, S. F. B. Tett, and P. D. Jones. 2006. "Uncertainty Estimates in Regional and Global Observed Temperature Changes: A New Data

Set from 1850." *Journal of Geophysical Research: Atmospheres* 111 (D12): D12106. doi:10.1029/2005JD006548.

Brown, Peter G., and Geoffrey Garver. 2009. *Right Relationship: Building a Whole Earth Economy.* San Francisco: Berrett-Koehler Publishers.

Costanza, Robert. 1992. "Toward an Operational Definition of Ecosystem Health." In *Ecosystem Health: New Goals for Environmental Management*, edited by Robert Costanza, Bryan G. Norton, and Benjamin D. Haskell, 239–256. Washington, DC: Island Press.

Costanza, Robert. 2000. "The Dynamics of the Ecological Footprint Concept." *Ecological Economics* 32: 341–345.

Costanza, Robert, Ralph d'Arge, Rudolf de Groot, Stephen Farber, Monica Grasso, Bruce Hannon, Karin Limburg, Shahid Naeem, Robert V. O'Neill, Jose Paruelo, Robert G. Raskin, Paul Sutton, and Marjan van den Belt. 1997. "The Value of the World's Ecosystem Services and Natural Capital." *Nature* 387 (6630): 253–260. doi: 10.1038/387253a0.

Costanza, Robert, and Michael Mageau. 1999. "What Is a Healthy Ecosystem?" *Aquatic Ecology* 33 (1): 105–115. doi: 10.1023/A:1009930313242.

Ehrlich, Paul R., and John P. Holdren. 1972. "Critique on 'the Closing Circle' (by Barry Commoner)." *Bulletin of the Atomic Scientists* 28 (5): 16, 18–27.

Henry, Mickaël, Maxime Béguin, Fabrice Requier, Orianne Rollin, Jean-François Odoux, Pierrick Aupinel, Jean Aptel, Sylvie Tchamitchian, and Axel Decourtye. 2012. "A Common Pesticide Decreases Foraging Success and Survival in Honey Bees." *Science* 336 (6079): 348–350. doi: 10.1126/science.1215039.

Intergovernmental Panel on Climate Change. 2007. *Fourth Assessment Report, Climate Change 2007: Synthesis Report.* Edited by The Core Writing Team, Rajendra K. Pachauri and Andy Reisinger. Geneva: Intergovernmental Panel on Climate Change.

Jollands, Nigel, Jonathan Lermit, and Murray Patterson. 2003. "The Usefulness of Aggregate Indicators in Policy Making and Evaluation: A Discussion with Application to Eco-Efficiency Indicators in New Zealand." *Australian National University Digital Collections, Open Access Research.* http://hdl.handle.net/1885/41033.

Jørgensen, Sven Erik. 2010a. "Eco-Exergy as Ecological Indicator." In *Handbook of Ecological Indicators for Assessment of Ecosystem Health*, edited by Sven Erik Jørgensen, Fu-Liu Xu, and Robert Costanza, 77–87. Boca Raton, FL: CRC Press.

Jørgensen, Sven Erik. 2010b. "Introduction." In *Handbook of Ecological Indicators for Assessment of Ecosystem Health*, edited by Sven Erik Jørgensen, Fu-Liu Xu, and Robert Costanza, 3–7. Boca Raton, FL: CRC Press.

Jørgensen, Sven E., Fu-Liu Xu, João C. Marques, and Fuensanta Salas. 2010. "Application of Indicators for the Assessment of Ecosystem Health." In *Handbook of Ecological Indicators for Assessment of Ecosystem Health*, edited by Sven Erik Jørgensen, Fu-Liu Xu, and Robert Costanza, 9–75. Boca Raton, FL: CRC Press.

Jørgensen, Sven Erik, Fu-Liu Xu, and Robert Costanza, eds. 2010. *Handbook of Ecological Indicators for Assessment of Ecosystem Health.* 2nd ed. Boca Raton, FL: CRC Press.

Koepsell, Thomas D., and Noel S. Weiss. 2004. "Measurement Error." In *Epidemiologic Methods: Studying the Occurrence of Illness*, 215–246. Oxford: Oxford University Press.

Losey, John E., and Mace Vaughan. 2006. "The Economic Value of Ecological Services Provided by Insects." *BioScience* 56 (4): 311–323. doi: 10.1641/0006-3568(2006)56[311:TEVOES]2.0.CO;2.

Lovett, Gary M., Timothy H. Tear, David C. Evers, Stuart E. G. Findlay, B. Jack Cosby, Judy K. Dunscomb, Charles T. Driscoll, and Kathleen C. Weathers. 2009. "Effects of Air Pollution on Ecosystems and Biological Diversity in the Eastern United States." *Annals of the New York Academy of Sciences* 1162 (1): 99–135. doi: 10.1111/j.1749-6632.2009.04153.x.

Martinez-Alier, Joan, Giorgos Kallis, Sandra Veuthey, Mariana Walter, and Leah Temper. 2010. "Social Metabolism, Ecological Distribution Conflicts, and Valuation Languages." *Ecological Economics* 70 (2): 153–158. doi: 10.1016/j.ecolecon.2010.09.024.

Opschoor, Hans. 2000. "The Ecological Footprint: Measuring Rod or Metaphor?" *Ecological Economics* 32 (3): 363–365. doi: 10.1016/S0921-8009(99)00155-X.

Rapport, David J. 1992. "Evaluating Ecosystem Health." *Journal of Aquatic Ecosystem Health* 1 (1): 15–24. doi: 10.1007/BF00044405.

Rockström, Johan, Will Steffen, Kevin Noone, Asa Persson, F. Stuart Chapin, III, Eric F. Lambin, Timothy M. Lenton, Marten Scheffer, Carl Folke, Hans Joachim Schellnhuber, Bjorn Nykvist, Cynthia A. de Wit, Terry Hughes, Sander van der Leeuw, Henning Rodhe, Sverker Sorlin, Peter K. Snyder, Robert Costanza, Uno Svedin, Malin Falkenmark, Louise Karlberg, Robert W. Corell, Victoria J. Fabry, James Hansen, Brian Walker, Diana Liverman, Katherine Richardson, Paul Crutzen, and Jonathan A. Foley. 2009a. "Planetary Boundaries: Exploring the Safe Operating Space for Humanity." *Ecology and Society* 14 (2): 32. http://www.ecologyandsociety.org/vol14/iss2/art32/.

Rockström, Johan, Will Steffen, Kevin Noone, Asa Persson, F. Stuart Chapin, III, Eric F. Lambin, Timothy M. Lenton, Marten Scheffer, Carl Folke, Hans Joachim Schellnhuber, Bjorn Nykvist, Cynthia A. de Wit, Terry Hughes, Sander van der Leeuw, Henning Rodhe, Sverker Sorlin, Peter K. Snyder, Robert Costanza, Uno Svedin, Malin Falkenmark, Louise Karlberg, Robert W. Corell, Victoria J. Fabry, James Hansen, Brian Walker, Diana Liverman, Katherine Richardson, Paul Crutzen, and Jonathan A. Foley. 2009b. "A Safe Operating Space for Humanity." *Nature* 461: 472–475. doi: 10.1038/461472a.

Roland, Martin, and Richard Morris. 1983a. "A Study of the Natural History of Back Pain. Part I: Development of a Reliable and Sensitive Measure of Disability in Low-Back Pain." *Spine* 8 (2): 141–144.

Roland, Martin, and Richard Morris. 1983b. "A Study of the Natural History of Low-Back Pain. Part II: Development of Guidelines for Trials of Treatment in Primary Care." *Spine* 8 (2): 145–150.

Stieb, David M., Marc Smith Doiron, Philip Blagden, and Richard T. Burnett. 2005. "Estimating the Public Health Burden Attributable to Air Pollution: An Illustration Using the Development of an Alternative Air Quality Index." *Journal of Toxicology and Environmental Health, Part A* 68 (13–14): 1275–1288. doi: 10.1080/15287390590936120.

Suter, Glenn W., II. 1993. "A Critique of Ecosystem Health Concepts and Indexes." *Environmental Toxicology and Chemistry* 12: 1533–1539. doi: 10.1002/etc.5620120903.

Ulén, B., M. Bechmann, J. Fölster, H. P. Jarvie, and H. Tunney. 2007. "Agriculture as a Phosphorus Source for Eutrophication in the North-West European Countries, Norway, Sweden, United Kingdom and Ireland: A Review." *Soil Use and Management* 23: 5–15. doi: 10.1111/j.1475-2743.2007.00115.x.

Whitehorn, Penelope R., Stephanie O'Connor, Felix L. Wackers, and Dave Goulson. 2012. "Neonicotinoid Pesticide Reduces Bumble Bee Colony Growth and Queen Production." *Science* 336 (6079): 351–352. doi: 10.1126/science.1215025.

Wright, David Hamilton. 1990. "Human Impacts on Energy Flow through Natural Ecosystems, and Implications for Species Endangerment." *Ambio* 19 (4): 189–194. doi: 10.2307/4313691.

CHAPTER FIVE

Boundaries and Indicators

Conceptualizing and Measuring Progress Toward an Economy of Right Relationship Constrained by Global Ecological Limits

GEOFFREY GARVER AND MARK S. GOLDBERG

What are the ecological boundaries and related parameters of the space in which the human economy must function to avoid overshoot, injustice, and collapse in interdependent ecological, economic, and social systems? What policy-oriented indicators will accurately and reliably show the extent to which the human enterprise is respecting those boundaries and parameters? In this chapter, these questions set the substantive framework for developing and continually improving high-quality economic indicators that can serve as beacons for reorienting the human economy toward a respectful and enduring relationship with Earth's life systems—one based on the notions of membership, householding, and entropic thrift introduced in chapter 2. Consistent with Herman Daly's conception of ecological economics (Daly 1996), these indicators should address the aggregate scale, distribution, and efficiency of the economy, with scale as the constraining envelope within which distribution and efficiency must be managed (Daly and Farley 2011). The notions of "right relationship" (Brown and Garver 2009) and "safe operating space" (Rockström et al. 2009) form the foundation of the proposed framework.

Three overarching considerations bear emphasis at the outset. First, the framework is built around the complex and often nonlinear and chaotic behavior of biogeochemical systems across a range of temporal and spatial scales. This systems-based framework poses significant challenges for dealing with uncertainty and measuring parameters suited to boundaries and indicators of the economy, as the previous chapter on measurement explained. Second, it requires a distinction between indicators of scale, for which a concept such as "planetary boundaries of safe operating space" is particularly well-suited, and indicators of distribution and efficiency, for which a concept such as "right relationship" provides criteria that overlap with but go beyond those related to scale and planetary boundaries. Consistent with safe operating space and right relationship, the concepts of membership, householding, and entropic thrift should guide the normative choices regarding distribution and efficiency, within ecological bounds. Third, the economic indicators envisioned are alternatives to prevailing or emerging indicators of the state of the human enterprise that are either explicitly or implicitly committed to economic growth as a necessary component of a sound economic trajectory. Contrary to the conventional notion of sustainable development, as reflected in the growth-driven Rio Principles, the Rio+20 outcome document *The Future We Want*, and the United Nations (UN) Environment Programme's green economy report,[1] this framework insists on the recognition of systems thresholds and other limits for key components of the human–Earth system as the nonnegotiable starting point for the development of indicators.

1. INTRODUCTION TO BOUNDARIES

In September 2009, an international team of researchers, led by Johan Rockström of the Stockholm Resilience Centre, proposed a novel ecological framework for guiding the human enterprise toward this vision (Rockström et al. 2009). This framework is built around the concept of "planetary boundaries" of the "safe operating space" for humanity. The boundaries are conceived as normative global ecological limits, beyond which humans face "the risk of deleterious or even catastrophic environmental change at continental to global scales" (Rockström et al. 2009:1). They are linked to estimates of thresholds of irreversible, nonlinear change or of undesirable levels of aggregate anthropogenic loading or disruption in key processes of the global ecosystem for which relatively stable conditions in the last 10,000 years have allowed expansion of human civilization. Rockström and his colleagues proposed nine planetary boundaries based on climate change, ocean acidification, stratospheric ozone depletion, atmospheric aerosol loading, land use, freshwater use, chemical pollution, biodiversity loss, and nutrient (i.e., nitrogen and phosphorous) cycles; they provided preliminary estimates for all of these except chemical pollution and atmospheric aerosol loading. Although keyed to human safety and survival—and therefore arguably at odds with the idea of membership of humans in the community of all living beings—the planetary boundaries are grounded on recognition of the essential reliance of humanity on Earth's life systems.

The proposed planetary boundaries may be considered as beacons, projected from the normative rubric of safe operating space, for guiding fundamental aspects of the human–Earth relationship—that is, the interactions of the human enterprise, consisting of the global entirety of human endeavors, with the ecosphere, the global entirety of the biotic and abiotic elements, and characteristics, that make up Earth and its atmosphere and support life on Earth (Huggett 1999). Other related boundaries are also possible, such as a limit on the global human ecological footprint that will adequately allow perpetual provisioning of the human enterprise while also maintaining acceptable levels of biodiversity (Ewing et al. 2010; Wackernagel and Rees 1996). A somewhat flawed attempt (Bishop, Amaratunga, and Rodriguez 2010) has been made to establish a similar boundary for human appropriation of net primary production (HANPP), a measure of the amount of biomass production that humans either harvest or appropriate by other means (Haberl et al. 2007).

Common features of all of these boundaries are that they accord primary importance to ecological considerations, and they reflect or suggest normative notions of the acceptable risk of catastrophic environmental consequences for humans and other life on the planet. In addition, the boundaries are related to thresholds in the interdynamic human–Earth system or to undesirable levels of aggregate ecological impacts due to human activity. These system thresholds are impossible to pinpoint with certainty because of the nonlinear dynamics of that system, in which change can occur chaotically in lurches, and the system can transcend points of no return long before feedback can manifest the consequent impacts. For many of the planetary boundaries, this uncertainty remains high enough to put their current usefulness in question. Likewise, human-induced loss of biodiversity and anthropogenic loadings of pollutants vary considerably from one locality or region to another, and the point at which their aggregate constitutes a catastrophic and globally significant disruption of ecological systems may be elusive. Indeed, the viability of some proposed boundaries, such as those related to nutrient loading and loss of biodiversity, has been questioned because they are difficult to conceptualize or model at the global level, as opposed to regionally or locally (Nordhaus, Shellenberger, and Blomqvist 2012). However, others are clearly of global significance, particularly the boundary for climate change. The boundary categories taken together reflect a comprehensive set of interacting systems features of the global ecosystem that frame the ecological contours of the human prospect.

Normative boundaries, which are linked to but distinct from thresholds or aggregate limits, are a human construct with implicit judgments as to the point beyond which the risk of exceeding system thresholds or aggregate pressures is unacceptable. Thus, boundaries establish a bridge between the vast complexities of the underlying science and the normative arenas in which people operate individually and collectively. For example, the proposed safe boundary for atmospheric levels of carbon dioxide is set with a margin of safety to avoid a point-of-no-return threshold that is based on solid scientific understanding of the myriad interrelated consequences of increasing temperatures across the planet, such as increases in sea level and extensive loss of biodiversity (Hansen et al. 2008). The threshold cannot be determined with certainty, but the boundary reflects a choice to operate safely within the threshold.

Where several boundaries are used to express the aggregate human ecological impact on Earth, as with the planetary boundaries, the boundaries' interrelatedness must be considered. Noting that "interactions among planetary boundaries may shift the safe level of one or several boundaries" (Rockström et al. 2009:24), the Rockström research team set each of their proposed boundaries on the assumption that no other boundary was transgressed. If this assumption does not hold, they suggest that the boundaries would most likely have to be adjusted so as to shrink humanity's safe operating space (Rockström et al. 2009). The relationship of boundaries to each other may reveal that some kind of ordering of the boundaries would make sense, particularly in view of concerns regarding the levels of uncertainty and the global viability of some of them. Even if some boundaries are conceptually more sound than others, their interactions have important implications. For example, concerns regarding land use are ultimately founded on the effects of land use on other boundaries, particularly biodiversity loss, climate change, and nitrogen loading (Nordhaus, Shellenberger, and Blomqvist 2012). As well, ocean acidification is directly related to increased levels of atmospheric carbon dioxide, for which the overarching consequence of concern is climate change. These relationships have an effect on the development of indicators, such as in determining relative priorities among impacts in regard to their contributions to the most urgent pressures on the global ecosystem.

The Rockström team's biophysically bounded space can be used to derive additional constraints focused more directly on the sociopolitical dimensions of the human sphere. Work dating back to the early 1970s (Ehrlich and Holdren 1972) has suggested that aggregate environmental impact is a function of the size of the human population, its affluence, and its technology; this is the well-known $I = P \times A \times T$, or IPAT, formulation, where I is impact, P is population, A is affluence or consumption, and T is technology. In theory, each of the planetary boundaries can be considered as a fixed limit of the I variable, which in turn constrains the P, A, and T variables; if P rises, A or T must fall if I is to remain unchanged.

If the focus is on population, the concept of carrying capacity is a variation of the boundaries concept. The sheer size of the human population already has an enormous impact on planetary processes: humans—a species representing approximately 5 percent of the total animal mass of the planet (McNeill 2000)—reached 7 billion in number in 2011 (United Nations Population Fund 2011). Yet, projections from the United Nations suggest that by 2050 the population of the planet will be between 8.34 and 10.9 billion (UN Population Division 2013). Daily and Ehrlich (1996) distinguished between two types of carrying capacities: one in which a maximum number of persons could be sustained biophysically under given technological capabilities and a second type, "social carrying capacity," whereby a maximum population could be sustained under various social systems and the associated patterns of resource consumption. Global carrying capacity, as defined by a maximum population size, is difficult to estimate, especially as it varies geographically and by associated regional socioeconomical systems. The planetary boundaries concept may help focus this effort by allowing for an analysis of population pressures on each of the boundaries, as well as of possible maximum populations that will allow the boundaries to be respected given feasible options for managing the A and T variables.

The concept of global limits or boundaries raises some noteworthy concerns. The global focus of the planetary boundaries concept should not preclude the possibility that for some (if not all) boundary categories, local or regional limits may be more important, or more readily determined and applied, than global ones. Expressing a global limit or boundary as evenly distributed would in almost all cases be misleading, because the local and regional contributions to the aggregate loading or impact, as well as the ecological consequences of that loading or impact, are inevitably spatially and temporally diverse and heterogeneous. In addition, the concept of a limit or boundary might imply that all is well until the limit or boundary is reached, whereas significant and undesirable deterioration in ecological or human well-being might occur below the limit or boundary.

2. INTRODUCTION TO INDICATORS

To be useful in governance and other contexts, boundaries must be expressed through measurable indicators. When determining the contextual framework of an indicator, it is important to consider *what* the indicator measures and *for* what purpose. The answers to these questions correspond to the underlying reasons for wanting to measure and gather information to which the indicator relates. For example, an economic indicator is only meaningful if it conveys information relevant to a stated objective for the economy. Thus, if the paramount objective of the global economy is to ensure continuous growth, then the combined gross domestic product (GDP) of the world's nations presumably would present meaningful and useful information. By contrast, Brown and Garver (2009) contemplated an economy with a purpose to preserve and enhance. the integrity, resilience, and beauty of the commonwealth of life. A set of indicators of the extent to which this objective is being attained, and with potential use in supporting policies for attaining it, would be quite different from GDP.

The premise of this chapter is that the notion of a bounded safe operating space for the human enterprise provides part of the overall context for indicators of whether the economy is functioning as it should. That is, indicators derived from the planetary boundaries, including subglobal indicators where the viability of global boundaries is doubtful or regional/local impacts may be more significant than aggregate global ones, may be helpful for determining whether the operating space for humanity is "safe." In addition to this human-oriented notion of safe operating space, for which the principle of membership might not be clearly respected, the notion of right relationship in Brown and Garver (2009) implies a need for additional objectives and additional indicators for the economy, such as objectives related to the flourishing of life and life systems and to interhuman, interspecies, and intergenerational fairness inherent in the notion of right relationship. Brown and Garver (2009) defined right relationship with reference to Aldo Leopold's land ethic, updated as follows: "A thing is right when it tends to preserve the integrity, resilience and beauty of the commonwealth of life" (Brown and Garver 2009:5). Right relationship refers to a "guidance system for functioning in harmony with scientific reality and enduring ethical traditions" (Brown and Garver 2009:4). In its full essence, it captures entirely the notions of membership, householding, and entropic thrift.

It is also essential to consider how an indicator will be used. Beyond creating an early-warning system based on the boundaries, the authors of the planetary boundaries concept clearly had in mind potential applications of the boundaries in governance systems, which could include their use in developing regulatory limits on, or incentive structures for, human activities. Yet, the lack of an institutional architecture suitable for implementing a governance approach based on interacting planetary boundaries, as well as other significant challenges, must be overcome before planetary boundaries can have a meaningful role in governance at the global or subglobal level (Galaz et al. 2012). Indicators might also be used to set agendas for research, including both research aimed at improving the rigor or reducing the uncertainty of the indicators themselves and research on application of the indicators in a specific context. An example of the first type of research is the work analyzing the Human Development Index (HDI) against a broad range of human development criteria to highlight areas of development that are not well captured by the HDI (Ranis, Stewart, and Samman 2006).

Examples of the second type of research include the calculation by Quebec's Commissioner of Sustainable Development of the ecological footprint for Quebec (Lachance 2007), a study comparing the ecological footprints of conventional and organic Italian wines (Niccolucci et al. 2008) and a study of the relationship between HANPP and bird diversity in Austria (Haberl et al. 2005). Indicators may also be used as means to communicate information, either about the indicators themselves or about the objectives that underlie them (Schiller et al. 2001). A prominent example of the use of an indicator for broad communication purposes is the ubiquitous—and often misleading—use of GDP to convey information about the state of the currently dominant growth-insistent version of the economy at local, national, and global levels. In applying the analytical framework described in this chapter, it is important to consider the adaptability of an indicator to uses in governance, research, communications, and other arenas.

The information conveyed by indicators of the relationship of the human enterprise with the ecosphere depends not only on the purpose and potential applications of the indicators, but also on the nature of the measurements, data, and assumptions that are needed to develop and make use of them. As explained further in chapter 4, all of these features of indicators can be analyzed in depth by examining their contextual framework, the spatial and temporal scales at which they apply, the scope of the information they attempt to convey, their commensurability in conveying information with respect to overarching value-laden objectives, and their usefulness in policy applications.

Also as discussed in chapter 4, the reliability and validity, or accuracy, of indicators must also be considered. If an indicator does not measure what it purports to measure (referred to as criterion validity), then it is of no value. If the indicator is not measured precisely, then again it cannot be used effectively to determine the importance of spatial or temporal trends, as its variability may be greater than the actual changes. In both regards, as explained in chapter 4, composite indicators have inherent limitations that complicate their interpretation. A classic example of a composite economic indicator is GDP, which measures the total dollar value of the throughput of goods and services of the economy, usually annually (Costanza et al. 2009). It is a composite indicator because it reduces different kinds of transactions involving a wide variety of goods and services to the single measure of monetary value. If confined to use as an indicator of total economic activity and nothing more, GDP might be relatively harmless. However, GDP is suspect when it is used as the predominant measure of national well-being, even though it ignores many nonmarket transactions such as housework, fails to distinguish energy and resource flows, and includes economic activity that adversely affects the environment as a positive contribution. For example, the costs associated with responding to the damage from the BP oil accident in the Gulf of Mexico may contribute more to GDP than the oil would have contributed had it entered the market once the legal transactions, cleanup costs, and other economic activity associated with the disaster are counted—as with the Exxon Valdez oil spill (Ogle 2000). The HDI and ecological footprint, mentioned previously, and the panoply of indicators subsumed in the paradigm of "ecosystem health" are also composite indicators that raise significant issues of interpretation in many contexts, as discussed extensively in the next chapter.

3. SAFE OPERATING SPACE AND RIGHT RELATIONSHIP AS CONTEXTUAL FRAMEWORKS FOR ECONOMIC INDICATORS: CONFRONTING SCALE, DISTRIBUTION, AND EFFICIENCY

The concepts of "safe operating space" and "right relationship," taken together, provide a context for developing indicators of the state of and trends in the economy. Although they are complementary and overlap to a considerable extent, they are not coextensive. Planetary boundaries associated with safe operating space for humanity provide a foundation for indicators primarily of scale (although the global distribution of pressures on the boundaries and of the resulting impacts is important). Right relationship not only provides guidance as to the ecological limits of human activities (i.e., to scale), but it also includes additional criteria that are more relevant to the development of indicators related to distribution and allocation, such as in regard to the fairness of distribution among humans and between humans and nonhuman life.

Rockström et al. (2009) developed a methodology that seeks to fix planetary boundaries so as to avoid a zone of uncertainty, within which "tipping points" (or changes of state) may be crossed; this may lead to irreversible deviation from the conditions that have allowed human civilization, along with other life, to flourish during the Holocene. The proposed boundaries, which the authors intended to set at the most cautious end of the zones of uncertainty identified for seven of the nine global indicators, are as follows (Rockström et al. 2009):

- For climate change, limiting carbon dioxide in the atmosphere to 350 ppm and net radiative forcing to +1 watt per meter squared (zone of uncertainty: 350–550 ppm and +1 to +1.5 watt per meter squared)
- For ocean acidification, maintaining at least 80 percent of the preindustrial level of aragonite in the surface waters of the oceans (zone of uncertainty: 70–80 percent)
- For depletion of stratospheric ozone, limiting the loss to no more than 5 percent of preindustrial levels (zone of uncertainty: 5–10 percent)
- For interference with nitrogen and phosphorus cycles, limiting the flow of phosphorus to the oceans to no more than 10 times greater than the flow due to phosphorus from natural weathering (zone of uncertainty: 10–100 times greater) and limiting the amount of nitrogen removed from the atmosphere by human means to no more than 35 megatons of nitrogen per year (25 percent of the amount naturally fixed by terrestrial ecosystems) (zone of uncertainty: 25–35 percent)
- For global freshwater use, limiting freshwater withdrawals to no more than 4,000 cubic kilometers per year (zone of uncertainty: 4,000–6,000 km^3 per year)
- For land use change, limiting the percentage of the global ice-free land surface converted to cropland to 15 percent (zone of uncertainty: 15–20 percent)
- For biodiversity, limiting the rate of extinction of species to no more than 10 extinctions per million species per year (zone of uncertainty: 10–100 extinctions)

The authors did not propose global boundaries for atmospheric aerosol (particulate) loading or chemical pollution, and none have become apparent since. The Rockström team was unable to set an initial boundary for atmospheric aerosols because of the complexity of the nature of aerosols and the uncertainty regarding the processes by which they affect climate and human health and cause other environmental damage. Likewise, chemical pollution has a wide range of direct impacts on human and ecosystemic functioning. It acts as a slow variable that influences the other boundaries, but the Rockström team did not estimate a boundary because of the huge number of chemicals (e.g., more than 80,000 in the United States alone) with widely variable impacts, and because of the uncertainty regarding the affect of chemical pollution on the other boundaries.

According to planetary boundaries researchers, proposed boundaries for atmospheric carbon dioxide concentration, biodiversity loss, and anthropogenic additions of nitrogen and phosphorus to the global ecosystem have already been crossed (Carpenter and Bennett 2011; Hansen et al. 2008; Intergovernmental Panel on Climate Change 2007b; Rockström et al. 2009; United Nations Environment Programme 2009). In 2013, it was broadly reported that the atmospheric concentration of carbon dioxide had reached 400 ppm, consistent with long-term trends (Dlugokencky and Tans 2013). Net radiative forcing was approximately 1.6 watts per square meter, the annual rate of species extinction was estimated to be between 100 and 1000 extinctions per million species, and the amount of nitrogen that humans fix annually was approximately 140 megatons (Rockström et al. 2009). Particular emphasis is given to the boundaries for climate change, nutrient cycling, and biodiversity in the analysis below, taking into account concerns that the planetary boundaries approach; some of the initial estimates that the Rockström research team proposed must reckon further with complexity of the temporal and geographical distribution of both the drivers that put pressure on different boundaries and the resulting impacts.

The zones of uncertainty associated with planetary boundaries reflect zones that the Rockström research team concluded encompass either a threshold representing an abrupt, nonlinear change in the human–Earth system, as with climate change due to atmospheric greenhouse gases, or an accumulation of regional impacts so as to cause a globally dangerous level of impact, as with land use change or nitrogen loading. These zones of uncertainty reflect incomplete scientific knowledge about the thresholds at which human activities will cause these changes, as well as the irreducible uncertainty discussed in the chapter 14. The authors explain that *boundaries*, in contrast to *thresholds*, reflect normative judgments as to the degree of risk that is tolerable in light of the consequences of crossing planetary thresholds in the global ecosystem, accounting for this uncertainty. The authors describe the conceptual framework for planetary boundaries as proposing "a strongly precautionary approach, by setting the discrete boundary value at the lower and more conservative bound of the uncertainty range" (Rockström et al. 2009, supplementary information:7).

The precautionary principle states that "when confronted with serious or irreversible environmental threats, the absence of absolute scientific certainty should not serve as a pretext for delaying the adoption of measures to prevent environmental degradation" (Saunier and Meganck 2009:229), which in the context of the planetary boundaries implies a conservative approach in setting the boundaries. Adhesion to this specific application of the precautionary principle is a fundamental feature of the economic indicators envisioned in this chapter, notwithstanding the considerable challenge that would be involved in reaching such a global consensus on the precautionary principle (Galaz et al. 2012).

An essential common feature of these boundaries is that they pertain to the scale of the aggregate impacts of human activities on key aspects of the global ecosystem, viewed holistically. The central concern behind the planetary boundaries concept is that human activity is cumulatively causing global limits to be transgressed so as to "destablise critical biophysical systems and trigger abrupt or irreversible environmental changes that would be deleterious or even catastrophic for human well-being" (Rockström et al. 2009:3). This feature distinguishes them from the reductionist approach usually taken in setting environmental limits under provisions of environmental law, whereby environmental controls are determined according to the technological and economic feasibility of reducing the emission or discharge of specific pollutants from specific sources or classes of sources—and not primarily according to ecologically derived limits of ecosystems.

At least one critique of the planetary boundaries concept warns that overreliance on thresholds in general can support justification of behavior right up to the threshold—the edge of the cliff—when other criteria might provide reasons for staying well back (Schlesinger 2009). "Right relationship" complements the boundaries concept so as to address this concern because it focuses on the integrity, resilience, and beauty of the commonwealth of life at all stages. The scientific reality reflected in right relationship is that Earth is essentially closed to material inputs but open to energy from the sun—characteristics that fundamentally define limits on Earth's life-support capacity. The ethical reality of right relationship is derived from the principles of membership, householding, and entropic thrift, which provide guidance for managing human affairs so as to maintain integrity, resilience, and beauty at any point within the boundaries.

The planetary boundaries concept is one expression of the ecological limits within which right relationship can exist. In this sense, right relationship and safe operating space are both largely consistent with principles from ecological economics, particularly the "biophysical constraints of the economic sub-system" (Rockström et al. 2009:6). Furthermore, both recognize the uncompromising nature of planetary thresholds of change to life-depleting conditions, such that "ecological and biophysical boundaries should be non-negotiable, and that social and economic develop[ment] (should) occur within the safe operating space provided by planetary boundaries" (Rockström et al. 2009, supplementary information:5–6). However, right relationship is not framed around the outer bounds of the acceptable levels of global environmental stresses that pose threats to human well-being; it seeks instead a positive, flourishing, mutually enhancing human–Earth relationship (Berry 1999). It is not clear that the situation that would exist if all of the parameters on which the planetary boundaries are based were at their "safe" limit is the same as the one in which the integrity, resilience, and beauty of the commonwealth of life is preserved. Indicators based on planetary boundaries must therefore leave room for additional refinements needed to reach a more nuanced set of ecological and societal outcomes.

If the planetary boundaries are mostly about the aggregate scale of impacts, what is missing? Criteria for distribution, including considerations of distributive justice, are not necessarily captured in indicators of aggregate scale. Right relationship has ethical foundations that provide normative criteria related to distribution, in that it "include[s] the fair sharing of the earth's life support capacities with all of life's commonwealth" (Brown and Garver 2009:17). In this sense, right relationship includes notions of interspecies, interhuman, and intergenerational fairness that are not as clearly implicit in the notion of safe operating space, even though a proposed boundary is set for loss of biodiversity. Indeed, right relationship implies a shared sense of well-being and capacity for fulfillment within the commonwealth of life that transcends fairness.

Also missing from the planetary boundaries concept are normative criteria for developing indicators of efficiency. Efficiency is about getting a certain amount of one thing with the least use of another. The concept of efficiency depends on the underlying idea of the person and related conceptions of the good. The principles of membership, householding, and entropic thrift presented in chapter 2 require a notion of efficiency that is different from the maximization of monetary gain at the least monetary cost, which is associated with efficiency in conventional economics—one that is bounded by the ecological limits within which the economy must operate. The normative goal that right relationship embraces emphasizes sufficiency (not maximum wealth or unlimited accumulation) for individual people and other living beings within a life-enhancing, flourishing system. From the perspective of right relationship, one measure of economic efficiency would be the ability of the economy to provision the human enterprise with the lowest contribution to aggregate scale measured against ecological boundaries (e.g., lower the greenhouse gas intensity of industrial production); another its ability to achieve fairness with the lowest impact measured in terms of provisioning (e.g., meeting basic needs of all life while avoiding excess or waste). Thus, for example, indicators of economic efficiency based on right relationship would account for both poverty and excess.

4. WHAT NEEDS TO BE MEASURED FOR SCALE-BASED, DISTRIBUTION-BASED, AND EFFICIENCY-BASED INDICATORS OF THE ECONOMY?

The three categories of indicators introduced in the preceding discussion—indicators of aggregate scale, indicators of distribution, and indicators of efficiency—provide the foundation for a determination of the relevant variables that need to be measured to compile the indicators. A detailed answer requires consideration of the contexts in which indicators are intended to be used. Section 5 presents this discussion with respect to the key context of governance.

In general, indicators of aggregate scale depend on a range of measurements that show the spatial and temporal distribution of the material and energy stocks and flows, along with land and water use, that support the economy. Human appropriation of these flows can be either direct, as with various forms of exploitation (of minerals, fossil fuels, water, and biomass) or indirect, as with land use change (Haberl et al. 2007). These material and energy flows must then be correlated with the planetary boundaries using a systems approach, an area where considerable research is still required (Galaz et al. 2012). As noted above, the system complexities underlying the planetary boundaries and related ecological parameters and the uncertainties involved in estimating them are considerable.

The measurements related to the aggregate scale of the economy can be seen as relating to the I variable in the IPAT or IPATE framework discussed earlier, with the I representing the boundaries individually or collectively. Ensuring sufficient (but not excessive) prosperity in an ecologically finite world means providing all present and future members of life's commonwealth with "bounded capabilities" to flourish, contingent on Earth's limited capacity to support life and on fair intragenerational, intergenerational, and interspecies sharing of that capacity (Brown and Garver 2009; Jackson 2009:45–47). Once the parameters that need to be measured for indicators of the global aggregates of scale are determined, indicators of distribution and efficiency come in. In reference to the IPAT or IPATE framework, questions of distribution or efficiency relate primarily to the combinations of the P, A, T, and E variables, which for a given value of I are infinite. As Rockström et al. (2009) pointed out, myriad pathways are possible for staying within the safe operating space.

Although right relationship can serve as a criterion for distribution and efficiency, it also transcends distribution and efficiency. For example, all life on Earth might get its fair share of an ecological pie that is sized not to exceed the planet's ecological capacity, but the relationships might still be dysfunctional in a way that falls short of right relationship. This may go primarily to the "how" of governance or other applications, where the indicators concept likely will get more and more elusive. Thus, the measurements that are needed to support indicators of distribution and efficiency are less clear-cut than those for aggregate scale because they depend on ethical criteria, such as those that underlie interhuman, interspecies, intragenerational, and intergenerational fairness associated with right relationship. Fairness and justice, however, do not necessarily mean equality; an unequal distribution may nonetheless fulfill the extremely diverse and reasonable needs and desires of all members of the commonwealth of life. Chapter 3 provides a rich terrain from which to draw out these questions of justice and fairness. Metrics and indicators that differentiate justice claims at the level of individual members of life's commonwealth may be impracticable to gather or apply. However, an attempt to combine the planetary boundaries with proposed social boundaries, so as to describe a safe and just space for humanity, revealed categories of social well-being that are relevant to questions of distribution and efficiency: specifically, food security, income, water and sanitation, health care, education, gender equality, equity, voice, jobs, and resilience (Raworth 2012).

5. GOVERNANCE AS A KEY CONTEXTUAL CRITERION FOR ECONOMIC INDICATORS BASED ON PLANETARY BOUNDARIES AND RIGHT RELATIONSHIP

A particularly intriguing feature of the proposed planetary boundaries is their potential application in governance arenas, particularly the possibility that governments at various levels could transform them into regulatory limits or other policies that would constrain the human enterprise within their boundaries. Rockström et al. (2009:28) explicitly noted this potential in stating that "the planetary boundary framework . . . suggests the need for novel and adaptive governance approaches at global, regional and local scales." In the supplementary information to that paper, Rockström et al. also compared the planetary boundaries approach to other normative frameworks that are primarily concerned with establishing limits on human impacts to the environment. These other approaches included the tolerable windows approach, which was developed in Germany as a means to frame greenhouse gas emissions strategies; the critical loads methodology, which has been used in Europe to set air pollution limits based on critical levels at which pollutants have adverse effects on receiving ecosystems; and the safe minimum standards approach, by which limits are set for environmental variables such as species population size, habitat, and water quality, taking into account nonlinearity and thresholds in the relevant ecosystems. The Montreal Protocol on Ozone Depleting Substances and the total maximum daily load program under the U.S. Clean Water Act are additional examples. Applications in governance are key aspects of the contextual framework for developing indicators based on the planetary boundaries concept.

The potential for planetary boundaries to be translated into regulatory or normative programs or standards faces significant obstacles (Galaz et al. 2012). The proposed planetary boundaries are uncompromising in a way that environmental standards reflecting the primacy of economic objectives and political constraints over environmental considerations are not (Garver 2013). The very nature of the boundaries, as defined, suggests that the failure to apply them in normative and proscriptive ways would raise an intolerable risk of catastrophe. Thus, the governance that the planetary boundaries concept implies is governance of necessity, not convenience, and it is governance that flips the hierarchy of factors to give ecological limits primacy over economic or sociopolitical factors. Socioeconomic goals—such as meeting basic human needs, along with more flexible social and psychological needs related to well-being (Jackson 2009), reducing sociopathic levels of inequality, promoting participation, and minimizing violent conflict—should be conditioned on the primary goal of limiting the aggregate scale of human activities and their ecological impacts.

Another challenge is that planetary boundaries imply the need for a much more comprehensive governance system—from the global to the local level—than has ever existed within the human community. For example, existing international environmental institutions with programs and mandates related to the planetary boundaries are fragmented and relatively inflexible, whereas meaningful implementation of the planetary boundaries approach in governance would require integrated research and policy development, as well as built-in flexibility and adaptability as planetary boundaries research evolves (Galaz et al. 2012). Varying challenges regarding measurement, described generally in chapter 4, confront indicators of all the boundaries, especially the ones that refer to land use (where satellite imagery can help), phosphorous and nitrogen cycles, freshwater use, and loss of biodiversity.[2] Typically, parameters related to these boundaries need to be measured accurately on small geographical scales and then aggregated to the global scale, with appropriate attention to tricky cross-scale dynamics (Dirnböck et al. 2008). For these boundaries, the most relevant and useful indicators, and the primary focus of governance, may be along local or regional scales, even if in the aggregate the boundaries are global (Nordhaus, Shellenberger, and Blomqvist 2012). Even with more homogenous global phenomena, such as climate change, ocean acidification, and stratospheric ozone depletion, different areas of the planet may be contributing or reacting to degradation in heterogeneous ways because of the distribution of ecosystems and the human population, geographic diversity, and other factors, again suggesting regional differences. Moreover, the boundaries relate to coupled human and biophysical systems that do not always follow linear, predictable patterns, and that involve tipping points, emergent properties, chaotic behavior, and stochastic events (Galaz et al. 2012;

Kotchen and Young 2007). Framing governance to confront these characteristics of the human–Earth relationship as reflected in the planetary boundaries will require an innovative, adaptive approach that transcends the normal desire for stable, predictable rules and policies.

Another challenge is to consider carefully the timescales at which applications of the boundaries in governance would take place. What do we do with estimates of planetary boundaries: rush toward the brink and slam on the brakes at the last minute, or cruise to a stop at a line well before the brink? The current economic framework essentially suggests that we need not worry about the brink because, as we approach the edge, prices of the relevant commodities will increase enough that we will find alternative technologies, slow down, and stop before plunging into the abyss. It is as if a force field will push us back harder the closer we get. A key problem with this framework is that prices reflect perceptions of certain kinds of scarcity, with a strong bias toward the present as opposed to the future; they are not based on the true scarcity of the life-support capacity that is reflected in a boundaries-based paradigm (Victor 2008). The conventional framework also tends to assume full knowledge and accounting of the costs behind prices—that all of the players have full knowledge and that everyone makes rational decisions.

Therefore, the ethical framework underlying the governance context for indicators based on planetary boundaries must take into account that knowledge underlying human transactions is incomplete and that knowledge is not shared equally among the "decision-makers" at various levels, from individuals to global institutions. Human beings also make "irrational" decisions because they do not have access to all of the relevant information, and some amount of uncertainty is inevitable. Thus, they must fill in blanks and act on assumptions—or, perhaps more often than we would like to think, on whimsy. Perhaps more importantly, irrational decisions arise because of conflict between what is perceived to be good for the individual in the moment and what is good for the broader community and the future.

Governance systems are needed to act as filters that "mediat[e] between human actions and biophysical processes" (Kotchen and Young 2007:150), as shown in figure 5.1. In this configuration, the governance filter "consists of the sets of rights, rules, and decision-making procedures that are created by humans to guide actions, including those that may have disruptive

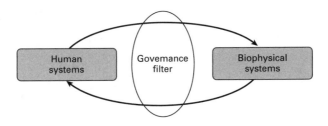

Figure 5.1. Governance filters in coupled human and biophysical systems. *Source*: Kotchen and Young (2007).

impacts on biophysical systems" (Kotchen and Young 2007:150), as well as mechanisms such as insurance programs that address adaptation to at least some impacts. Whereas natural resource management has traditionally focused on regulating biophysical systems so as to ensure human welfare (protecting humans from the volatility of biophysical systems), the arrival of the Anthropocene[3] heightens the need to apply this governance filter more to human actions so as to reduce their impacts on biophysical systems (protecting biophysical systems from humans; Kotchen and Young 2007). The planetary boundaries concept falls neatly into this depiction of a governance filter.

As currently proposed, the planetary boundaries are an initial estimate in need of further refinement (Rockström et al. 2009). For the two global variables without initial estimates (chemical pollution and atmospheric aerosols), the boundaries concept appears to be very difficult, if not impossible, to apply. The boundaries are not "a roadmap for sustainable development," but rather an overall envelope "within which humanity has the flexibility to choose a myriad pathways for human well-being and development" (Rockström et al. 2009:7). In this sense, as noted earlier in this chapter, their proposal is similar to framing governance choices using the $I = f(PATE)$ framework, by which a fixed limit on the value for human impacts, I (which could represent the planetary boundaries individually or collectively), allows a range of flexibility for decisions regarding population (P), affluence and consumption (A), technology (T), and ethics (E) (Brown and Garver 2009).

The Rockström research team also acknowledged the need for further research on the application of risk assessment and the precautionary principle in setting standards, in order to sort out apparent discrepancies that typically exist between the generally low risks deemed acceptable in human health and welfare (particularly from invisible, carcinogenic, low-risk but high-magnitude, inequitable, and involuntary risks) and the generally higher risks deemed acceptable in environmental decision-making (Rockström et al. 2009, supplementary information). Nonetheless, their initial estimates provide a basis for assessing, at least preliminarily, the further development of the governance context for the boundaries, individually and collectively. Here, particular attention is given to the climate change, nitrogen cycle, and biodiversity boundaries.

Table 5.1 provides a framework for translating the planetary boundaries into policy applications, with examples of key applications of the climate change, nitrogen cycle, and biodiversity boundaries in the context of governance. In making this translation, parameters that are readily useful in policy or regulatory settings must be identified for indicators at the global, regional/national, and local scales. Moreover, indicators must be established at different geographic scales with awareness of interscale dynamics, both from a governance perspective and an ecological perspective. As noted earlier, where the viability of global-scale boundaries is in doubt (Nordhaus, Shellenberger, and Blomqvist 2012) or where regional/local boundaries or indicators are deemed more appropriate for other reasons, a focus on subglobal scales will be more important. It is also important to identify the key policy arenas or options and the economic sectors that need to be targeted at different scales. Table 5.1 lists models or existing frameworks that may serve as models for governance options.

Ten key criteria are important to bear in mind in considering governance applications as central to the contextual framework for indicators of an ecologically bounded and ethical economy. Given the current overshoot of Earth's ecological capacity, the economic and ecological governance for which indicators are needed should include the ten following mutually reinforcing features (Garver 2013):

1. It should recognize that humans are part of Earth's life systems, not separate from it—hence, the notion of membership articulated in chapter 2.
2. Legal and policy regimes must be constrained by ecological considerations necessary to avoid catastrophic outcomes and promote the

TABLE 5.1

Trends in boundaries and ranges of uncertainty from the preindustrial era for global ecological limits for the human economy

Process or factor	Boundary	Range of uncertainty	Preindustrial	Current (2012)
Climate change: CO_2 /radiative forcing	350 ppm, +1 w/m^2	350–550 ppm, +1–1.5 w/m^2	280 ppm, 0 w/m^2	390 ppm, +1 w/m^2
Land use: global ice-free land surface converted to cropland	15%	15–20%	NA	12%
Acidification of the ocean (Ω_{arag})	2.75 (80% preindustrial)	2.41–2.75 (70–80%)	3.44	2.9
Stratospheric ozone	~278 Dobson Units (5% less than preindustrial, extrapolar stratospheric ozone)	5–10% less than preindustrial	290 Dobson Units	280–290 Dobson Units
Biogeochemical nitrogen: amount of nitrogen removed from the atmosphere by humans	35 Mt/year	25–35 Mt/year	Negligible	Anthropogenic: 150 Mt/year
Biogeochemical phosphorous: flow of phosphorus to the oceans from human use	10 times natural flow (10 MT/year), 24 in freshwater lakes, 160 mg per m^3 in rivers	10–100 times natural flow (~10–100 Mt/year)	1.1 Mt/year	9 Mt/year
Global freshwater use (human withdrawals/year)	4000 km^3 per year	4000–6000 km^3 per year	NA	~4000 km^3 per year
Loss of biological diversity	10 E/MSY	10–100 E/MSY	0.1–1 E/MSY (fossil record)	>100 E/MSY

Note: E/MSY, extinctions per million species per year; NA, not available.
Sources: Carpenter and Bennett (2011), Dlugokencky and Tans (2013), Douglass and Fioletov (2010), Rockström et al. (2009).

flourishing of life, with the socioeconomic sphere fully contained within these ecological constraints and ecosystems restored where necessary.

3. Boundaries-based laws and policies must permeate legal and policy regimes in a systemic, integrated way, not be seen as a specialty area of law or policy.

4. Because the human enterprise has already surpassed global ecological limits, legal and policy regimes should be radically refocused on reduction of the throughput of material and energy in the economy.

5. Boundaries-based governance must be global but distributed fairly using principles of proportionality and subsidiarity, with protection of the global commons and public goods paramount and constraints on property rights and individual choice implemented as needed to keep the economy within ecological limits.

6. Legal and policy regimes must ensure fair sharing of resources among present and future generations of humans and other life.

7. Boundaries-based laws and policies must be binding and supranational, with supremacy over subglobal legal regimes as necessary, and with rights of enforcement for nonstate actors.

8. A greatly expanded program of research and monitoring for improved understanding and continual adjustment of ecological boundaries and means for respecting them is needed to support boundaries-based governance approaches from the global to the local level.

9. Boundaries-based governance approaches require precaution about crossing planetary boundaries, with margins of safety to ensure both that the boundaries are respected from the global to the local level and that Earth's life systems have the capacity to flourish.

10. Boundaries-based governance must be adaptive because ecosystems evolve constantly and because we need to get started on a comprehensive effort to constrain the economy within ecological limits despite uncertainty.

5.1. Climate Change

An immediately apparent characteristic of the proposed climate change boundary is that the variables chosen to establish it—concentration of carbon dioxide in the atmosphere and net radiative forcing—are global, indivisible variables that are not directly applicable in a regulatory or other system of governance designed to apportion the right to assert a limited level of environmental impact globally, regionally, and locally. For this reason, the climate change boundary resists the criticism of being nonviable or meaningless at the global level (Nordhaus, Shellenberger, and Blomqvist 2012). In many senses, the climate change boundary is the soundest and most important one, given its relation to most of the others.

To fit more comfortably in governance contexts, the climate change boundary needs to be translated into net emissions of carbon dioxide and other greenhouse gases, taking into account factors such as their relative contributions to climate change, long-term trends in feedbacks and the behavior of terrestrial and marine sources and sinks of carbon, and the time frame in which stabilization at 350 ppm carbon dioxide in the atmosphere and net radiative forcing of 1 watt per square meter is sought. These boundaries are presumed to correlate roughly with a limit on average global warming of about 2 degrees Celsius (Rockström et al. 2009)—the temperature rise agreed to internationally as a target, as in the Copenhagen and Cancun climate accords in 2009 and 2010. Ultimately, the climate change boundary boils down to the amount of fossil fuels—coal, oil, and natural gas—that humans can ethically extract consistent with the principles of membership, householding, and entropic thrift that underlie right relationship and safe operating space.

These are complicated matters—the 2007 Fourth Assessment Report of the Intergovernmental Panel on Climate Change (IPCC) addresses them in more than two thousand pages. The Rockström research team noted that the IPCC's stabilization scenarios, which are built around achieving long-term stabilization at various atmospheric concentrations of carbon dioxide, are similar to boundaries, although the IPCC does not frame those scenarios around thresholds involving abrupt, nonlinear, and irreversible system changes. Nonetheless, the stabilization scenarios of the IPCC and others provide a starting point for translating the Rockström research team's climate change boundary into a form that is more readily adapted to the governance context. Essentially, the exercise is simple. First, a stabilization curve needs to be developed, based on the relationship of net greenhouse gas emissions to atmospheric greenhouse gas concentrations and the resulting change in average global temperature. The stabilization curve will yield a long-term set of annual limits on greenhouse gas emissions that must be respected to keep the average global temperature from exceeding unacceptable limits—roundly accepted as 2 degrees Celsius above the pre-industrial average. Next, the allowable sources of those annual greenhouse gas emissions must be determined, which in turn indicates the sources and amounts of fossil fuels that can be burned within the annual limit taking into account carbon sinks—an assessment that requires continual adjustment as sinks trend toward saturation.

An illustration based on the information available in the IPCC's Fourth Assessment Report puts this approach in perspective. The IPCC estimated that stabilizing carbon dioxide equivalent concentrations in the atmosphere at 450 ppm, which is roughly equivalent to 350 ppm carbon dioxide,[4] in the long term would require developed countries to reduce their carbon dioxide emissions from levels in 1990 by 25 to 40 percent by 2020, and by 80 to 95 percent by 2050 (Intergovernmental Panel on Climate Change 2007a). The shaded band in figure 5.2 shows the total global reductions in carbon dioxide emissions that the IPCC estimated would be needed for long-term stabilization at 350 to 400 ppm of carbon dioxide in the atmosphere. The eventual dipping of those pathways below zero emissions accounts for the possible development of technologies that

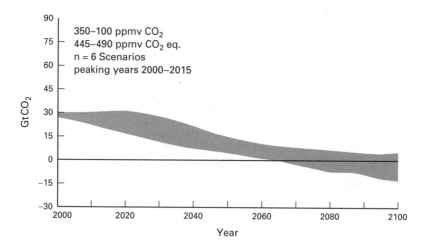

Figure 5.2. Global emissions pathways to stabilization at 350–400 ppm CO_2 in the atmosphere. *Source*: Intergovernmental Panel on Climate Change (2007b).

remove carbon dioxide from the atmosphere. Some would include geoengineering options here, but the potential for extremely negative unforeseen consequences with most geoengineering options, not to mention the technological challenges, is highly problematic (Cao and Caldeira 2010; Robock 2008).

Translating the climate change boundary into targets for reductions of emissions, for nations, firms, or individuals, is an essential step for implementing the climate change boundary in well-founded proscriptive or other forms of governance. For example, one proposal is to establish a long-term climate stabilization curve to determine the total amount of greenhouse gases that can safely be emitted by the end of a fossil-fuel phase-out period and to divide that total equally among developed and developing countries, as represented by inclusion or exclusion from annex B of the Kyoto Protocol (Schlesinger, Ring, and Cross 2012). The policy options in various sectors in table 5.2 (Intergovernmental Panel on Climate Change 2007a) provide examples of the specific contexts of governance in which the climate change boundary could be applied. Working toward a particular stabilization target allows flexibility, both in terms of the different kinds of policy options and because the same stabilization target can be met, on the same schedule, following either a relatively high peak in emissions followed by a sharp decrease or a relatively low peak in emissions followed by a more gradual decrease, where the cumulative emissions are the same (Anderson and Bows 2008).

However a stabilization curve keyed to the climate change boundary is derived, the choices drawn from the possible menus of options must add up to the reductions needed to meet the climate change boundary. In this sense, the primacy of the boundary must be underscored. The climate change boundary, unlike the IPCC's six emission scenarios (Intergovernmental Panel on Climate Change 2007b), is not an output scenario based on assumptions about the rate of economic growth, human development and lifestyles, technological change, population, and other factors accounted for in the IPCC's scenarios. Rather, inherent in the boundaries concept is the notion that the climate change and other boundaries compel a set of governance options that, while flexible, are ecologically bounded— meaning that patterns of economic and human development, technological change, demographics, and so on must be adapted to stay within the boundary.

TABLE 5.2
Framework for developing policy-relevant indicators based on planetary boundaries

Planetary boundary	Geographic scale	What to measure	Key policy arenas and/or options	Target sectors	Existing models or starting points
Climate change	Global	Net GHG emissions (sources minus sinks); global ecosystem effects of climate change	Post-Kyoto GHG and climate change agreement	National governments	UNFCCC, Kyoto Protocol
	Regional, national	National accounts of net GHG emissions; regional ecosystem effects of climate change	Regional or national GHG emissions control, or other climate policies	Subnational governments or specific sectors or sources (e.g., power plants, motor vehicles)	European carbon market; U.S. Clean Air Act
	Subnational, local	Source-specific GHG emissions or carbon uptake; local ecosystem effects of climate change	Regulatory limits on specific sources; measures to protect carbon sinks; carbon capture and storage; carbon taxes; etc.	All economic sectors	State programs in California, Northeastern states in the U.S.
Nitrogen flux	Global	Total nitrogen loading	Agreement on nitrogen loading	National governments	None directly on topic
	Regional, national	National or regional N loads from fertilizers, fossil fuels and other sources; ecosystem effects of N loading	Water and air pollution laws, non-point source pollution controls	Subnational governments, key sectors such as agricultural and fossil-fuel powered power plants	Convention on Long-range Transport of Air Pollutants; U.S. Clean Water Act total maximum daily load and nonpoint source pollution programs
	Subnational, local	Local N loading, and ecosystem effects of N loading	State/provincial or local water and air pollution laws and policies	Agriculture, power plants	State/provincial or local nutrient management or air pollution programs

Biodiversity loss	Global	Extinction rates; status of and changes in species abundance and diversity; status of and changes in habitat; status of and changes in drivers of biodiversity loss (habitat loss, overexploitation, pollution, invasive species, climate change), e.g. relationship of ecological, footprint or HANPP to biodiversity metrics; ecosystem effects of biodiversity loss	Global agreement on habitat change or loss, overexploitation, pollution, invasive alien species, climate change	National governments	Convention on Trade in Endangered Species
	Regional, national	Regional, national	National species and habitat protection legislation	Subnational or local governments, all economic sectors, and landusers ad owners	U.S. Endangered Species Act
	Subnational, local	Subnational, local	Local land use controls and property restrictions; habitat reserves; ecological restoration programs	Land users and owners, all economic sectors	State/provincial and local species or habitat protection laws; ecosystem conservation programs; state/provincial or local parks and reserves

Note: GHG, greenhouse gas; UNFCCC, United Nations Framework Convention on Climate Change.

The promise of the planetary boundaries concept is that it frames the options in terms of biophysical realities, allowing a margin of safety to account for risk and uncertainty but with no softening of ecological thresholds to accommodate socioeconomic or political concerns. The dark side of this promise is the international community's collective inability to date to respond to a concept like planetary boundaries. The global community has recognized the need for, but not committed to, "scaled-up overall mitigation efforts that allow for the achievement of desired stabilization levels [which] are necessary, with developed country Parties showing leadership by undertaking ambitious emission reductions" (UN Framework Convention on Climate Change 2010, point 2[a]). The world's nations agree on the need to respect a target of 2 degree Celsius and have made reference to the emissions reductions that the IPCC's Fourth Assessment Report said were necessary to reach that target—an implicit incorporation of reductions beyond the decrease in emissions in developed countries of 25 to 40 percent from 1990 levels by 2020 associated with an atmospheric concentration of carbon dioxide of 450 ppm. However, the most recent policies of the United States, whose nonbinding target is to reduce its carbon dioxide emissions in 2020 by only approximately 4 percent compared to its emissions in 1990, and of Canada, whose nonbinding target is to *increase* its carbon dioxide emissions in 2020 by about 2 percent compared to its emissions in 1990, lend credence to the pessimism of Anderson and Bows (2008). Yet, those policies represent not simply a compromise of economic and environmental concerns, but a dangerous departure from the emissions reductions required to avoid the risk of catastrophic systemic changes in the climate system.

5.2. Nitrogen Loading

According to Rockström et al. (2009), humans now add nearly four times the amount of nitrogen to global ecosystems than the level of the proposed planetary boundary for nitrogen. Through conversion of atmospheric nitrogen to ammonia using the Haber-Bosch process (approximately 80 megatons of nitrogen per year), agricultural fixation of nitrogen from leguminous crops (approximately 40 megatons of nitrogen per year), combustion of fossil fuel (approximately 20 megatons of nitrogen per year), and burning of biomass (approximately 10 megatons of nitrogen per year),

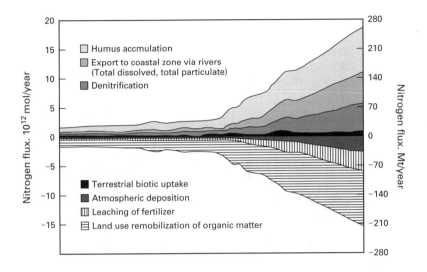

Figure 5.3. Modeled flux of human-induced nitrogen flux in global land systems. *Source:* Mackenzie, Ver, and Lerman (2002).

humans now incorporate into global ecosystems about the same amount of nitrogen as is fixed by nonanthropogenic processes (Rockström et al. 2009). Prior to the industrial revolution, human fixation of atmospheric nitrogen was negligible (Rockström et al. 2009). Figures 5.3 and 5.4 show a modeled time series of the human-induced changes in the partitioning of the nitrogen released and stored in global land (figure 5.3) and coastal margin (figure 5.4) systems, with projections out to the year 2030 assuming a business-as-usual scenario (Mackenzie, Ver, and Lerman 2002). Positive values on those figures represent storage mechanisms, and negative values represent release mechanisms.

Along with human-caused additions of phosphorus and sulfur, this rapid expansion in the addition of nitrogen to biogeochemical systems contributes to climate change, increases in smog and ground-level ozone levels, eutrophication of aquatic systems, and acid deposition (Mackenzie, Ver, and Lerman 2002). At the same time, the accumulation of nitrogen and phosphorus in the environment can enhance the ability of terrestrial ecosystems to capture atmospheric carbon dioxide (Mackenzie, Ver, and Lerman 2002). These various impacts are significant both locally or regionally,

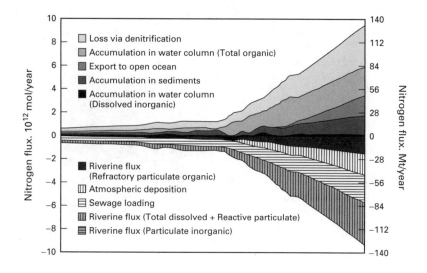

Figure 5.4. Modeled flux of human-induced nitrogen flux in global coastal margin. *Source*: Mackenzie, Ver, and Lerman (2002).

where they contribute to urban and other locally significant air pollution and to eutrophication of aquatic systems, and globally, where nitrogen acts as a "slow variable, eroding the resilience of important sub-systems of the Earth System" (Rockström et al. 2009:15). As well, nitrous oxide contributes to climate change. Thus, the planetary boundary for nitrogen is relevant to governance at all levels, from local to global. The units for the nitrogen boundary—megatons of nitrogen per year—could simplify its direct application in governance contexts, such as by setting a limit on the amount of nitrogen fixed through the Haber-Bosch process as ammonia. However, the end use of this ammonia as fertilizer used in food production greatly complicates such a simple application of the boundary, even if the Rockström researchers see the nitrogen boundary as a potential impetus for reducing synthetic fertilizer production. Furthermore, the nitrogen boundary is a "first guess" (Rockström et al. 2009) and in need of further development, and others have found it to be somewhat arbitrary (Schlesinger 2009) if not virtually meaningless (Nordhaus, Shellenberger, and Blomqvist 2012).

In the short term, attempts to limit smog and to protect local or regional aquatic systems are likely to be the most active areas of governance, as in places where excessive nutrients have caused outbreaks of cyanobacteria or accelerated eutrophication. These local and regional efforts are likely to vary considerably, depending on the pressures from various sectors that contribute to nitrogen pollution and on the sensitivity of local and regional ecosystems. Bouwman et al. (2002) provided an estimate of the global distribution of critical loads of nitrogen eutrophication (i.e., the estimated maximum ecosystem tolerance of added nitrogen) and of the extent to which those critical loads have been exceeded. Matched with information on sectoral contributions, this distribution can support the development of priorities for regional and local controls on nitrogen pollution. This is essentially the approach taken in Europe using the critical loads and levels methodology, where "critical load" is defined as a "quantitative estimate of an exposure to one or more pollutants below which significant harmful effects on specified sensitive elements of the environment do not occur according to present knowledge" (Umweltbundesamt 2004:V-1). A similar approach is taken with the total maximum daily load (TMDL) program under the U.S. Clean Water Act, which requires plans for reducing the input of pollutants from point sources and nonpoint sources (the most important source of nitrogen-based pollutants, especially from agriculture) to water bodies that do not meet federal water quality standards. However, controlling eutrophication due to nutrient pollution has been a particularly difficult challenge in both the critical loads and levels approach and the TMDL program at the regional level. At the global level, further research on the biogeochemistry of nitrogen and on the impact of local and regional controls is needed to determine the most effective way to refine and implement the nitrogen boundary in global governance contexts.

5.3. Biodiversity

Of the seven boundaries that the Rockström research team proposed, the boundary for biodiversity perhaps best illustrates the challenge of translating planetary boundaries into a form suitable for governance applications. Although climate change is already exacerbating pressures on biodiversity globally and is expected to continue to do so for quite some time (Butchart et al. 2010; Ehrlich and Pringle 2008), the most pressing threats to biodiversity arise primarily at the local level, through direct takings of species, habitat destruction or degradation, toxic pollution and invasive species (Ehrlich and Pringle 2008). As the Rockström team acknowledged, considerable uncertainty impedes piecing together these diverse local situations to derive a global boundary. Yet, even in jurisdictions with apparent policies of zero-tolerance for species extinctions, in which endangered and threatened species are protected either strictly, as in the United States, or less so, as in Quebec and the rest of Canada, threats to biodiversity persist. Furthermore, studies indicate not only that global goals for reducing biodiversity losses are not being met but also that declines are worsening (Butchart et al. 2010; Ehrlich and Pringle 2008). In Quebec, threatened and endangered species are protected under the Act Respecting Threatened or Vulnerable Species (1999), but the government resources for conducting comprehensive inventories and for ensuring protection of species and their habitat on private property are limited. Thus, the Western chorus frog, listed as threatened in Quebec, faces ongoing and incremental fragmentation of its habitat from suburban and urban development in the Montérégie region. Each increment is a small diminution of the species' chance of long-term survival in Quebec, but the cumulative effect poses a significant threat (Équipe de rétablissement de la rainette faux-grillon de l'ouest 2000; Quebec Ministry of Natural Resources and Wildlife 2010).

A planetary boundary based on the rate of extinction does not easily translate into effective regulations and policies at the local level. Extinction is the end point of a generally long process of decline, and preservation of biodiversity therefore requires mechanisms of governance that apply at earlier points in the process. These mechanisms must be fashioned so as to be applicable at every scale of importance—in this case many species, down to the landowner level. Furthermore, a complete set of complementary, mutually reinforcing mechanisms is necessary. A restriction on land uses to protect species of concern so as to maintain biodiversity will be effective only if the responsible authorities also have the capacity to conduct comprehensive inventories of the occurrence and abundance of species and to obtain comprehensive information on the ecological requirements for the species' long-term survival. Furthermore, they must have the ability to apply a precautionary approach that does not place the burden of uncertainty regarding factors that affect species' survival on the species. It is safe to say that jurisdictions in which all of these conditions are satisfied are few and far between—if they exist at all.

The planetary boundary for biodiversity based on extinction rates clearly requires some form of translation to make it useful in regulatory applications. However, as with the nitrogen boundary, "it remains very difficult to define a boundary level of biodiversity loss that, if transgressed for long periods of time, could result in undesired, non-linear Earth System change at regional to global scales" (Rockström et al. 2009:18). Thus, a boundary based on extinction rates is proposed as an interim indicator. Ultimately, a different boundary measure or measures is needed, given that data on the abundance and distribution of species is limited and unequal among species, the relationship between extinction and global environmental change is complex and not well understood, and rates of speciation and extinction vary widely among groups of organisms and habitats (Samper 2009). Some doubt that this can ever be done meaningfully (Nordhaus, Shellenberger, and Blomqvist 2012).

Accordingly, enforcement of laws and implementation of programs that protect species and habitats at the local and regional level should continue and be strengthened. In the context of global governance, in the short term the primary agenda with respect to biodiversity should be an aggressive, well-funded research and monitoring agenda. A key area of research is to develop better knowledge of species around the globe that are facing extinction, which can be used to identify hotspots for priorities for governance related to preservation of biodiversity. Research is also needed to identify bioindicator species that can be monitored and serve as the focal points of governance action aimed at preserving biodiversity (Bestelmeyer and Wiens 2001; Rossi and van Halder 2010; Stuart et al. 2010).

Another potentially promising area is research on the relationship between species extinctions and human use of ecosystems on which they depend. For example, using species–energy curves, which relate the number of species to the total production of available energy in a given area, the effect of HANPP on the total energy available has yielded field-verified estimates of the percentage of species expected to be extinct or endangered (Wright 1990). More recent work has expanded the research on the relationship between HANPP, or of activities that contribute to it, and species diversity in various contexts (Haberl et al. 2007). Specific examples include the relationship of livestock grazing (Bestelmeyer and Wiens 2001), agricultural production more generally (Haberl et al. 2004b) and land-use decision-making (Bestelmeyer and Wiens 2001) to biodiversity, using HANPP and other measures of human impact. Bishop, Amaratunga, and Rodriguez (2010) proposed a "sustainable" threshold for HANPP at the level necessary to satisfy basic human food needs while preserving ecosystem functions that will limit biodiversity loss; they made a preliminary estimate of 9.72 Pg C/yr, a 60 percent decrease from actual HANPP in 2005. However, establishing such a threshold is difficult because the nature and quality of primary production are highly variable spatially; intensification of land use with pesticides, fertilizers, and irrigation involve sustainability tradeoffs and can raise net primary production while lowering HANPP; and HANPP does not account well for the waste absorption function of ecosystems (Erb et al. 2009; Foley et al. 2007; Haberl et al. 2004a).

Further refinement of methodologies for relating HANPP and biodiversity, especially in a spatially explicit manner, would be useful for better identifying or forecasting—and perhaps controlling—threats to biodiversity. More broadly, tracking the movement of HANPP and other material and energy sources in the global economy at high resolution, in conjunction with geographically explicit data on the drivers and impacts related to biodiversity—as well as to climate change, biogeochemical fluxes, and other key systemic processes—will allow for better governance of impending sustainability challenges at an ecosystem-specific scale. Ongoing extension of the analysis of HANPP and its relationship to biodiversity and other planetary boundaries can particularly help with challenges related to feeding a rising human population that is trending toward greater consumption of meat; deciding the future role of biofuels in the global energy picture; contending with the trend toward increasing metabolic rift in an increasingly globalized economy; and controlling the impacts of intensive biomass production that depends on fertilization, genetically modified organisms, pesticides and irrigation (Bringezu, O'Brien, and Schütz 2012; Erb et al. 2009; Foley et al. 2011).

6. AN ITERATIVE APPROACH TO DEVELOPING AND USING INDICATORS BASED ON PLANETARY BOUNDARIES

The general approach of the Rockström research team to estimating planetary boundaries is based on precaution, where being cautious about crossing the boundaries takes precedence over economic or political risks associated with managing the economy to stay within the boundaries. This implies a need to apply the indicators in governance contexts, even though some degree of uncertainty may remain regarding the boundaries or the thresholds to which they relate. For some planetary boundaries, such as the biodiversity and nitrogen-loading boundaries, the current level of uncertainty precludes effective implementation of governance-ready indicators based on the global boundary, although knowledge of regional and local impacts is sufficient to improve governance mechanisms to address concerns at those scales. For others, such as the stratospheric ozone and climate change boundaries, current levels of uncertainty are insufficient to preclude applying indicators based on the boundaries from being implemented in global governance contexts. However, the Rockström research team explained that the boundaries were each set on the understanding that all of the other boundaries were not exceeded for significant time periods. Because they interrelate, if that assumption is not true, then the boundaries need to be adjusted. Under the analytical framework proposed for the assessment of indicators based on planetary boundaries and complementary principles such as right relationship, the governance context needs to be dynamic and fluid, so as to allow for an iterative process in which the indicators and the objectives for which they are designed to provide information may be refined over time in light of additional understandings gained through research and applications. The intention is that this iterative approach will lead to continual improvement of indicators and an increase in their value and potential to be applied in a variety of contexts over time.

7. CONCLUSION

The premises of membership, householding, and entropic thrift that undergird the approach to ecological economics in this book can be broken down further into criteria that allow for normative assessment of how humanity is faring in relation to them. In chapter 2, virtues such as courage, epistemological humility, atonement, fairness, and respect were presented as indicators of the economy's functioning. In addition to those qualitative indicators, this chapter presented a framework for developing some quantitative indicators for ecological economics. These quantitative indicators are built on the overarching concepts of safe operating space for humanity and right relationship; they take into account the challenges associated with measurement and irreducible uncertainty in the complex global ecosystem.

A concerted effort, from the global to the local level, will be required to develop a robust set of qualitative and quantitative indicators that can be used to establish rules of decision for respecting the essential premises of ecological economics.

NOTES

1. Rio Principle 12 states in part: "States should cooperate to promote a supportive and open international economic system that would lead to economic growth and sustainable development in all countries, to better address the problems of environmental degradation." *The Future We Want*, the outcome document adopted at the Rio+20 conference in June 2012, makes over twenty references to the international community's commitment to sustained economic growth.

2. As noted above, the challenges with respect to chemical pollution and atmospheric aerosol loading are such that boundaries have not even been proposed.

3. The term *Anthropocene* was coined in 2000 to describe the era beginning at least at the start of the industrial revolution in the eighteenth century in which large-scale and long-lasting impacts of humans on Earth's ecology and geology became globally important (Crutzen and Stoermer 2000). The proposed planetary boundaries were developed in large part with reference to the "relatively stable environment of the Holocene, the current interglacial period that began about 10,000 years ago, allow[ing] agriculture and complex societies, including the present, to develop and flourish" (Rockström et al. 2009:3).

4. Rockström et al.'s (2009) proposed climate change boundary is based on carbon dioxide concentrations—not carbon dioxide equivalent concentrations, which accounts for the contribution of other greenhouse gases—on the theory that, at least for now, the contributions of other greenhouse gases to global warming are roughly compensated by the cooling effect of atmospheric aerosols from human activities.

REFERENCES

Anderson, Kevin, and Alice Bows. 2008. "Reframing the Climate Change Challenge in Light of Post-2000 Emission Trends." *Philosophical Transactions of the Royal Society A: Mathematical, Physical and Engineering Sciences* 366 (1882): 3863–3882. doi: 10.1098/rsta.2008.0138.

Berry, Thomas. 1999. *The Great Work: Our Way into the Future*. New York: Three Rivers Press.

Bestelmeyer, Brandon T., and John A. Wiens. 2001. "Ant Biodiversity in Semiarid Landscape Mosaics: The Consequences of Grazing Vs. Natural Heterogeneity." *Ecological Applications* 11 (4): 1123–1140. doi: 10.1890/1051-0761(2001)011[1123:ABISLM]2.0.CO;2.

Bishop, Justin D. K., Gehan A. J. Amaratunga, and Cuauhtemoc Rodriguez. 2010. "Quantifying the Limits of HANPP and Carbon Emissions Which Prolong Total Species Well-Being." *Environment, Development and Sustainability* 12 (2): 213–231. doi: 10.1007/s10668-009-9190-7.

Bouwman, A. F., D. P. Van Vuuren, R. G. Derwent, and M. Posch. 2002. "A Global Analysis of Acidification and Eutrophication of Terrestrial Ecosystems." *Water, Air, and Soil Pollution* 141 (1–4): 349–382. doi: 10.1023/A:1021398008726.

Bringezu, Stefan, Meghan O'Brien, and Helmut Schütz. 2012. "Beyond Biofuels: Assessing Global Land Use for Domestic Consumption of Biomass: A Conceptual and Empirical Contribution to Sustainable Management of Global Resources." *Land Use Policy* 29 (1): 224–232. doi: 10.1016/j.landusepol.2011.06.010.

Brown, Peter G., and Geoffrey Garver. 2009. *Right Relationship: Building a Whole Earth Economy.* San Francisco: Berrett-Koehler Publishers.

Butchart, Stuart H. M., Matt Walpole, Ben Collen, Arco van Strien, Jörn P. W. Scharlemann, Rosamunde E. A. Almond, Jonathan E. M. Baillie, Bastian Bomhard, Claire Brown, John Bruno, Kent E. Carpenter, Geneviève M. Carr, Janice Chanson, Anna M. Chenery, Jorge Csirke, Nick C. Davidson, Frank Dentener, Matt Foster, Alessandro Galli, James N. Galloway, Piero Genovesi, Richard D. Gregory, Marc Hockings, Valerie Kapos, Jean-Francois Lamarque, Fiona Leverington, Jonathan Loh, Melodie A. McGeoch, Louise McRae, Anahit Minasyan, Monica Hernández Morcillo, Thomasina E. E. Oldfield, Daniel Pauly, Suhel Quader, Carmen Revenga, John R. Sauer, Benjamin Skolnik, Dian Spear, Damon Stanwell-Smith, Simon N. Stuart, Andy Symes, Megan Tierney, Tristan D. Tyrrell, Jean-Christophe Vié, and Reg Watson. 2010. "Global Biodiversity: Indicators of Recent Declines." *Science* 328 (5982): 1164–1168. doi: 10.1126/science.1187512.

Cao, Long, and Ken Caldeira. 2010. "Atmospheric Carbon Dioxide Removal: Long-Term Consequences and Commitment." *Environmental Research Letters* 5 (2): 024011. doi: 10.1088/1748-9326/5/2/024011.

Carpenter, Stephen R., and Elena M. Bennett. 2011. "Reconsideration of the Planetary Boundary for Phosphorus." *Environmental Research Letters* 6 (1): 014009. doi: 10.1088/1748-9326/6/1/014009.

Costanza, Robert, Maureen Hart, Stephen Posner, and John Talberth. 2009. "Beyond GDP: The Need for New Measures of Progress." In *The Pardee Papers, No. 4.* Boston: The Frederick S. Pardee Center for the Study of the Longer-Range Future, Boston University.

Crutzen, Paul, and E. Stoermer. 2000. "The Anthropocene." *Global Change Newsletter* 41 (1): 17–18.

Daily, Gretchen C., and Paul R. Ehrlich. 1996. "Socioeconomic Equity, Sustainability, and Earth's Carrying Capacity." *Ecological Applications* 6 (4): 991–1001. doi: 10.2307/2269582.

Daly, Herman E. 1996. *Beyond Growth: The Economics of Sustainable Development.* Boston: Beacon Press.

Daly, Herman E., and Joshua C. Farley. 2011. *Ecological Economics: Principles and Applications.* 2nd ed. Washington, DC: Island Press.

Dirnböck, Thomas, Peter Bezák, Stefan Dullinger, Helmut Haberl, Hermann Lotze-Campen, Michael Mirtl, Johannes Peterseil, Steve Redpath, Simron Jit Singh, Justin Travis, and Sander M.J. Wijdeven. 2008. "Scaling Issues in Long-Term Socio-Ecological Biodiversity Research: A Review of European Cases." In *Social Ecology Working Paper 100.* Vienna: Institute of Social Ecology.

Dlugokencky, Ed, and Pieter Tans. 2013. "Trends in Atmospheric Carbon Dioxide—Global." National Oceanic & Atmospheric Administration/Earth System Research Laboratory. Accessed September 12, 2013. http://www.esrl.noaa.gov/gmd/ccgg/trends/global.html.

Douglass, A., and V. Fioletov. 2010. "Chapter 2: Stratospheric Ozone and Surface Ultraviolet Radiation." In *Scientific Assessment of Ozone Depletion: 2010*. Geneva: World Meteorological Organization. http://ozone.unep.org/Assessment_Panels/SAP/Scientific_Assessment_2010/04-Chapter_2.pdf.

Ehrlich, Paul R., and John P. Holdren. 1972. "Critique on 'the Closing Circle' (by Barry Commoner)." *Bulletin of the Atomic Scientists* 28 (5): 16, 18–27.

Ehrlich, Paul R., and Robert M. Pringle. 2008. "Where Does Biodiversity Go from Here? A Grim Business-as-Usual Forecast and a Hopeful Portfolio of Partial Solutions." *Proceedings of the National Academy of Sciences of the United States of America* 105 (Supplement 1): 11579–11586. doi: 10.2307/25463380.

Équipe de rétablissement de la rainette faux-grillon de l'ouest. 2000. *Plan De Rétablissement De La Rainette Faux-Grillon De L'ouest (Pseudacris Triseriata) Au Québec*. Edited by J. Jutras. Québec: Société de la faune et des parcs du Québec.

Erb, Karl-Heinz, Fridolin Krausmann, Veronika Gaube, Simone Gingrich, Alberte Bondeau, Marina Fischer-Kowalski, and Helmut Haberl. 2009. "Analyzing the Global Human Appropriation of Net Primary Production—Processes, Trajectories, Implications. An Introduction." *Ecological Economics* 69 (2): 250–259. doi: 10.1016/j.ecolecon.2009.07.001.

Ewing, Brad, David Moore, Steven Goldfinger, Anna Oursler, Anders Reed, and Mathis Wackernagel. 2010. *The Ecological Footprint Atlas 2010*. Oakland, CA: Global Footprint Network. http://www.footprintnetwork.org/images/uploads/Ecological_Footprint_Atlas_2010.pdf.

Foley, Jonathan A., Chad Monfreda, Navin Ramankutty, and David Zaks. 2007. "Our Share of the Planetary Pie." *Proceedings of the National Academy of Sciences of the United States of America* 104 (31): 12585–12586. doi: 10.2307/25436347.

Foley, Jonathan A., Navin Ramankutty, Kate A. Brauman, Emily S. Cassidy, James S. Gerber, Matt Johnston, Nathaniel D. Mueller, Christine O'Connell, Deepak K. Ray, Paul C. West, Christian Balzer, Elena M. Bennett, Stephen R. Carpenter, Jason Hill, Chad Monfreda, Stephen Polasky, Johan Rockstrom, John Sheehan, Stefan Siebert, David Tilman, and David P.M. Zaks. 2011. "Solutions for a Cultivated Planet." *Nature* 478: 337–342. doi: 10.1038/nature10452.

Galaz, Victor, Frank Biermann, Beatrice Crona, Derk Loorbach, Carl Folke, Per Olsson, Måns Nilsson, Jeremy Allouche, Åsa Persson, and Gunilla Reischl. 2012. "'Planetary Boundaries'—Exploring the Challenges for Global Environmental Governance." *Current Opinion in Environmental Sustainability* 4 (1): 80–87. doi: 10.1016/j.cosust.2012.01.006.

Garver, Geoffrey. 2013. "The Rule of Ecological Law: The Legal Complement to Degrowth Economics." *Sustainability* 5 (1): 316–337. doi: 10.3390/su5010316.

Haberl, Helmut, K. Heinz Erb, Fridolin Krausmann, Veronika Gaube, Alberte Bondeau, Christoph Plutzar, Simone Gingrich, Wolfgang Lucht, and Marina Fischer-Kowalski. 2007. "Quantifying and Mapping the Human Appropriation of Net Primary

Production in Earth's Terrestrial Ecosystems." *Proceedings of the National Academy of Sciences of the United States of America* 104 (31): 12942–12947. http://www.jstor .org/stable/25436409.

Haberl, Helmut, Christoph Plutzar, Karl-Heinz Erb, Veronika Gaube, Martin Poll-heimer, and Niels B. Schulz. 2005. "Human Appropriation of Net Primary Production as Determinant of Avifauna Diversity in Austria." *Agriculture, Ecosystems & Environment* 110 (3–4): 119–131. doi: 10.1016/j.agee.2005.03.009.

Haberl, Helmut, Mathis Wackernagel, Fridolin Krausmann, Karl-Heinz Erb, and Chad Monfreda. 2004a. "Ecological Footprints and Human Appropriation of Net Primary Production: A Comparison." *Land Use Policy* 21 (3): 279–288. doi: 10.1016/j .landusepol.2003.10.008.

Haberl, Helmut, Niels B. Schulz, Christoph Plutzar, Karl Heinz Erb, Fridolin Kraus-mann, Wolfgang Loibl, Dietmar Moser, Norbert Sauberer, Helga Weisz, Harald G. Zechmeister, and Peter Zulka. 2004b. "Human Appropriation of Net Primary Production and Species Diversity in Agricultural Landscapes." *Agriculture, Ecosystems & Environment* 102 (2): 213–218. doi: 10.1016/j.agee.2003.07.004.

Hansen, James, Makiko Sato, Pushker Kharecha, David Beerling, Robert Berner, Valerie Masson-Delmotte, Mark Pagani, Maureen Raymo, Dana L. Royer, and James C. Zachos. 2008. "Target Atmospheric CO_2: Where Should Humanity Aim?" *The Open Atmospheric Science Journal* 2: 217–231. doi: 10.2174/1874282300802010217.

Huggett, R. J. 1999. "Ecosphere, Biosphere, or Gaia? What to Call the Global Ecosystem." *Global Ecology and Biogeography* 8 (6): 425–431. doi: 10.1046/j.1365–2699.1999.00158.x.

Intergovernmental Panel on Climate Change. 2007a. *Fourth Assessment Report, Climate Change 2007: Mitigation of Climate Change, Contribution of Working Group III.* Edited by Bert Metz, Ogunlade Davidson, Peter Bosch, Rutu Dave and Leo Meyer. Geneva. Accessed September 12, 2014. http://www.ipcc.ch/publications_and_data /ar4/wg3/en/contents.html.

Intergovernmental Panel on Climate Change. 2007b. *Fourth Assessment Report, Climate Change 2007: The Physical Science Basis, Contribution of Working Group I.* Edited by Susan Solomon, Dahe Qin, Martin Manning, Melinda Marquis, Kristen Averyt, Melinda M. B. Tignor, Henry LeRoy Miller, Jr. and Zhenlin Chen. Geneva. Accessed September 12, 2014. http://www.ipcc.ch/publications_and_data/ar4/wg1/en/contents .html.

Jackson, Tim. 2009. *Prosperity without Growth: Economics for a Finite Planet.* London: Earthscan.

Kotchen, Matthew J., and Oran R. Young. 2007. "Meeting the Challenges of the Anthropocene: Towards a Science of Coupled Human–Biophysical Systems." *Global Environmental Change* 17 (2): 149–151. doi: 10.1016/j.gloenvcha.2007.01.001.

Lachance, Renaud. 2007. *Rapport Du Vérificateur Général Du Québec À L'assemblée Nationale Pour L'année 2007-2008.* Quebec City: Vérificateur général.

Mackenzie, Fred T., Leah May Ver, and Abraham Lerman. 2002. "Century-Scale Nitrogen and Phosphorus Controls of the Carbon Cycle." *Chemical Geology* 190 (1–4): 13–32. doi: 10.1016/S0009-2541(02)00108-0.

McNeill, John R. 2000. *Something New Under the Sun: An Environmental History of the Twentieth-Century World.* New York: W. W. Norton.

Niccolucci, Valentina, Alessandro Galli, Justin Kitzes, Riccardo M. Pulselli, Stefano Borsa, and Nadia Marchettini. 2008. "Ecological Footprint Analysis Applied to the Production of Two Italian Wines." *Agriculture, Ecosystems & Environment* 128 (3): 162–166. doi: 10.1016/j.agee.2008.05.015.

Nordhaus, Ted, Michael Shellenberger, and Linus Blomqvist. 2012. *The Planetary Boundaries Hypothesis: A Review of the Evidence*. Oakland, CA: Breakthrough Institute.

Ogle, Greg. 2000. "Accounting for Economic Welfare: Politics, Problems and Potentials." *Environmental Politics* 9 (3): 109–128. doi: 10.1080/09644010008414540.

Quebec Ministry of Natural Resources and Wildlife. 2010. *Rainette Faux-Grillon De L'ouest. Fiches Descriptives Des Espèces Menacées Ou Vulnerables*: Government of Quebec.

Ranis, Gustav, Frances Stewart, and Emma Samman. 2006. "Human Development: Beyond the Human Development Index." *Journal of Human Development* 7 (3): 323–358. doi: 10.1080/14649880600815917.

Raworth, Kate. 2012. *A Safe and Just Space for Humanity: Can We Live Within the Donut?* Oxford, UK: Oxfam International.

Robock, Alan. 2008. "20 Reasons Why Geoengineering May Be a Bad Idea." *Bulletin of the Atomic Scientists* 64 (2): 14–18. doi: 10.2968/064002006.

Rockström, Johan, Will Steffen, Kevin Noone, Asa Persson, F. Stuart Chapin, III, Eric F. Lambin, Timothy M. Lenton, Marten Scheffer, Carl Folke, Hans Joachim Schellnhuber, Bjorn Nykvist, Cynthia A. de Wit, Terry Hughes, Sander van der Leeuw, Henning Rodhe, Sverker Sorlin, Peter K. Snyder, Robert Costanza, Uno Svedin, Malin Falkenmark, Louise Karlberg, Robert W. Corell, Victoria J. Fabry, James Hansen, Brian Walker, Diana Liverman, Katherine Richardson, Paul Crutzen, and Jonathan A. Foley. 2009. "Planetary Boundaries: Exploring the Safe Operating Space for Humanity." *Ecology and Society* 14 (2): 32. http://www.ecologyandsociety.org/vol14/iss2/art32/.

Rossi, J. P., and I. van Halder. 2010. "Towards Indicators of Butterfly Biodiversity Based on a Multiscale Landscape Description." *Ecological Indicators* 10 (2): 452–458. doi: 10.1016/j.ecolind.2009.07.016.

Samper, Cristián. 2009. "Planetary Boundaries: Rethinking Biodiversity." *Nature Reports Climate Change* 3: 118–119. doi: 10.1038/climate.2009.99.

Saunier, Richard E., and Richard Albert Meganck. 2009. *Dictionary and Introduction to Global Environmental Governance*. 2nd ed. London: Earthscan.

Schiller, Andrew, Carolyn T. Hunsaker, Michael A. Kane, Amy K. Wolfe, Virginia H. Dale, Glenn W. Suter, Clifford S. Russell, Georgine Pion, Molly H. Jensen, and Victoria C. Konar. 2001. "Communicating Ecological Indicators to Decision Makers and the Public." *Conservation Ecology* 5 (1): 19. http://www.ecologyandsociety.org/vol5/iss1/art19/.

Schlesinger, Michael E., Michael J. Ring, and Emily F. Cross. 2012. "A Revised Fair Plan to Safeguard Earth's Climate." *Journal of Environmental Protection* 3: 1330–1335.

Schlesinger, William H. 2009. "Planetary Boundaries: Thresholds Risk Prolonged Degradation." *Nature Reports Climate Change* 3: 112–113. doi:10.1038/climate.2009.93.

Stuart, S. N., E. O. Wilson, J. A. McNeely, R. A. Mittermeier, and J. P. Rodríguez. 2010. "The Barometer of Life." *Science* 328 (5975): 177. doi: 10.1126/science.1188606.

Umweltbundesamt. 2004. *Manual on Methodologies and Criteria for Modelling and Mapping Critical Loads & Levels and Air Pollution Effects, Risks and Trends.* Berlin: Federal Environmental Agency (Umweltbundesamt).

United Nations Framework Convention on Climate Change. 2010. *Draft Decision -/ Cp.16: Outcome of the Work of the Ad Hoc Working Group on Long-Term Cooperative Action under the Convention.* Accessed September 12, 2014. http://unfccc.int/files /meetings/cop_16/application/pdf/cop16_lca.pdf.

United Nations Environment Programme. 2009. *Climate Change Science Compendium 2009.* Edited by Catherine P. McMullen and Jason Jabbour. Nairobi: United Nations Environment Programme.

United Nations Population Division. 2013. *World Population Prospects: The 2012 Revision.* Accessed September 12, 2014. http://esa.un.org/wpp/Excel-Data/population. htm.

United Nations Population Fund. 2011. *State of World Population 2011.* New York: United Nations Population Fund.

Victor, Peter A. 2008. *Managing Without Growth: Slower by Design, Not Disaster.* Cheltenham, UK: Edward Elgar.

Wackernagel, Mathis, and William E. Rees. 1996. *Our Ecological Footprint: Reducing Human Impact on the Earth.* Gabriola Island, BC, Canada: New Society Publishers.

Wright, David Hamilton. 1990. "Human Impacts on Energy Flow through Natural Ecosystems, and Implications for Species Endangerment." *Ambio* 19 (4): 189–194. doi: 10.2307/4313691.

CHAPTER SIX

Revisiting the Metaphor of Human Health for Assessing Ecological Systems and Its Application to Ecological Economics

MARK S. GOLDBERG, GEOFFREY GARVER, AND NANCY E. MAYO

1. INTRODUCTION

Since the 1990s, the notion of "ecosystem health" has become prominent in the ecological literature (Aguirre et al. 2002; Costanza, Norton, and Haskell 1992; Jørgensen, Xu, and Costanza 2010; Rapport et al. 1998; Scow et al. 2000). A main motivation for the definition and use of the concept of "ecosystem health" was to provide a uniform framework for assessing the state and function of ecosystems and for managing ecosystems using knowledge that can be integrated from various disciplines (Rapport 1998b). Its relevance to ecological economics is clear: essential indicators of ecological function and their states are required in order to monitor unbridled human activities and to ensure that severe stress and destruction of ecosystems is monitored, avoided, and (where necessary) mitigated.

Rapport (1998a) discussed the strengths and limitations of using the human health metaphor. He indicated that ecosystem health can be defined operationally and quantitatively. It can be used for a systematic diagnosis of systems that are in stress, especially for the understanding of etiology; it also can be used both for the purposes of remediation and prevention. Nonetheless, he recognized that this metaphor cannot be used simply and would need to include multiple determinants of stress and disease. Indeed, Jørgensen (2010:3) indicated that "it is clear today that it is not possible to find one indicator or even a few indicators that can be used generally, as some naively thought when ecosystem health assessment . . . was introduced."

Although the metaphor of ecological health is derived from its application to individual humans and human populations, it does not appear that an in-depth exploration of the paradigm of human health has been discussed in the ecological literature, although VanLeeuwen et al. (1999) provided a model whereby human health is embedded within the planet. This chapter emerged from an inquiry into how the health metaphor might be used to facilitate discussion of regional and global ecological boundaries that must be used to limit human activities. Even a superficial exploration of either the human health or ecological health literature shows that there has been much debate over the use of the metaphor. Such a review raises the question of whether the term "health" is truly useful, and, if so, under what circumstances it can be applied appropriately.

It appears that this metaphor or analogy was taken originally from human medicine. Some of the earliest writings on the subject suggest that the essential aspect relates to a physician diagnosing a disease in an individual using a set of measurements, mostly at the physiological level. For example, Jørgensen (2010) stated that "we go to a doctor to get a diagnosis . . . and hopefully initiate a cure to bring us back to normal (or healthy conditions). Doctors will apply several indicators/examinations (pulse, blood pressure, sugar in the blood, and urine, etc.) before they come up with a diagnosis and proper cure." The purpose is practical and very important—a serious problem is identified, a clear diagnosis is made, the causes are identified, and remediation and subsequent management are affected.

A relatively simple case in ecology can be used to illustrate the parallels to human health. Consider the ramifications of treating agricultural runoff in a lake. Specific measurements are made to identify the extent of the eutrophication of the lake. The investigation reveals that the excess nitrogen and phosphorous entered the lake from runoff of adjoining farmlands and other sources. A straightforward solution is to reduce the use of fertilizers and associated runoff, although this is not trivial to implement.

Räike et al. (2003) in Finland showed the effectiveness of control measures, finding that decreased loadings were observed after control measures were taken to reduce emissions from wastewater treatment plants and pulp and paper operations, as well as to reduce the use of nitrogen and phosphorous fertilizers among farmers. Other examples may involve less obvious signs of damage and may have multiple interacting factors operating through different pathways and with effects of different magnitude. Diagnosis and remediation are thus complex. The hope of using the metaphor of ecological health is that it will stimulate an interdisciplinary approach, making "diagnosis" and "treatment" more effective.

We show below that human health is far more than the absence of specific diseases, and that sole reliance on physiological indicators does not provide a means for its definition or its measurement. Indeed, human health is an elusive concept that defies a simple, comprehensive definition; it comprises so many dimensions that to define it, especially in quantitative terms, is essentially impossible. Measurement in humans of the vast panoply of indicators is problematic: in almost no situations that we are aware of would all aspects of health be measured in a particular individual. We discuss briefly some of the means by which various aspects of human health are measured, and some of the essential characteristics of these measurements. We thus argue that applying the analogy to ecosystems—which are of course not organisms but complex, hierarchical systems—has significant difficulties. In particular, the field of inquiry that is concerned with the measurement of indicators of function in an ecosystem defines the field of "ecosystem health," but we conclude that measurement of selected indicators cannot be used to imply health, although they may imply lack of health. We also make some suggestions, which have been made previously by others, regarding the measurement of "states" in ecosystems.

2. A COMMENTARY ON THE METAPHOR OF ECOSYSTEM HEALTH AS DERIVED FROM HUMAN HEALTH: MEASUREMENT OF DISEASE VERSUS CAPACITY TO FUNCTION

2.1. The Role of Physicians

The primary role of physicians is to "find it and fix it." They are trained to identify and treat pathologies that require attention and remediation. For example, in the context of a routine visit to a general practitioner, often a standard physical examination will be conducted, evaluating such things as enlarged lymphatic glands and obvious breathing and heart abnormalities. The physician may ask questions regarding health and complaints, which may assist in uncovering specific problems. The history of the patient is important because previous illnesses will factor into the interpretation of the examination; results from previous tests are often reviewed.

In addition, a number of essential physiological measurements may be taken, including such things as height, weight, pulse rate, pupil size, blood pressure, and body temperature; blood tests also may be ordered. Many of the blood tests can serve as markers for specific bodily functions, such as liver function (through assays that measure liver-related enzymes). Most indicators have "normal" ranges that are defined from observations of distributions in "healthy" populations. Temperature, blood pressure, pulse rate, and respiration rate are essential vital signs. Within the context of a specific investigation, other tests may be ordered. There are many indicators that are not measured routinely, especially those that may predict future health, such as plaque in the blood vessels, arterial elasticity, and lung function. Lung function is a strong predictor of mortality in the distant future (Sabia et al. 2010); however, in routine clinical practice, only people showing frank lung dysfunction (e.g., chronic bronchitis, emphysema, or asthma) would be tested and usually only by a respirologist.

In "healthy" individuals, it would be expected that some prime physiological indicators, such as blood pressure, resting heart rate, lipids in blood, and sugar in urine, would all be in the "normal" range. However, this only tells part of the story. For example, people with back pain whose activities of daily life are constrained may also have normal values of essential biomarkers. Some other conditions are episodic; for example, people with asthma may look normal on pulmonary function tests when they are not experiencing an exacerbation, but they will appear abnormal during a crisis. In addition, there are conditions that show exacerbations and remissions during the course of life but may also be life-shortening (e.g., multiple sclerosis).

Human health certainly is affected by exogenous influences or contaminants, such as air pollution, excess sunlight, and *Escherichia coli* in a food product. However, many human illnesses are endogenous, arising from failure at the cellular, tissue, or organ level—often for unknown reasons.

An ecosystem also has exogenous and potentially endogenous influences, such as acid rain (exogenous) and overpopulation of a species (endogenous).

Physicians have little to do with general matters of individual health, other than possibly suggesting to their patients that they should exercise or eat properly. Other indicators of consequence that are investigated rarely by general practitioners include the ability to conduct activities of daily living, levels of stress, and mental health. There are some medical specialties that deal with larger issues; for example, sports medicine physicians are concerned about musculoskeletal health and are quite attuned to providing support for rehabilitation and prevention of injury through an interaction with other health professionals, such as physiotherapists and athletic therapists. Occupational physicians are concerned with causes of disease and are trained to identify those that have origins in the workplace.

Physicians have little to do with the development of population health, which is a matter of public and societal issues.[1] For example, the increase of diabetes in North America and elsewhere (Lipscombe and Hux 2007) is related in part to the fragmentation of our cities, so that transportation is mostly via the automobile, as well as to our consumer-based society that leads to overeating and general sedentariness, especially in children (Gortmaker et al. 1996). Physicians come late in the process in trying to deal with the complications of obesity, which include diabetes and heart disease. Other examples abound, including exposure to air pollution that causes manifold health problems (Chen, Goldberg, and Villeneuve 2008), ultraviolet light causing skin cancer (Leiter and Garbe 2008), and disinfection byproducts in water associated with the incidence of cancer (Richardson et al. 2007). The instruments that have been developed to measure various components of "health" come not from medicine but from allied fields that deal with outcomes, such as epidemiology.

2.2. The Role of Allied Health Fields in Measuring Indicators of Health

A number of indicators are used by nonphysicians to go beyond the measurement of specific physiological and pathological biomarkers. Indicators of health include freedom from symptoms of pain, fatigue, and emotional distress; the ability to carry out activities that are necessary for daily life (e.g., dressing, personal hygiene, moving around the home or community, purchasing and preparing food, maintaining a home); and engagement in family, social, and economic roles. Many tests and questionnaires exist to measure these aspects of health; some of these are tailored for specific subgroups of the population, such as the elderly or children.

Many of these other measures also have reference values that are derived from population-based distributions or from benchmarks for safety. For example, a common measure of community mobility required for basic activities of daily living is how far a person can walk in six minutes (American Thoracic Society 2002). The value will indicate whether the person is performing as expected for their age and sex, but the value does not provide any indication as to why a person walks only that distance or whether the limitation is temporary or permanent. There are also self-reported measures of overall and specific mental health (e.g., happiness, anxiety, depression) as well other measures of activities of daily life (e.g., simple things such as being able to tie one's shoes, recreational activities).

Many of these indices were developed to assess health in specific populations or situations, whereas others are of a more global nature that is applicable in many situations. In addition to having population norms, their validity (i.e., they measure what they purport to measure) and reliability (repeatability) are also assessed before they are put into widespread use.

There are also composite indicators of an individual's health relative to a broader population of humans that are based on a number of constructs, with each construct defined by a series of questions. Two of the most widespread instruments are the Short-Form Health Survey (SF-36; Ware and Sherbourne 1992) and the Euroqol (Kind 1996). Questions are assigned numerical values and the calculation of an overall score is achieved by summing, or otherwise, the values of each item. There has been considerable progress in developing and validating these indicators so that their meaning has a grounded interpretation. They are often evaluated in "healthy populations" and the range of scores helps define what is defined as "normal," providing a baseline by which to make comparisons. In addition, in formal studies comparisons between a group with some pathology, for example, will be compared to an otherwise comparable "healthy" one (e.g., comparing activities of daily life amongst those people with chronic back pain to those without). Thus, a complex numerical scale becomes meaningful. Nevertheless, as mentioned in chapter 4, acting on a disability in an individual requires specific knowledge of the problem, and it is only in evaluating the individual items in a complex indicator that the specific problem can be identified and treated.

Despite over a half a century of advancement in the measurement of health, markers of disease predominate the measures used in clinical practice. Often, the mismatch between what is measured clinically and how the person feels leads to suboptimal treatment and outcome. A person may show abnormally high levels of blood lipids but may feel very well; hence, he or she may not be motivated to start or continue with pharmacological therapy. Another person may feel very unwell but may not have markedly abnormal values on physiological parameters (a finding not uncommon with a condition such as fibromyalgia), and hence is dissatisfied with the lack of treatment or concern by the medical community.

In the health field, methods of measurement are now incorporating multiple indicators of health (most often measured on ordinal scales), calibrating the critical ordinal categories in a hierarchical manner, and creating measures that have interval-like properties. In this way, change in different aspects of health can be quantified and related to changes going at a cellular or organ level or at an environmental level (Andrich 2011).

2.3. Definitions of Human Health

The previous discussion indicates that physicians have a limited role in health: they can identify pathologies and some can be treated. As Richard Smith, a past editor of the *British Medical Journal*, wrote in a blog for that journal:

> But what is health? For most doctors that's an uninteresting question. Doctors are interested in disease, not health. Medical textbooks are a massive catalogue of diseases. There are thousands of ways for the body and mind to go wrong, which is why disease is so interesting. We've put huge energy into classifying disease, and even psychiatrists have identified over 4000 ways in which our minds may malfunction. Health for doctors is a negative state—the absence of disease. In fact, health is an illusion. If you let doctors get to work with their genetic analysis, blood tests, and advanced imaging techniques then everybody will be found to be defective—"diseased." . . . So I'm not happy with health being defined as the absence of disease. Nor am I keen on the World Health Organisation's definition of "Complete physical, psychological, and social

wellbeing," a state reached only at the moment of mutual orgasm, joked Peter Skrabanek. It's a ludicrous definition that would leave most of us unhealthy most of the time. My favourite definition of the moment is Sigmund Freud's definition, which was never written down by him, of "the capacity to love and work."
(Smith 2008)

2.3.1. World Health Organization definition of health

In 1948, the World Health Organization (WHO) defined health as "a state of complete physical, mental and social well being, not merely the absence of disease or infirmity" (World Health Organization 1948). In 1986, WHO defined health as "a resource for everyday life, not the objective of living. Health is a positive concept emphasizing social and personal resources, as well as physical capacities" (World Health Organization 2006).

Richard Smith was not the only one to have issues with the WHO definition of health. Rodolpho Saracci argued that a serious pitfall with the WHO definition is that it does not discriminate between health and happiness (Saracci 1997). The connection with social well-being brings in the notion of happiness, so that any deviation from happiness, however defined (and like health, happiness is not trivial to define), implies a state of imperfect health. Larson (1999) discussed other problems with the WHO definition, including that there is no consensus on the meaning of well-being and social well-being; health is defined differently in different cultures; there is no clear ranking of health states; we all suffer from time to time from various complaints, so essentially no individual fits the definition; and the definition is so vague as to admit myriad ways to operationalize health. Larson (1999:123) also quoted other possible definitions, including that "health may be defined as the absence of illness, having strength and robustness, and high quality of life. It may be defined as quantity (length of life) and quality, or it may be defined as consisting of the absence of physical disability, psychological disability, and pain" (Brown et al. 1984; Feinstein 1992).

Larson (1999) discussed the conceptualization of health in terms of other models. Of interest is the wellness model that is "health promotion and progress toward higher functioning, energy, comfort, and integration of mind, body, and spirit" (Larson 1999:125, table 1). In this model, there is an integration with spiritual and mental functioning, having reserves to overcome illness, striving toward higher levels of functioning. VanLeeuwen et al. (1999) also discussed a number of models of human health, some of which were also discussed by Larson, that included ecological factors but also some of the concepts related to hierarchies found in nature. Their "butterfly model" embeds human health within the ecosystem (similar to ecological economics) and attempts to incorporate broad interacting links between the biophysical and socioeconomic environments.

In response to these and other criticisms, the WHO developed a framework for defining constructs important to health: the International Classification of Functioning, Disability, and Health (World Health Organization 2014). The International Classification of Functioning, Disability, and Health is based on what is called a biopsychosocial model, which attempts an integration of the medical and social aspects of health. Figure 6.1 shows a conceptual model based on World Health Organization (2002).

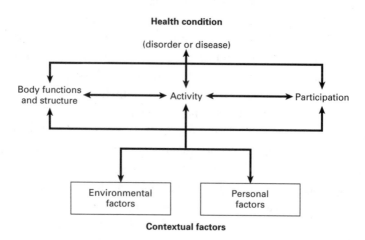

Figure 6.1. Conceptualization of the model of disability used in the International Classification of Functioning, Disability, and Health. The figure implies that disability and functioning result from the interaction of health conditions (diseases, disorders and injuries) and "contextual factors." The contextual factors are distinguished by external factors, referred to as "environmental factors" that include social contexts and geographical and climatic ones and "personal factors" that relate to the individual (e.g., age, gender, education). These factors have a bearing on how individuals perceive disability, or lack of disability. The middle part of the box shows the main components of human function, and dysfunction relates to impairments (problems in body function or structure), limitations of physical activities, and limitations in participating in life activities. *Source*: World Health Organization (2002).

The International Classification of Functioning, Disability, and Health provides a unified and standard language and framework for the description of health and health-related states. It defines components of health, both positive (function) and negative (disability), and some health-relevant components of well-being. Constructs for disabilities, impairments, limitations, and restrictions are the objective and exteriorized signs of the individual, but what people "feel" about their function is considered akin to well-being. Measuring only well-being without measuring function, and vice versa, misses an essential dimension of the health of an individual, especially given that people adapt to poor function and a high degree of expressed well-being does not indicate need of care or services.

To summarize, although the definition of health by the WHO is cited widely, there is no singular definition of health. (Also, the model by which "one recognizes health when one sees it" is highly flawed; consider a person with latent lung cancer, who appears healthy one day and dies three months later.) Indeed, many models have been developed over time to attempt to capture various elements of human health. There have also been a number of instruments geared toward measuring different components of health, including the International Classification of Functioning, Disability, and Health, which is based mostly on the WHO definition; however, other instruments are also of considerable value (e.g., the SF-36 [Ware and Sherbourne 1992] and the Euroqol [Kind 1996]). Indeed, it is clear that the full spectrum of health goes beyond the internal signs, as reflected by physiological and pathological parameters measured by physicians; it even goes beyond the exteriorized signs of disability and includes the concept of "well-being"—or how a person "feels" about their health condition or its consequences. Moreover, health is an evolving process, and individuals change in different ways through time. In short, human health in its entirety cannot be measured in a specific individual and certainly not in groups.

3. DEFINITIONS OF ECOSYSTEM HEALTH

The original motivation was to understand the etiology of "stressed" ecosystems and to provide a global method of assessment. As described previously, the literature on ecosystem health has its roots in the paradigm of the physician diagnosing illness and disease in a patient (Jørgensen 2010;

Rapport 1992; Rapport, Regier, and Hutchinson 1985). Indeed, Schaeffer (1996) wrote that "the four-step sequential process that must be accomplished in making a diagnosis of a human or animal disease . . . also applies to ecosystem diseases." Rapport (1992) discussed the need to superimpose human values on the assessment of ecosystem health; he cited the example of indigenous forests in New Zealand being converted to plantations of radiata pine for the purposes of human exploitation, with conservationists arguing against this transformation. For the purposes of humans evaluating their effects on the biosphere, some sort of ethics is required. Perhaps, the "land ethic" of Aldo Leopold (1949) could be a starting point (see chapter 7).

Regardless of the ethics, there is no debate about the importance of identifying problems with ecosystems, uncovering their etiology, and carrying out remediation. Essential to this purpose is the ability to measure characteristics or states of ecosystems and making comparisons to "normal circumstances," as we do with humans, or to map out temporal trends. Presence of disease, dysfunction, out of the "normal range" of key indicators, and statistical correlations showing relationships with metrics of ill-health (e.g., acid rain and the loss of fish stock in lakes) would indicate an ecosystem's ill health; perhaps some cure could be prescribed to make the system right.

Some investigators have attempted to operationalize the concept of ecosystem health. For example, Costanza (1992) suggested that ecosystem health is "a comprehensive, multiscale, hierarchical measure of system resilience (Holling 1986), organization, and vigor." He suggested that health is "a measure of the overall performance of a complex system that is built up from the behavior of its parts." A definition, developed at a conference, was proposed: "An ecological system is healthy and free from 'distress syndrome' if it is stable and sustainable—that is, if it is active and maintains its organization and autonomy over time and is resilient to stress." From this, Costanza defined an ecosystem health index that was the product of indices measuring vigor, organization, and resilience. In a follow-up to this paper, Costanza and Mageau (1999) described a definition of ecosystem health as follows: homeostasis, absence of disease, diversity or complexity, stability or resilience, vigor or scope for growth, and balance between system components. They extended these previous definitions to propose "ecosystem health as a comprehensive, multiscale, dynamic, hierarchical measure of system resilience, organization, and vigor"[2] (Costanza and Mageau 1999:106).

This is somewhat analogous to the original WHO definition, with "well-being" replaced by resilience, organization, and vigor. Costanza and Mageau (1999:106) indicated that "system health implies a weighted summation or a more complex operation over the component parts, where the weighting factors incorporate an assessment of the relative importance of each component to the functioning of the whole. . . . In the practice of human medicine, these weighting factors or values are contained in the body of knowledge and experience embodied in the medical practitioner." Unfortunately, this last statement is patently false, as clinicians do not provide a judgment as to the health of an individual but rather to the presence or absence of specific diseases and the possibility of cure or management.

Moreover, as discussed in chapter 4, complex indicators are in fact difficult to interpret because their values can be ambiguous.[3] Although benchmarking (validation) can overcome these drawbacks, they would still need to be disaggregated for specific remedial actions. Also, combining indices in either a multiplicative or additive framework is fraught with difficulty: there is the unverifiable assumption that one can develop a meaningful complex index by some mathematical formula that combines disparate indices. Validation of such an index is required and, as in studies of human health, must be benchmarked. Lastly, such indices are exceedingly difficult to interpret.

In 1993, Suter (1993) criticized the notion of ecosystem health as follows:

1. It misrepresents health. He noted that the original WHO definition of health was indeed subjective and that physicians do not in fact measure health, as we have shown previously.
2. It misrepresents ecology. Ecosystems are not organisms (and certainly they do not have central nervous systems).
3. Definitions of ecosystem health, such Karr et al.'s (1986) Index of Biotic Integrity, combine indicators that are heterogeneous and, because they cannot be benchmarked, have no intrinsic meaning. To quote Suter (1993:1533, Abstract): "They cannot be predicted, so that they are not applicable to most regulatory problems; they have no diagnostic power; effects on one component are eclipsed by responses of other components [this is the main issue of combining in an arbitrary fashion indices that measure different constructs and have different units]; and the

reason for a high or low value is unknown [the issue of variability from a statistical framework]."

Wicklum and Davies (1995:997, Abstract) were also critical of the concept:

"The phrase ecosystem health is based on an invalid analogy with human health requiring acceptance of an optimum condition and homeostatic processes maintaining the ecosystem at a definable optimum state. Similarly, ecosystem integrity is not an objective, quantifiable property of an ecosystem. Health and integrity are not inherent properties of an ecosystem and are not supported by either empirical evidence or ecological theory."

As with human health, assessments of ecosystem function and states do not require a clear definition of ecosystem health. Indeed, Rice (2003:236) used the term "indicators of ecosystem status" rather than "indicators of environmental health." He stated that "a review prepared three years ago . . . found over two-hundred different indicators of ecosystem status, without being exhaustive. The challenge is not to find an indicator of ecosystem status to use. It is to choose the set that will serve the users' needs best."

Jørgensen, Xu, and Costanza (2010) discussed criteria by which indicators should be selected from management and scientific points of view, including being simple to apply and understand, relevant, scientifically justifiable, quantitative, reasonable cost, easy to use, sensitive to small variations (i.e., in terms of the language used in our chapter on measurement, they should be valid and reliable), independent of reference states, and widely applicable. These authors indicated that few indicators meet all of these criteria.

The notion of developing orthogonal qualitative descriptors, or perhaps "properties," to define various components of health is indeed analogous to what is done in human health. For example, we speak of mental health, physical health, quality of life, etc. As in human health, it is unlikely that resilience, organization, and vigor are the only set of descriptors. Jørgensen, Xu, and Costanza (2010) classified ecosystem indicators into eight "levels": presence and absence of species, ratio of populations of classes of organisms, concentrations of chemical compounds, populations or concentrations in trophic levels, rates of essential processes, composite indicators (e.g., complex indicators such as Odum's [1971] set of indicators), holistic indicators (these are also complex and refer to concepts such as resistance and resilience), and thermodynamic variables (e.g., exergy).

4. DISCUSSION AND CONCLUSIONS

We have shown that a definition of human health is elusive, and the various definitions that have been attempted have serious shortcomings. As well, the myriad attributes and domains that make up the concept of human health cannot be measured uniquely in any individual or in populations. It is clear that human health goes beyond the diagnosis of disease and that there are many inherent properties of human health, and most of these are never measured by physicians. In fact, physicians play almost no role in assessing health; they deal with pathology, remediation, and possible cure. Health professionals working in various fields have defined many indices to capture the various dimensions inherent in human health. One of the essential components in defining and developing these indices is to measure their validity and reliability. Complex indicators in human health have no meaning unless they are benchmarked.

The current concept of ecosystem health is plagued by the incorrect analogy of physicians as the lead players in defining human health, although ecologists acting as physicians to diagnose and correct pathology is of course correct and essential. Moreover, as indicated by Suter (1993) and others, ecosystems are not organisms. An implication to ecology is that complex indices that have been defined to measure ecosystem health do not and cannot measure all of the dimensions in complex ecosystems. This has been acknowledged by a number of other ecologists (e.g., Jørgensen, Xu, and Costanza 2010).

Because we are not ecologists, we would be rather bold to presume that we have any expertise in developing relevant indicators. Surely, however, as in any scientific discipline, the assessment of states of ecosystems must be based on defining clearly the objectives of the research or the issues related to management. The job is then to make measurements of all relevant parameters that reflect these objectives accurately, as well as to be wary of combining these parameters into complex indices that have no intrinsic meaning. De Leo and Levin (1997) suggested "that it is much more useful to characterize in detail the functional and structural aspects of ecosystems to provide a conceptual framework for assessing the impact of human activity on biological systems and to identify practical consequences stemming from this framework."

Human populations have relatively little variability in their genomic constitution as compared to the constitutions of the myriad types and numbers of ecosystems on the planet. This means that humans are much more homogeneous than ecosystems, and the task of measurement and assessment in humans should be much easier than with ecosystems. Nevertheless, perhaps we can speculate that the model of human functioning, disability, and health could be used by analogy in assessing ecosystem function: each ecosystem can be considered to have a capacity to function, which may or may not be altered; given a specific level of capacity, each component of the ecosystem may or may not perform "optimally" in achieving the activities and roles for which they have evolved to do. Each component in the system is interdependent and forms an integral part of the environmental cloth for the whole ecosystem. This would provide a measurement framework for ecosystem health that would go beyond only the physiological. On the other hand, perhaps the classification system for indicators of ecosystems by Jørgensen, Xu, and Costanza (2010) may be sufficient.

The above functional assessments probably are more closely related to concepts of the health of ecosystems and less generally to planetary systems, although it is likely that one would still need an array of "physiological" parameters to fully characterize health. Moreover, the monitoring of these parameters in time and space would be needed. It is thus unlikely that a proper measure of ecosystem health could ever be achieved using one complex, integrated index.

NOTES

1. Unless they are working as epidemiologists or in public or occupational health.

2. The resilience of a system refers to its ability to maintain its structure and pattern of behavior in the presence of stress. The vigor of a system is simply a measure of its activity, metabolism, or primary productivity. The organization of a system refers to the number and diversity of interactions between the components of the system.

3. For example, let us consider a complex indicator obtained by multiplying two variables (A, B) together. Let us assume for simplicity that each are coded as 0, 1, 2, 3. The interpretation of this complex indicator is ambiguous because different combinations yield the same product: for example, $A(=1) \times B(=2) = 2$ and $A(=2) \times B(=1)$ also equals 2. See chapter 4.

REFERENCES

Aguirre, A. Alonso, Richard S. Ostfeld, Gary M. Tabor, Carol House, and Mary C. Pearl, eds. 2002. *Conservation Medicine: Ecological Health in Practice*. Oxford: Oxford University Press.

American Thoracic Society. 2002. "ATS Statement: Guidelines for the Six-Minute Walk Test." *American Journal of Respiratory and Critical Care Medicine* 166 (1): 111–117. doi: 10.1164/rccm.166/1/111.

Andrich, David. 2011. "Rating Scales and Rasch Measurement." *Expert Review of Pharmacoeconomics & Outcomes Research* 11 (5): 571–585. doi: 10.1586/erp.11.59.

Brown, J. H., L. E. Kazis, P. W. Spitz, P. Gertman, J. F. Fries, and R. F. Meenan. 1984. "The Dimensions of Health Outcomes: A Cross-Validated Examination of Health Status Measurement." *American Journal of Public Health* 74 (2): 159–161. doi: 10.2105/AJPH.74.2.159.

Chen, Hong, Mark Goldberg, and Paul J. Villeneuve. 2008. "A Systematic Review of the Relation between Long-Term Exposure to Ambient Air Pollution and Chronic Diseases." *Reviews on Environmental Health* 23 (4): 243–298. doi: 10.1515/REVEH.2008.23.4.243.

Costanza, Robert. 1992. "Toward an Operational Definition of Ecosystem Health." In *Ecosystem Health: New Goals for Environmental Management*, edited by Robert Costanza, Bryan G. Norton and Benjamin D. Haskell, 239–256. Washington, DC: Island Press.

Costanza, Robert, and Michael Mageau. 1999. "What Is a Healthy Ecosystem?" *Aquatic Ecology* 33 (1): 105–115. doi: 10.1023/A:1009930313242.

Costanza, Robert, Bryan G. Norton, and Benjamin D. Haskell, eds. 1992. *Ecosystem Health: New Goals for Environmental Management*. Washington, DC: Island Press.

De Leo, G. A., and S. Levin. 1997. "The Multifaceted Aspects of Ecosystem Integrity." *Conservation Ecology* 1 (1): 3. http://www.consecol.org/vol1/iss1/art3/.

Feinstein, Alvan R. 1992. "Benefits and Obstacles for Development of Health Status Assessment Measures in Clinical Settings." *Medical Care* 30 (5): MS50–MS56. doi: 10.2307/3766229.

Gortmaker, S. L., A. Must, A.M. Sobol, K. Peterson, G. A. Colditz, and W. H. Dietz. 1996. "Television Viewing as a Cause of Increasing Obesity among Children in the United States, 1986–1990." *Archives of Pediatrics & Adolescent Medicine* 150 (4): 356–362. doi: 10.1001/archpedi.1996.02170290022003.

Holling, C. S. 1986. "The Resilience of Terrestrial Ecosystems." In *Sustainable Development of the Biosphere*, edited by W. C. Clark and R. E. Munn, 292–320. Cambridge, UK: Cambridge University Press.

Jørgensen, Sven Erik. 2010. "Introduction." In *Handbook of Ecological Indicators for Assessment of Ecosystem Health*, edited by Sven Erik Jørgensen, Fu-Liu Xu, and Robert Costanza, 3–7. Boca Raton, FL: CRC Press.

Jørgensen, Sven Erik, Fu-Liu Xu, and Robert Costanza, eds. 2010. *Handbook of Ecological Indicators for Assessment of Ecosystem Health*. 2nd ed. Boca Raton, FL: CRC Press.

Karr, James R, Kurt D Fausch, Paul L Angermeier, Philip R Yant, and Isaac J Schlosser. 1986. *Assessing Biological Integrity in Running Waters: A Method and Its Rationale*. Champaign, IL: Illinois Natural History Survey.

Kind, Paul. 1996. "The Euroqol Instrument: An Index of Health-Related Quality of Life." In *Quality of Life and Pharmacoeconomics in Clinical Trials*, edited by Bert Spilker, 191–201. Philadelphia: Lippincott-Raven.

Larson, James S. 1999. "The Conceptualization of Health." *Medical Care Research and Review* 56 (2): 123–136. doi: 10.1177/107755879905600201.

Leiter, Ulrike, and Claus Garbe. 2008. "Epidemiology of Melanoma and Nonmelanoma Skin Cancer—the Role of Sunlight." In *Sunlight, Vitamin D and Skin Cancer*, edited by Jörg Reichrath, 89–103. New York: Springer. doi:10.1007/978-0-387-77574-6_8.

Leopold, Aldo. 1949. *A Sand County Almanac and Sketches Here and There*. New York: Oxford University Press.

Lipscombe, Lorraine L., and Janet E. Hux. 2007. "Trends in Diabetes Prevalence, Incidence, and Mortality in Ontario, Canada 1995–2005: A Population-Based Study." *The Lancet* 369 (9563): 750–756. doi: 10.1016/S0140-6736(07)60361-4.

Odum, Eugene. 1971. *Fundamentals of Ecology*. Philadelphia: Saunders.

Räike, A., O. P. Pietiläinen, S. Rekolainen, P. Kauppila, H. Pitkänen, J. Niemi, A. Raateland, and J. Vuorenmaa. 2003. "Trends of Phosphorus, Nitrogen and Chlorophyll a Concentrations in Finnish Rivers and Lakes in 1975–2000." *Science of The Total Environment* 310 (1–3): 47–59. doi: 10.1016/S0048-9697(02)00622-8.

Rapport, David J. 1992. "Evaluating Ecosystem Health." *Journal of Aquatic Ecosystem Health* 1 (1): 15–24. doi: 10.1007/BF00044405.

Rapport, David J. 1998a. "Defining Ecosystem Health." In *Ecosystem Health*, edited by David J. Rapport, Robert Costanza, Paul R. Epstein, Connie Gaudet, and Richard Levins, 18–33. Oxford: Blackwell Science.

Rapport, David J. 1998b. "Need for a New Paradigm." In *Ecosystem Health*, edited by David J. Rapport, Robert Costanza, Paul R. Epstein, Connie Gaudet, and Richard Levins, 3–17. Oxford: Blackwell Science.

Rapport, David J., Robert Costanza, Paul R. Epstein, Connie Gaudet, and Richard Levins, eds. 1998. *Ecosystem Health*. Oxford: Blackwell Science.

Rapport, David J., H. A. Regier, and T. C. Hutchinson. 1985. "Ecosystem Behavior Under Stress." *The American Naturalist* 125 (5): 617–640. doi: 10.2307/2461475.

Rice, Jake. 2003. "Environmental Health Indicators." *Ocean & Coastal Management* 46 (3–4): 235–259. doi: 10.1016/S0964-5691(03)00006-1.

Richardson, Susan D., Michael J. Plewa, Elizabeth D. Wagner, Rita Schoeny, and David M. DeMarini. 2007. "Occurrence, Genotoxicity, and Carcinogenicity of Regulated and Emerging Disinfection by-Products in Drinking Water: A Review and Roadmap for Research." *Mutation Research/Reviews in Mutation Research* 636 (1–3): 178–242. doi: 10.1016/j.mrrev.2007.09.001.

Sabia, Séverine, Martin Shipley, Alexis Elbaz, Michael Marmot, Mika Kivimaki, Francine Kauffmann, and Archana Singh-Manoux. 2010. "Why Does Lung Function Predict Mortality? Results from the Whitehall Ii Cohort Study." *American Journal of Epidemiology* 172 (12): 1415–1423. doi: 10.1093/aje/kwq294.

Saracci, Rodolfo. 1997. "The World Health Organisation Needs to Reconsider Its Definition of Health." *BMJ: British Medical Journal* 314 (7091): 1409–1410. doi: 10.2307/25174539.

Schaeffer, David J. 1996. "Diagnosing Ecosystem Health." *Ecotoxicology and Environmental Safety* 34 (1): 18–34. doi: 10.1006/eesa.1996.0041.

Scow, K. M., G. E. Fogg, D. E. Hinton, and M. L. Johnson. 2000. *Integrated Assessment of Ecosystem Health.* Boca Raton, FL: Lewis Publishers.

Smith, Richard. 2008. "The End of Disease and the Beginning of Health." Accessed January 17, 2015. http://blogs.bmj.com/bmj/2008/07/08/richard-smith-the-end-of -disease-and-the-beginning-of-health/.

Suter, Glenn W., II. 1993. "A Critique of Ecosystem Health Concepts and Indexes." *Environmental Toxicology and Chemistry* 12: 1533–1539. doi: 10.1002/etc.5620120903.

VanLeeuwen, J. A., D. Waltner-Toews, T. Abernathy, and B. Smit. 1999. "Evolving Models of Human Health toward an Ecosystem Context." *Ecosystem Health* 5 (3): 204–219. doi: 10.1046/j.1526-0992.1999.09931.x.

Ware, John E., Jr., and Cathy Donald Sherbourne. 1992. "The MOS 36-Item Short-Form Health Survey (Sf-36): I. Conceptual Framework and Item Selection." *Medical Care* 30 (6): 473–483. doi: 10.2307/3765916.

Wicklum, D., and Ronald W. Davies. 1995. "Ecosystem Health and Integrity?" *Canadian Journal of Botany* 73 (7): 997–1000. doi: 10.1139/b95-108.

World Health Organization. 1948. *Preamble to the Constitution of the World Health Organization as Adopted by the International Health Conference, New York, 19–22 June 1946.* Accessed January 17, 2015. http://www.who.int/about/definition/en /print.html.

World Health Organization. 2002. *Towards a Common Language for Functioning, Disability and Health: ICF, the International Classification of Functioning, Disability and Health.* Accessed January 17, 2015. http://www.who.int/classifications/icf/training /icfbeginnersguide.pdf.

World Health Organization. 2006. *Constitution of the World Health Organization: Basic Documents.* 45th ed. Accessed January 17, 2015. http://www.who.int/governance/eb /who_constitution_en.pdf.

World Health Organization. 2014. "International Classification of Functioning, Disability and Health (ICF)." Accessed January 17, 2015. http://www.who.int /classifications/icf/en/.

CHAPTER SEVEN

Following in Aldo Leopold's Footsteps

Humans-in-Ecosystem and Implications for Ecosystem Health

QI FENG LIN AND JAMES W. FYLES

1. INTRODUCTION

Our goal in this chapter is to propose and explore two related changes in our current thinking on the relationship between humans and the environment, which we feel are crucial to this effort. The first is the idea of humans as being part of the ecosystem as opposed to the common assumption that humans are separate from it. The second is a reinterpretation of the concept of ecosystem health in view of this humans-in-ecosystem perspective.

We begin by discussing the work of Aldo Leopold (1887–1948), the American forester, conservationist, and author of *A Sand County Almanac* (1949). This chapter examines how Leopold's land ethic was based on his sound understanding of the land, as well as how he used the concept of land health to refer to the intrinsic characteristic state of land that humans should strive to preserve. We also discuss the concept of the ecosystem, which gained prominence in the academic community toward the end of Leopold's life and has since become a common scientific term. We then consider Tim Ingold's description of the physical environment as a "domain of entanglement," which we feel is a compelling way to help people redefine themselves as being part of the environment. We discuss how this new self-image in turn forces us to rethink concepts such as ecosystem health, which has been used by the scientific community to refer to the holistic, well-functioning state of ecosystems. We note (in reference to other chapters of this book) that the word "health" may lack a definitive and precise definition, but this is compensated for by its connotative quality, which provides an interpretative space for users. Finally, we discuss an essay by Leopold entitled "A Mighty Fortress," which conveys the multiplicity and richness of the health concept as applied to land, before concluding with a brief remark on the importance of metaphor in constructing an ethics for the Anthropocene.

2. ALDO LEOPOLD'S LAND ETHIC AND LAND HEALTH

The writings of Aldo Leopold were a milestone in the discourse on the relationship between humans and the environment (Leopold 2013). Leopold's work is particularly relevant to our present discourse in ecological economics because he was grappling with an earlier version of the same issue as we are now: a profound deterioration of the environment as a result of economic thinking that had yet to be informed by ecological principles. After decades of promoting conservation within the existent economic system, Leopold eventually turned his mind toward challenging the underlying assumptions of society on the role of human beings in the environment and began to develop the correlative concepts of land ethic and land health.

2.1. Personal History

Leopold's views concerning humans and land evolved gradually over his long career in conservation, first as a forester in the U.S. Forest Service and then as professor of game and later wildlife management at the University of Wisconsin. He studied at the Yale Forest School, which was established in 1900 with funding from Gifford Pinchot's family, and he was influenced by Pinchot's utilitarian views of the human–forest relationship. Upon graduating in 1909, Leopold moved to the southwest territories of Arizona and New Mexico to begin his career with the Forest Service. His early responsibilities included forest management, grazing and recreation policy, and a nascent game and fish program. During this period, he advocated policies such as the extermination of large predators and the draining of wetlands, which went against the ecological wisdom he would develop later (Leopold [1915] 1991, [1945] 1991; Meine 2010).

Leopold rose quickly through the ranks of the Forest Service. In 1924, he moved to the Forest Products Laboratory of the Forest Service in Madison, Wisconsin to serve as assistant and later associate director. He grew increasingly uncomfortable there, in part because his maturing ideas, which were always concerned with the broader subject of human–land relations, could not be reconciled with "the industrial motif of this otherwise admirable institution" (Leopold [1947] 1987). He left the Forest Service in 1928.[1]

In 1933, Leopold joined the University of Wisconsin at Madison as professor of game management in the Department of Agricultural Economics.[2] In 1935, he purchased a worn-out farm, nicknamed "the shack," near Baraboo, Wisconsin, which became his family's weekend retreat. This gave him an opportunity to spend time in a rural landscape where he could observe the ecological drama of wild plants and animals, try his hand at land stewardship, and reflect upon the role of humans on the land.

Leopold encountered an array of conservation issues during his career. These issues, set against a background of rapid social and economic transformation beginning in the 1910s, through the Great Depression and the Dust Bowl of the 1930s, and finally the catastrophic Second World War and subsequent demobilization of the 1940s, prompted Leopold to reflect upon the nature of the relationship between humans and the environment.

2.2. The Land Ethic

Although Leopold had been concerned with conservation during his employment at the Forest Service, it was during his professorship at Wisconsin that he devoted more attention and energy toward the challenge of promoting conservation of "the land," a term he used in his writings to refer to the collective whole of soil, water, plants, animals, and people (Leopold [1944] 1991). This led him to elucidate and propose solutions to the various impacts of society on the land, which are generated and mediated through government policy and private economic production and consumption (Leopold [1933] 1991, [1934] 1991b). However, the impediments to conservation, as Leopold saw them, were overwhelming: the economic growth imperative; the resulting competitive pressure, leading to relentless resource utilization and urban development; the general lack of an aesthetic sense in society; a tension between private and public interests; the rapid industrialization of the economy; and the resulting drift in social consciousness away from the biophysical processes that sustain human communities and toward a fascination with gadgetry.

Eventually, Leopold concluded that conservation of the land would need to be based on an ethic. In an essay entitled "The Land Ethic," published in his collection of essays *A Sand County Almanac* (1949), he articulated his land ethic.[3] Leopold felt that insights from ecology bear important implications for ethics. The field of ecology, then relatively new, revealed a web of connections among the different components of land. These interconnections complicate matters because one can no longer devote ethical or economic attention to only a subset of species or resources chosen for their utility to humankind; ecology requires that land be treated as an integral whole. To respect this ecological reality, Leopold proposed that the sphere of ethical concern be expanded to include the land. This is accompanied by a transformation of consciousness of human beings from the dominant mentality of "conqueror of the land-community to plain member and citizen of it" (Leopold 1949:204).

In the view of most readers, the substance of Leopold's land ethic is stated in the following[4]:

> Quit thinking about decent land-use as solely an economic problem. Examine each question in terms of what is ethically and aesthetically right, as well as what is economically expedient. A thing is right when it tends to preserve the integrity, stability, and beauty of the biotic community. It is wrong when it tends otherwise.
> (Leopold 1949:224–225)

Leopold's dictum spoke most directly to land-use decision makers: foresters, farmers, government bureaucrats, and the like. It also spoke to the layperson whose consumption patterns and mentality toward the land affect land-use decisions in myriad and sometimes imperceptible ways. Indeed, during Leopold's lifetime, the percentage of the U.S. population living in rural areas gradually decreased from more than 64 percent in the 1880s and 1890s to approximately 44 percent in the 1930s and 1940s; it has continued to decrease to approximately 19 percent in 2010 (U.S. Census Bureau 2012:13–14). Leopold's land ethic, appropriate to its time, now needs to be accompanied by a more explicit "consumption ethic" for it to gain traction in our present highly urbanized and high-consumption society (MacCleery 2000).[5]

Leopold's use of the concepts of integrity and stability in the biotic community reflected his understanding of the land mechanism. For Leopold, the integrity of the land referred to its functional integrity, "a state of vigorous self-renewal" synonymous with health, which needs to be preserved (Leopold [1944] 1991). By stability, Leopold was not implying a static state of the land. Instead, he was referring to a dynamic equilibrium in the land mechanism, in contrast to the "self-accelerating rather than self-compensating departures from normal functioning" brought about by exploitative human land use, which he observed in the American landscape during the 1920s and 1930s (Leopold [1935] 1991; Meine 2004). His ideas were informed by his observations on how the stability of various ecosystem processes depends on the species diversity of that ecosystem. The more the ecosystem is able to keep its original species diversity and population levels, Leopold argued, the more stable the various processes of the ecosystem would be (Leopold [1944] 1991).[6] It follows that human modification of the land should be "as gentle and as little as possible" (Leopold [1944] 1991:315).

Leopold's criterion of beauty was based on his thinking that human interaction with the land ought to include both utility and beauty, and that recognizing and preserving the beauty in the landscape could serve to counterbalance the hard-headed and callous character of economic thinking. In an essay to encourage conservation on farmlands, Leopold wrote that "the landscape of any farm is the owner's portrait of himself" (Leopold [1939] 1991b).[7] According to Callicott (2008), Leopold's aesthetic appreciation of nature, which could be described as his "land aesthetic," is based not only on the physical appearance of natural objects and places, but also on the knowledge of their evolutionary history and ecological relationships. The end result is an extension of the "traditional criteria of natural beauty to the point where they essentially merged with his sense of long-term utility based on land health" (Meine 2004:112). Indeed, Newton (2006:347) suggested that in Leopold's view beauty was an attribute of lands that possessed stability and integrity. Leopold felt that the appreciation of nature should be accessible to the layperson and that improving the public sensitivity of landscapes and the underlying biophysical processes was crucial to maintaining the health of the land (Meine 2004:112).

Finally, if humans are part of the land, then Leopold's call for integrity, stability, and beauty in the land also applies to the human subcommunity of the land—to our lives, both individually and as a society.

The three principles articulated by Peter Brown in chapter 2—membership, householding, and entropic thrift—are similar to the concepts in Leopold's land ethic. The first, membership, is a reference to Leopold's idea of human beings as plain members and citizens of the biotic community. The second principle, householding, echoes Leopold's proposal of substituting ecology—the study of the household, from Greek oikos and logos—for engineering as the guiding principle of civilization (Leopold [1938] 1991). Finally, the concept of entropic thrift reminds one of Leopold's call for maintaining the complexity and diversity of food chains so as to preserve or restore the network of energy flow in the land (Leopold 1949:214–218), as articulated in his "biotic view of land."

2.3. "Land Health" as Rationale for and Result of the Land Ethic

Leopold's land ethic, which is essentially a reimagining of the role of human beings in the land, was grounded in his ecological understanding of how the land works. He drew upon the latest contemporary research in ecology to articulate his biotic view of land (Leopold [1939] 1991a). He understood land as having a highly complex and organized structure, being a network of soil, water, plants, and animals through which energy flows. It is "a slowly augmented revolving fund of life" open to energy from the sun and storing some of that energy away in soils, peats, and long-lived forests (Leopold [1939] 1991a). Leopold's comprehension of the structure and processes of the natural environment remains ecologically sound for the most part from today's perspective.[8] Although the field of landscape ecology had not emerged at the time, Leopold's view of a landscape in which various land types (e.g., woodlands, fields, and wetlands) are tied together by the movements of animals, water, and nutrients is fully consistent with current thinking in landscape ecology (Turner 2005).[9] Leopold also used the words "community" and "organism" to describe the land, which facilitated his use of the concept of land health.

Two characteristics of Leopold's ecological understanding of land, which informed his land ethic, are worth noting here. First, in Leopold's thinking humans are not separated from the land, even when land is represented in a scientific form such as a land pyramid or a food web (see figure 7.1). He considered humans to be part of the pyramid, sharing "an intermediate layer with the bears, raccoons, and squirrels, which eat both meat and vegetables" (Leopold 1949:215). However, he understood that human modification of the land is of a different order compared to the other organisms: the use of scientific tools and technology was causing severe degradation of the land. Leopold concluded that "man-made changes are of a different order than evolutionary changes, and have effects more comprehensive than is intended or foreseen" (Leopold 1949:218).

Second, Leopold recognized the limits of science in understanding how the land works: "The ordinary citizen today assumes that science knows what makes the [biotic] community clock tick; the scientist is equally sure that he does not. He knows that the biotic mechanism is so complex that its workings may never be fully understood" (Leopold 1949:205). The irreducible complexity and uncertainty in the workings of the land call for a certain measure of refrain and humility from humans when interacting with the land. Leopold thus imagined his land ethic as a guide for human behavior in complex ecological situations (Leopold 1949:203).

Leopold's "land health," which he had introduced in his published work but eventually explored in more depth in his unpublished manuscripts,[10] plays an important role in his overall land ethic: "A land ethic . . . reflects the existence of an ecological conscience, and this in turn reflects a conviction of individual responsibility for the health of the land. Health is the capacity of the land for self-renewal" (Leopold 1949:221). Leopold was thus using land health, which is a normative concept, as a guide for human activity. Leopold used the term "unity" to describe this land-health attribute of the original native state of land, and called for "unified conservation" (instead of lopsided, single-track conservation that focuses only on a single resource) so that human activities are configured to promote and maintain land health instead of causing damage to it (Leopold [1944] 1991). The goal here is to discern and maintain the intrinsic characteristics of the land and to prevent or reverse the derangement of the land or "land-sickness," which is caused by excessive or misguided human activity in industry and agriculture (Leopold [1941] 1991). Examples of

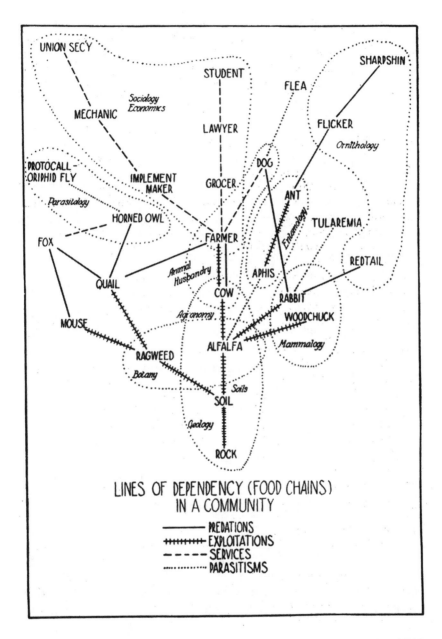

Figure 7.1. The figure that accompanied Leopold's "The Role of Wildlife in a Liberal Education" as it appeared in *Transactions of the Seventh North American Wildlife Conference.* It "traces some of the lines of dependency (or food chains, so called) in an ordinary community" (Leopold [1942] 1991). *Source:* The Aldo Leopold Foundation.

symptoms of land-sickness that Leopold mentioned include loss of soil fertility, soil erosion, abnormal floods and shortages in water systems, and the sudden disappearance or irruption of plants and animal species (Leopold 1949:194, 221). Moreover, because Leopold viewed humans as part of land, his concept of land health encompassed the health of natural systems, humans included (Meine 2004:100).[11] Interestingly enough, one of Leopold's earliest uses of the phrase "land ethic" occurred in a pithy 1935 essay entitled "Land Pathology," in which he discussed the problems of relying simply on the profit motive in conservation; the social, historical, and cultural reasons for the limited success of conservation so far; and the importance of ethics and aesthetics in tempering economic activity on the land (Leopold [1935] 1991).

Leopold's ecological knowledge and land health concepts are therefore an integral component of his land ethic, providing a normative framework for understanding how the land works and helping him to establish a "mental image of land" in relation to which "we can be ethical" in the land ethic essay (Flader and Callicott 1991; Leopold 1949:214). He proposed two yardsticks of land health: soil fertility and diversity of fauna and flora (Leopold [c.1942] 1999). In a manuscript entitled "The Land-Health Concept and Conservation," Leopold articulated four broad guidelines to achieve land health: (1) preserve native species; (2) avoid violence in land use, such as large-scale earth moving or application of chemical pesticides; (3) inculcate among landowners a relationship with the land that goes beyond economics; and (4) limit the size of human population (Leopold [1946] 1999).

3. ECOSYSTEMS AND ECOSYSTEM HEALTH

The early twentieth century saw rapid development in the field of ecology. Leopold himself was a significant contributor to the field of wildlife ecology, beginning with his pioneering textbook, *Game Management* (Leopold 1933), and his writings on how conservation needs to be implemented according to ecological principles.[12] Although Leopold used the familiar word "land" to refer to soil, water, plants, and animals, other scholars in ecology were coining new phrases or words to refer to similar concepts.

One of the important concepts that was founded during this period was the "ecosystem," a term that was coined by Arthur Roy Clapham (1904–1990) after he was asked by Arthur Tansley (1871–1955) for a suitable word to describe the "biological and physical components of an environment in relation to each other as a unit" (Willis 1994). Although Tansley subsequently introduced the term to the scientific community in a seminal paper (Tansley 1935), the term did not gain traction until Raymond Lindeman (1942) applied the term in his analysis of trophic dynamics of Cedar Bog Lake in Minnesota. This groundbreaking work helped establish the concept of ecosystem in ecology and marked the advent of ecosystem ecology (Cook 1977; McIntosh 1985:196–198; Worster 1994:306–311). The ecosystem concept was given a further boost when Eugene Odum organized his influential 1953 textbook *Fundamentals of Ecology* around ecosystems and their structure and function. Derivative concepts that had emerged include ecosystem management (Grumbine 1994, 1997), ecosystem health (Jørgensen, Xu, and Costanza 2010), and ecosystem services (Daily 1997; de Groot, Wilson, and Boumans 2002; Millennium Ecosystem Assessment 2003)—the last of which has become a key focus of ecological economics (Gómez-Baggethun et al. 2010).

In the next section, we follow Leopold's lead by exploring how human beings could be viewed as being part of the ecosystem and by characterizing ecosystem health accordingly.[13] We then highlight the complexity of ecosystems and the importance of judgment when assessing ecosystem health by discussing Leopold's essay "A Mighty Fortress" in *A Sand County Almanac*, in which he describes how the presence of tree diseases contributed to a flourishing wildlife in the woodlot on his farm.

3.1. The Place of Humans in Ecosystems

An ecosystem is an assemblage of life-forms, geological features, and climate factors, drawn together by biogeochemical and physical processes. Perhaps because the term "ecosystem" was invented to refer to the biophysical environment as an object of scientific study, most people would not consider humans as part of the ecosystem, although some may be aware of the interactions between humans and the biophysical environment. This dualistic mindset of viewing humans as being alienated from the natural environment, which became dominant during the modern age, has enabled us to scientifically investigate the myriad biophysical characteristics of an ecosystem. Unfortunately, the same mindset has also facilitated human exploitation of the environment to the present day.

A possible way out of our environmental predicament is to consider ourselves as another life-form of the ecosystem, or, as Leopold put it, "plain member and citizen" of the "land-community" (Leopold 1949:204). This is admittedly easier for the terms Leopold used—"land" and the less well-known ecological concept of "biotic community"—than for ecosystems, which is a widely used scientific concept often interpreted as excluding humans. However, we need to be cognizant of the fact that humans are utterly dependent on the biophysical environment and that "the unit of survival is the organism and its environment" (Bateson 1972:457). Humans are fundamentally embedded in the ecosystem, not only physically but also through the myriad interrelationships that regulate our well-being. For example, humans may not be living in a lake or a rainforest, but their living conditions are coupled to these ecosystems. The physical landscape is modified according to human minds and hands, responding to feedback from the environment and the needs of human beings; the result is that conditions of the human mind and body are reflected in the landscape.

This view of seeing the relationship between humans and the rest of the biophysical environment as being mutually constitutive is supported by studies on the microbial communities residing on or within the human body, or "human-associated microbiota" (Robinson, Bohannan, and Young 2010). It is estimated that the microbiota associated with a human body consists of trillions of microorganisms but constitutes only 1–3 percent of body mass due to its smaller mass. This microbiota plays an important role in human body function. For example, the gut microbiota helps the body to absorb energy and nutrients from food and contributes to regular immune function. Its composition is influenced by age, diet, environment (nationality), and disease and medication (Lozupone et al. 2012). These findings suggest that the human body is a superorganism—a veritable ecosystem that is in constant interaction with other, larger ecosystems. This has profound implications for our sense of self and definition of health (Pollan 2013).[14] Indeed, "human health can be thought of as a collective property of the human-associated microbiota" (Robinson, Bohannan, and Young 2010:468). These insights on how human existence is conditioned by the ecology of the external biophysical environment and our (mostly internal) microbiota undermine the individualistic and self-interested model of the human being on which modern economics is based.

What we need now are mental images and descriptions to help remind us of how inescapably embedded we humans are within our physical and social context. Terms such as "ecosystem" and "environment" do not explicitly reflect this embeddedness. Tim Ingold (2006) proposed thinking of an organism (e.g., a human) as a "relational constitution of being" where an organism would be imagined as an ever-ramifying web of lines of growth, with multiple lines branching out from a single source. Each organism is enmeshed within a "domain of entanglement" and the effect of its behavior ramifies through this entanglement to varying degrees.[15] Indeed, we can see that everything on the planet is part of a global domain of entanglement and constituted by relationships.

This depiction takes on greater significance when applied to humans, who, "being organisms, likewise extend along the multiple pathways of their involvement in the world" (Ingold 2006:13). The lines of growth originating from different sources become entangled with one another, such that what we refer to as "the environment"—not referring merely to the biophysical, but also the social, intellectual, spiritual, etc.—would here be depicted as a domain of entanglement. We are constantly and inextricably entangled with other humans and nonhuman organisms—past, present, and future—and the geology and climate of our planet through our lifestyle and culture. Indeed, even the most remote places on the planet can be severely affected by human activity.

If we accept that humans are embedded within an ecosystem, the next step is to think holistically and reflect upon what the intrinsic character of an ecosystem that includes human presence ought to be. As mentioned earlier, our thinking on this matter evolves alongside our concept of the human self and our relationship to the environment. To restate this idea using the terms introduced previously, our understanding of the intrinsic character of an ecosystem, when considered as the domain of entanglement of our relational constitution of being, coevolves with our thinking on the desired nature of the entanglement and our understanding of the nature of our own being, both of which are influenced by culture.

This discussion on the normative state of human being is admittedly going beyond the domain of science and reminds one of Leopold's idea of humans as "plain member and citizen" of the "land-community" (Leopold 1949:204) living on "the land" and his correlated concepts of land ethic and land health. One possible way of studying and reflecting upon the intrinsic character of an ecosystem, for guiding human beings, is through the concept of health.

3.2. Health of an Ecosystem that Includes Humans

Health is a complex concept, whether it is applied to humans or to ecosystems. It is difficult to grasp the concept solely from a scientific perspective: its meaning cannot be fully encapsulated by a definitive description; it cannot be assessed using only quantitative tools such as metrics or indices; and it does not refer merely to a state of absence of disease (Goldberg and Garver, chapter 4 this volume).

Perhaps it is more fruitful to interpret health as a connotative concept where assessment of the condition of an organism is based not only on certain key principles, but also by taking into account the information and relationships that are relevant to its particular context. This is not a revolutionary idea; it merely makes more salient the importance of human judgment and interpretation in assessing health. Indeed, assessment of health is not a strictly scientific endeavor: our assessment of our own health is based partly on our sense of feeling about how we are doing, and we could make similar assessments of others (e.g., the people and animals who are close to us) by observing their mien (although we submit that scientific investigation requires human judgment and skills as well, more than usually recognized). This expanded, connotative sense of health enables one to recognize the complex interrelationships of being in this world, as made explicit by the terms "relational constitution of being" and "domain of entanglement" mentioned previously.

Leopold's understanding of land health—and the land ethic that emerged from it—were the product of a lifetime of observation and interpretation. Leopold was a keen observer of the relationships that played out on the land and of the participation of humanity in them. Much of his writing is an invitation to us to "see" more in the land. However, Leopold passed away in 1948, on the threshold of the exponential rise in the ability of science to quantify. Many methodologies that emerged after Leopold's time sought to validate human observation; however, in doing so, they devalued the knowledgeable and experienced observer relative to the well-calibrated instrument. Leopold's ability to "read the land" would not gain traction in today's scientific grant and publishing establishment.

Decades of methodological innovation and careful measurement have allowed us to see deeply into the genome of individuals and into the diversity of any square meter of the land. Although informing us as never before of the land and its functions, one of the products of this expanding knowledge has been an incapacitating awareness of degree of complexity at different scales of the land. The emergence of large-scale data analytics or "big data" notwithstanding, our ability to provide an integrated assessment of an abused tract of land in Wisconsin or anywhere else, although potentially informed by better data, has not advanced much beyond what was apparent to the tuned senses of Leopold. This is particularly true when we recognize that our assessment on steps to be taken next on the land must extend beyond our scientific measurements to the ethical and dimensions in which the data reside. When faced with potentially overwhelming information, although not devaluing the data, we may be wise to seek out the modern-day Leopolds—the foresters, farmers, Aboriginal elders who have spent lives in observation—and ask them what they see in the land. As scientists, we should view our science with humility, recognizing the irreducible complexity in which we are entangled and the limitations it imposes in our application of scientific knowledge to a specific problem or place. We should be open to all sources of insight, whether they are molecular measurements of gene function, remote images from space, or the understanding of a knowledgeable and experienced observer of the land.

The concept of health, which is already rich in meaning and interpretation, becomes metaphorical when applied to an ecosystem. Indeed, as discussed previously, research on human-associated microbiota suggests that the human body could be considered as an ecosystem, although humans differ from ecosystems in that humans possess consciousness while ecosystems are commonly considered otherwise by modern Western thought. However, an ecosystem can display a dynamic equilibrium and possess the potential to develop in novel ways. In this context, ecosystem health refers to the "proper" characteristics of all that is involved in the ecosystem. The concept of ecosystem health is therefore a heuristic and metaphorical device to help us deal as best we can with the complex task of characterizing and understanding ecosystems and our place in them. This interpretation may locate ecosystem health some distance away from its scientific mooring. However, because health is a connotative concept that most people can understand, applying the metaphor of health to ecosystems can yield a fruitful, if ambiguous, interpretive space to work with, which is commensurate with and suggestive of the immense potential of an ecosystem for its own evolution and development. If we extend the concept of human health broadly enough to include ecosystems, then ecosystem health may be viewed as an extension of human health and not as a metaphor. However, this view is based solely on the consideration of human well-being and neglects the multiplicity of values and perspectives in an ecosystem, as we will see in the next subsection.

The thinking that humans are part of ecosystems forces us to rethink ecosystem health and raises the question of what a healthy landscape that is inhabited by plant and animal species, including homo sapiens, would look like. In other words, given our knowledge of the intrinsic character of the nonhuman aspects of an ecosystem and our needs as a species, our present goal is to establish and maintain a healthy joint condition of both humans and the rest of the ecosystem. This ultimately leads us to confront the profound question of the relationship between humans and the other life-forms and what our role in the landscape should be. If we participate in the ecosystem—our biophysical domain of entanglement—in a respectful and mindful way, we not only minimize our impact on its natural workings, but we also preserve the long-term life-supporting mechanism for all organisms (including ourselves). In our studies of ecosystem health, we should bear in mind that we are an integral component of the ecosystem we are studying; our membership is established through our physical presence in the environment and through our consciousness (i.e., mental presence) when perceiving it.

Although Leopold's land health concept has been credited for contributing to the emergence of the concept of ecosystem health (Callicott 1992; Kolb, Wagner, and Covington 1994; Rapport 1995), the concepts are different in terms of the contexts in which they were used. Leopold's land health was not a stand-alone concept, but it was derived as part of his broader effort to rethink the human-land relationship with an eye toward establishing a land ethic and promoting conservation of the land. Ecosystem health, on the other hand, was conceived as a scientific concept that emerged out of the development of the field of ecology and ecosystem in the decades after Leopold's death.

3.3. A Mighty Fortress for Life

In one of his essays in *A Sand County Almanac*, "A Mighty Fortress," Leopold (1949) suggested the notion of adopting a less anthropocentric position in relation to our domain of entanglement in order to allow the domain more space for its natural workings and its myriad ecological dramas to unfold. He described how disease-afflicted trees in the woodlot on his shack property enriched the woodlot with higher levels of wildlife activity. Trees that were infected by diseases and insect pests became protective habitats and rich sources of food for wildlife. Diseased and dying oak trees which became windfalls found "new life" as habitat for grouse and wild bees. Trees that were infested with termites became a rich mine of food for chickadees. The diseases in his woodlot taught Leopold that wildlife depends on diseased trees for habitat. Commenting on how a raccoon had taken refuge against a hunter in a half-uprooted maple that was weakened in the roots by a fungus disease, Leopold (1949) observed that the tree "offers an impregnable fortress for coondom [*sic*]"[16] (see figure 7.2).

Figure 7.2. While Leopold espoused leaving diseased and dying trees in its place to promote habitat for wildlife, the opposite had occurred here. This former "raccoon tree" near the Leopold shack, with Leopold standing before it to serve as a scale, was apparently chopped down by (presumably ecologically uninformed) raccoon hunters long before this photo was taken in April 1937 (McCabe 1987:96). *Source:* The Aldo Leopold Foundation.

In this remarkable section, Leopold uses the metaphor from Martin Luther's hymn, "A Mighty Fortress Is Our God." However, Leopold shifts its understanding dramatically, effectively substituting the word "evolution" for God. It is evolution that is a "bulwark never failing."

These observations made on a Wisconsin farm in the early twentieth century have been explored, reiterated, and confirmed by generations of research ecologists concerned with conserving species and ecosystems. Disturbance agents such as insects, fire, and flooding are now viewed as integral components of the ecosystem (Hobbs and Huenneke 1992). Dead wood, once an anathema to the forester, is now regarded as a key resource to be valued and managed (Harmon et al. 2004). In embracing the complexity of the ecosystem and the entanglement of its entire species, we have been forced to change our perspective on what constitutes a valued component.

The implication of Leopold's observation—that the health of an ecosystem entails the presence of what have been commonly considered as diseases—forces us to expand and reflect on our concept of health. Death and disease are part and parcel of nature and life. The health of the land is not reducible to the health of its individual constituent organisms. The labelling of species and conditions as "disease," "pests," or "weeds" reflects the inconvenience they cause to humans. This does not mean that we willfully tolerate infestations or plagues. Instead, we need to recognize, as Leopold pointed out, that "no species are inherently a pest": whether a species is considered expedient for human activity depends on its population size and distribution and the interpretation of what is good by humans (Leopold [1943] 1991). This forces us to reexamine the meaning of freedom and liberty for human individuals, concepts that have come to characterize the modern age.[17] Again, and going back to Leopold's land ethic, we need to cease using our anthropocentric, economic criteria as primary measures when assessing the good of living organisms and habitats with which we are entangled.

Of course, Leopold could afford to let nature have a free hand in his woodlot because he was not hard-pressed to extract economic value from it. Indeed, from society's point of view, the economic value of Leopold's farm property, which was abandoned by its previous tax-delinquent owner, had already been depleted. However, this is not the current reality in most parts of the world, where the cultural narrative of progress and the economic growth imperative are causing humans to profoundly disrupt the developmental possibilities of ecosystems.

The question that confronts us humans is how our consciousness and actions will affect the long-term potential of the natural world. At the heart of this issue is whether we consider ourselves to be part of nature, and hence as agents of the forces of nature, or whether we consider ourselves to be separate from nature and hence constitute a force on this world that is distinct from that of nature. To use Leopold's metaphors, we could adopt a conqueror's mentality in the environment and use it as a habitat for humans only, ignoring the fact that this might undermine our own long-term viability. Or, we could view human society as inescapably embedded in the ecosystems of Earth, learning to live in a way that recognizes and supports the collective health of the ecosystems, including our own.

4. FINAL THOUGHTS

The history of civilization has been one of constant innovation, development, and evolution of human societies, as well as conflicts and wars. Our current work to reconcile the human economy with ecological principles, essentially a re-envisioning project, represents an opportunity to continue this process in a manner that is more conscious and mindful of the life-sustaining processes that our human society depends on than had been generally demonstrated in the past century or so.

Progress toward a holistic melding of ecological and economic principles would require a sound understanding of the intrinsic character of the ecosystem. Basing human action on untenable assumptions about the relationship between humans and ecosystems—from the smallest local plot of land to the scale of the entire planet—is manifest folly. Normative guides for human action with respect to the ecosystem—at the largest scale, the entire planet—must be based on a deep and critical knowledge of the ecosystem, coupled, as Leopold would hasten to add, with deep humility. This should be achieved through not only the scientific mode of thinking but also other forms of knowledge and perception, such as the humanities and the arts, which supply rich and essential metaphors for understanding human existence (Midgley 2001). We need to develop a scientifically realistic and enlightened conception of the nature of human being and our place in the ecosystem—indeed, in the universe—if we are to establish a healthy existence within it.

Recognizing the embeddedness of humans within the domain of entanglement would hopefully lead to a general respect and appreciation of the myriad life-forms and life-supporting biophysical processes of this planet. Leopold's land ethic was the result of his understanding of the ecology of the land and his reflection of the place of humans in it. Moreover, he recognized that an ethic is constantly evolving with society (Leopold 1949:225). It is perhaps time for society—beginning with small groups, such as communities and towns—to reflect upon its place in the domain of entanglement and develop its own ethic toward it.

Such an ethic would also enable us to deal better with the complexity and uncertainty that is inherent in our domain of entanglement. Once we begin to recognize the extent of our entanglement with one another and the biophysical environment, the multiplicity of interactions would make it impossible for us to grasp every detail from an analytical perspective. The value of an ethic here is to provide an alternative way of guiding our conduct under such immensely complex situations. Leopold's land ethic, which calls for the preservation of integrity, stability, and beauty of the biophysical environment, does not provide any specific prescriptions on human behavior. However, it provides a thought-provoking and solemn ethic against which we can evaluate our possible actions.

Human thought is inescapably anthropocentric in the sense that our thinking is biased toward ensuring human well-being in its various conceptions. Such thinking is necessary and perhaps unavoidable for our survival. However, in Western culture, this mentality has intensified in the past century or so to the point of privileging human material well-being over that of the rest of the environment, resulting in profound transformations of the planet that are detrimental to the long-term well-being of all life-forms. Our challenge, henceforth, is to consider humans and the biophysical environment as a unified whole and revise human actions according to this perspective. We need to ground human consciousness and activity in a foundation of being alive and open to a world that is always coming into being, just like ourselves. We need to learn how to cooperate with our biophysical environment instead of leading an existence that has largely been in contention with it. For this purpose, metaphors such as health can help us to cognitively and emotionally grasp what is at stake in the Anthropocene.

NOTES

We are grateful to the editors, Peter Brown and Peter Timmerman, for their patience and helpful suggestions; to the other authors of the book for thought-provoking discussions; to Curt Meine and Peter Whitehouse for commenting on the manuscript; and to the Aldo Leopold Foundation for permission to use the figures in this chapter.

1. In 1924, William B. Greeley, then head of the Forest Service and Leopold's instructor at the Yale Forest School, expressed his desire to have Leopold assume the position of assistant director at the Forest Products Laboratory with the thinking that he would succeed then-director Carlile P. Winslow; Winslow was expected to step down within a year. Leopold never fully explained his decision for the move (Meine 2010:225). However, his decision to leave in 1928 was wise in retrospect: Winslow did not step down as director until 1946 (Havlick 2009).

2. Between 1928 and 1933, Leopold conducted game surveys for Sporting Arms and Ammunitions Manufacturers' Institute in eight north-central states (Newton 2006:107–114).

3. Following Leopold's abrupt death in 1948 due to a heart attack, his son Luna headed a team comprising former students and friends to prepare the manuscript for publication. Perhaps the most important change was to shift "The Land Ethic" from its original first position in Part III to its present final position. See Ribbens (1987) and Meine (2010:523–524).

4. The term "biotic community" was one of the terms that was used in ecological discourse during Leopold's time and was used in Charles Elton's *Animal Ecology* (1927). Leopold was using its metaphorical aspect to describe how "the plants, animals, men, and soil are a community of interdependent parts" (Leopold [1934] 1991a:209).

5. See the chapter by Janice Harvey in this volume (chapter 11) for a discussion on culture, consumerism, and possible ways of implementing cultural change.

6. Leopold was assuming the idea that biological diversity enhances ecological stability, the "diversity-stability hypothesis," which was challenged by studies conducted by Robert May and others. However, these studies, which were based on mathematical models, had shortcomings, including a focus on individual species population. Beginning in the mid-1990s, the hypothesis was rehabilitated by an experimentally driven research program and a shift in focus from population to ecosystem stability. See Mikkelson (2009).

7. Leopold was influenced by the work of the regional artist John Steuart Curry. From 1936 until his death in 1946, Curry was the initial appointee of the artist-in-residence program at the University of Wisconsin's College of Agriculture, the first such program in the United States (Cronon and Jenkins 1994:783). Curry was appointed to the College's Department of Agricultural Economics in 1943 (Glover 1952:338–339), which strongly suggests that he would have crossed path with Leopold. In the context of the sentence quoted here, Leopold cited Curry and the other two great artists of the regionalist movement in the Midwest: "What is the meaning of John Steuart Curry, Grant Wood, Thomas Benton? They are showing us drama in the red barn, the stark silo, the team heaving over the hill, the country store, black against the sunset. All I am saying is that there is also drama in every bush, if you can see it. . . . The landscape of any farm is the owner's portrait of himself. Conservation implies self-expression in

that landscape, rather than blind compliance with economic dogma" (Leopold [1939] 1991b: 263).

8. In Leopold's model of the land pyramid, energy was thought to return to the soil through death and decay. This model of energy cycling through the food chain was proven wrong when Lindeman (1942) established from his study of trophic dynamics of a lake that energy flows through and out of an ecosystem (as degraded heat in accordance with the second law of thermodynamics). Thus, the energetic aspect of Leopold's "land pyramid" was already outdated when in 1947 he incorporated it into "The Land Ethic" essay of *A Sand County Almanac* (Leopold 1949; Meine 2010: 501).

9. For discussion of Leopold's thinking on ecology in light of present-day ecological findings, see, for example, Wolfe and von Berg (1988), Silbernagel (2003), Ripple and Beschta (2005), and Binkley et al. (2006).

10. These manuscripts have since been published: "Conservation: In Whole or in Part?" (Leopold [1944] 1991); "Biotic Land-Use" (Leopold [c.1942] 1999); and "The Land-Health Concept and Conservation" (Leopold [1946] 1999).

11. For a detailed discussion of Leopold's concept of land health, see Newton (2006: 316–351).

12. At the time of his death in 1948, Leopold was thinking of revising *Game Management* (Meine 2010: 523).

13. For discussions on integrating humans into ecosystems, see, for example, Berkes, Colding, and Folke (2003) and Bechtold et al. (2013). The fact that the latter considered this concept as on the "frontier" of ecosystem science suggests that ecosystems are commonly perceived to be without humans.

14. Lozupone et al. (2012) used "macroecosystem" to refer to the physical ecosystem.

15. Ingold articulated these concepts as part of his critique of the conventional understanding of animism.

16. Leopold coined the word "coondom" to refer to "raccoon-dom," the world of raccoons.

17. See the discussion on these concepts in relation to ecological economics by Bruce Jennings in chapter 10.

REFERENCES

Bateson, Gregory. 1972. "Form, Substance, and Difference." In *Steps to an Ecology of Mind*, 454–471. Chicago: University of Chicago Press.

Bechtold, Heather A., Jorge Durán, David L. Strayer, Kathleen C. Weathers, Angelica P. Alvarado, Neil D. Bettez, Michelle A. Hersh, Robert C. Johnson, Eric G. Keeling, Jennifer L. Morse, Andrea M. Previtali, and Alexandra Rodríguez. 2013. "Frontiers in Ecosystem Science." In *Fundamentals of Ecosystem Science*, edited by Kathie Weathers, David L. Strayer, and Gene E. Likens, 279–296. Waltham, MA: Academic Press.

Berkes, Fikret, Johan Colding, and Carl Folke, eds. 2003. *Navigating Social-Ecological Systems: Building Resilience for Complexity and Change*. Cambridge, UK: Cambridge University Press.

Binkley, Dan, Margaret Moore, William Romme, and Peter Brown. 2006. "Was Aldo Leopold Right About the Kaibab Deer Herd?" *Ecosystems* 9 (2): 227–241. doi: 10.1007/s10021-005-0100-z.

Callicott, J. Baird. 1992. "Aldo Leopold's Metaphor." In *Ecosystem Health: New Goals for Environmental Management*, edited by Robert Costanza, Bryan G. Norton and Benjamin D. Haskell, 42–56. Washington, DC: Island Press.

Callicott, J. Baird. 2008. "Leopold's Land Aesthetic." In *Nature, Aesthetics, and Environmentalism: From Beauty to Duty*, edited by Allen Carlson and Sheila Lintott, 105–118. New York: Columbia University Press.

Cook, Robert Edward. 1977. "Raymond Lindeman and the Trophic-Dynamic Concept in Ecology." *Science* 198 (4312): 22–26.

Cronon, E. David, and John W. Jenkins. 1994. *The University of Wisconsin: A History. Vol. 3, Politics, Depression, and War (1925–1945)*. Madison: University of Wisconsin Press. http://digital.library.wisc.edu/1711.dl/UW.UWHist19251945v3.

Daily, Gretchen C., ed. 1997. *Nature's Services: Societal Dependence on Natural Ecosystems*. Washington, DC: Island Press.

de Groot, Rudolf S., Matthew A. Wilson, and Roelof M. J. Boumans. 2002. "A Typology for the Classification, Description and Valuation of Ecosystem Functions, Goods and Services." *Ecological Economics* 41 (3): 393–408. doi: 10.1016/S0921-8009(02)00089-7.

Elton, Charles. 1927. *Animal Ecology*. London: Sidgwick and Jackson.

Flader, Susan L., and J. Baird Callicott. 1991. "Introduction." In *The River of the Mother of God and Other Essays*, edited by S. L. Flader and J. B. Callicott, 3–31. Madison: University of Wisconsin Press.

Glover, Wilbur H. 1952. *Farm and College: The College of Agriculture of the University of Wisconsin; A History*. Madison: University of Wisconsin Press.

Gómez-Baggethun, Erik, Rudolf de Groot, Pedro L. Lomas, and Carlos Montes. 2010. "The History of Ecosystem Services in Economic Theory and Practice: From Early Notions to Markets and Payment Schemes." *Ecological Economics* 69 (6): 1209–1218. doi: 10.1016/j.ecolecon.2009.11.007.

Grumbine, R. Edward. 1994. "What Is Ecosystem Management?" *Conservation Biology* 8: 27–38.

Grumbine, R. Edward. 1997. "Reflections on 'What Is Ecosystem Management?'" *Conservation Biology* 11: 41–47. doi: 10.1046/j.1523-1739.1997.95479.x.

Harmon, M. E., J. F. Franklin, F. J. Swanson, P. Sollins, S. V. Gregory, J. D. Lattin, N. H. Anderson, S. P. Cline, N. G. Aumen, J. R. Sedell, G. W. Lienkaemper, K. Cromack, Jr., and K. W. Cummins. 2004. "Ecology of Coarse Woody Debris in Temperate Ecosystems." In *Advances in Ecological Research*, edited by H. Caswell, 59–234. Waltham, MA: Academic Press.

Havlick, David G. 2009. *US Forest Service History: Carlile P. 'Cap' Winslow*. Forest History Society. Accessed January 20, 2015. http://www.foresthistory.org/ASPNET/People/Scientists/Winslow.aspx.

Hobbs, Richard J., and Laura F. Huenneke. 1992. "Disturbance, Diversity, and Invasion: Implications for Conservation." *Conservation Biology* 6 (3): 324–337. doi: 10.2307/2386033.

Ingold, Tim. 2006. "Rethinking the Animate, Re-Animating Thought." *Ethnos* 71 (1): 9–20. doi: 10.1080/00141840600603111.

Jørgensen, Sven Erik, Fu-Liu Xu, and Robert Costanza, eds. 2010. *Handbook of Ecological Indicators for Assessment of Ecosystem Health*. 2nd ed. Boca Raton, FL: CRC Press.

Kolb, T. E., M. R. Wagner, and W. W. Covington. 1994. "Concepts of Forest Health: Utilitarian and Ecosystem Perspectives." *Journal of Forestry* 92 (7): 10–15.

Leopold, Aldo. (1915) 1991. "The Varmint Question." In *The River of the Mother of God and Other Essays*, edited by Susan L. Flader and J. Baird Callicott, 47–48. Madison: University of Wisconsin Press.

Leopold, Aldo. 1933. *Game Management*. New York: Charles Scribner's Sons.

Leopold, Aldo. (1933) 1991. "The Conservation Ethic." In *The River of the Mother of God and Other Essays*, edited by Susan L. Flader and J. Baird Callicott, 181–192. Madison: University of Wisconsin Press.

Leopold, Aldo. (1934) 1991a. "The Arboretum and the University " In *The River of the Mother of God and Other Essays*, edited by Susan L. Flader and J. Baird Callicott, 209–211. Madison: University of Wisconsin Press.

Leopold, Aldo. (1934) 1991b. "Conservation Economics." In *The River of the Mother of God and Other Essays*, edited by Susan L. Flader and J. Baird Callicott, 193–202. Madison: University of Wisconsin Press.

Leopold, Aldo. (1935) 1991. "Land Pathology." In *The River of the Mother of God and Other Essays*, edited by Susan L. Flader and J. Baird Callicott, 212–217. Madison: University of Wisconsin Press.

Leopold, Aldo. (1938) 1991. "Engineering and Conservation." In *The River of the Mother of God and Other Essays*, edited by Susan L. Flader and J. Baird Callicott, 249–254. Madison: University of Wisconsin Press.

Leopold, Aldo. (1939) 1991a. "A Biotic View of Land." In *The River of the Mother of God and Other Essays*, edited by Susan L. Flader and J. Baird Callicott, 266–273. Madison: University of Wisconsin Press.

Leopold, Aldo. (1939) 1991b. "The Farmer as a Conservationist." In *The River of the Mother of God and Other Essays*, edited by Susan L. Flader and J. Baird Callicott, 255–265. Madison: University of Wisconsin Press.

Leopold, Aldo. (1941) 1991. "Wilderness as a Land Laboratory." In *The River of the Mother of God and Other Essays*, edited by Susan L. Flader and J. Baird Callicott, 287–289. Madison: University of Wisconsin Press.

Leopold, Aldo. (1942) 1991. "The Role of Wildlife in a Liberal Education." In *The River of the Mother of God and Other Essays*, edited by Susan L. Flader and J. Baird Callicott, 301–305. Madison: University of Wisconsin Press.

Leopold, Aldo. (1942) 1999. "Biotic Land-Use." In *For the Health of the Land: Previously Unpublished Essays and Other Writings*, edited by J. Baird Callicott and Eric T. Freyfogle, 198–207. Washington, DC: Island Press.

Leopold, Aldo. (1943) 1991. "What Is a Weed?" In *The River of the Mother of God and Other Essays*, edited by Susan L. Flader and J. Baird Callicott, 306–309. Madison: University of Wisconsin Press.

Leopold, Aldo. (1944) 1991. "Conservation: In Whole or in Part?" In *The River of the Mother of God and Other Essays*, edited by Susan L. Flader and J. Baird Callicott, 310–319. Madison: University of Wisconsin Press.

Leopold, Aldo. (1945) 1991. "Review of Young and Goldman, the Wolves of North America." In *The River of the Mother of God and Other Essays*, edited by Susan L. Flader and J. Baird Callicott, 320–322. Madison: University of Wisconsin Press.

Leopold, Aldo. (1946) 1999. "The Land-Health Concept and Conservation." In *For the Health of the Land: Previously Unpublished Essays and Other Writings*, edited by J. Baird Callicott and Eric T. Freyfogle, 218–226. Washington, DC: Island Press.

Leopold, Aldo. (1947) 1987. "Foreword." In *Companion to a Sand County Almanac: Interpretive & Critical Essays*, edited by J. Baird Callicott, 281–288. Madison: University of Wisconsin Press.

Leopold, Aldo. 1949. *A Sand County Almanac and Sketches Here and There*. New York: Oxford University Press.

Leopold, Aldo. 2013. *A Sand County Almanac & Other Writings on Ecology and Conservation*. Edited by Curt Meine. New York: Library of America.

Lindeman, Raymond L. 1942. "The Trophic-Dynamic Aspect of Ecology." *Ecology* 23 (4): 399–417. doi: 10.2307/1930126.

Lozupone, Catherine A., Jesse I. Stombaugh, Jeffrey I. Gordon, Janet K. Jansson, and Rob Knight. 2012. "Diversity, Stability and Resilience of the Human Gut Microbiota." *Nature* 489 (7415): 220–230. doi: 10.1038/nature11550.

MacCleery, Douglas W. 2000. "Aldo Leopold's Land Ethic: Is It Only Half a Loaf?" *Journal of Forestry* 98 (10): 5–7.

McCabe, Robert A. 1987. *Aldo Leopold: The Professor*. Madison, WI: Rusty Rock Press.

McIntosh, Robert P. 1985. *The Background of Ecology: Concept and Theory*. Cambridge, UK: Cambridge University Press.

Meine, Curt. 2004. *Correction Lines: Essays on Land, Leopold, and Conservation*. Washington, DC: Island Press.

Meine, Curt. 2010. *Aldo Leopold: His Life and Work*. Madison: University of Wisconsin Press.

Midgley, Mary. 2001. *Science and Poetry*. London: Routledge.

Mikkelson, Gregory M. 2009. "Diversity-Stability Hypothesis." In *Encyclopedia of Environmental Ethics and Philosophy*, edited by J. Baird Callicott and Robert Frodeman, 255–256. Detroit: Macmillan.

Millennium Ecosystem Assessment. 2003. *Ecosystems and Human Well-Being: A Framework for Assessment*. Washington, DC: Island Press.

Newton, Julianne Lutz. 2006. *Aldo Leopold's Odyssey*. Washington, DC: Island Press.

Odum, Eugene P. 1953. *Fundamentals of Ecology*. Philadelphia: Saunders.

Pollan, Michael. 2013. "Say Hello to the 100 Trillion Bacteria That Make up Your Microbiome." *The New York Times Magazine*, May 15. Accessed January 20, 2015. http://www.nytimes.com/2013/05/19/magazine/say-hello-to-the-100-trillion-bacteria-that-make-up-your-microbiome.html.

Rapport, David J. 1995. "Ecosystem Health: More Than a Metaphor?" *Environmental Values* 4 (4): 287–309.

Ribbens, Dennis. 1987. "The Making of *a Sand County Almanac*." In *Companion to a Sand County Almanac*, edited by J. Baird Callicott, 91–109. Madison: University of Wisconsin Press.

Ripple, W. J., and R. L. Beschta. 2005. "Linking Wolves and Plants: Aldo Leopold on Trophic Cascades." *Bioscience* 55: 613–621. doi: 10.1641/0006-3568(2005)055[0613:LWAPAL]2.0.C.

Robinson, Courtney J., Brendan J. M. Bohannan, and Vincent B. Young. 2010. "From Structure to Function: The Ecology of Host-Associated Microbial Communities." *Microbiology and Molecular Biology Reviews* 74 (3): 453–476. doi: 10.1128 /mmbr.00014-10.

Silbernagel, Janet. 2003. "Spatial Theory in Early Conservation Design: Examples from Aldo Leopold's Work." *Landscape Ecology* 18 (7): 635–646. doi: 10.1023/B:LAND .0000004458.18101.4d.

Tansley, A. G. 1935. "The Use and Abuse of Vegetational Concepts and Terms." *Ecology* 16: 284–307.

Turner, Monica G. 2005. "Landscape Ecology: What Is the State of the Science?" *Annual Review of Ecology, Evolution, and Systematics* 36: 319–344. doi: 10.1146/annurev .ecolsys.36.102003.152614.

U.S. Census Bureau. 2012. *2010 Census of Population and Housing: Population and Housing Unit Counts. CPH-2-1, United States Summary*. Washington, DC: US Government Printing Office. Accessed January 20, 2015. http://www.census.gov/prod /cen2010/cph-2-1.pdf.

Willis, A. J. 1994. "Arthur Roy Clapham. 24 May 1904–18 December 1990." *Biographical Memoirs of Fellows of the Royal Society* 39: 72–90. doi: 10.1098/rsbm.1994.0005.

Wolfe, Michael L., and Friedrich-Christian von Berg. 1988. "Deer and Forestry in Germany: Half a Century after Aldo Leopold." *Journal of Forestry* 86 (5): 25–31.

Worster, Donald. 1994. *Nature's Economy: A History of Ecological Ideas*. 2nd ed. Cambridge, UK: Cambridge University Press.

PART III

IMPLICATIONS:
STEPS TOWARD REALIZING
AN ECOLOGICAL ECONOMY

INTRODUCTION AND CHAPTER SUMMARIES

The last part of this volume follows up on the previous two parts by exploring how an ecological economy can be implemented, at the micro- and macro-levels. As pointed out in part 1, our current environmental crisis is a result of a peculiar type of thinking and relating to the environment, which resulted in a dominant approach that shows little concern for its environmental context, as well as the related pressing problems of social and ecological justice. Recognizing this myopia calls for changes in our current patterns of thought and behavior. Needed changes in micro- and macroeconomics, and in our overall cultural ethos, are detailed in the next three chapters.

SUMMARY OF CHAPTERS

"Toward an Ecological Macroeconomics," by Peter A. Victor and Tim Jackson

In this chapter, the authors share the results of their research in developing an ecological macroeconomics. They begin with a discussion on the dilemma of the current economic system, noting that the natural dynamics of the system push it toward either perpetual growth or collapse. The business-as-usual scenario of relentless economic growth is needed for continual and stable functioning of the system, even though it undermines the life-support system of the planet. To resist growth is to risk economic and social collapse. What is needed is an ecological macroeconomics that would be based on some key existing macroeconomic variables (e.g., investment, government expenditure) while incorporating some new elements (e.g., a transformation of the money system, a different national accounting system that takes into account environmental and resource variables).

Three models are presented in this chapter. The first is Victor's LowGrow model of the Canadian economy, which showed that slower growth in net investment, productivity, government expenditure, and cessation of growth in population, besides other conditions, will lead to a form of growth of the gross domestic product that eventually levels off. The second is a simulation model of the UK economy, created to examine the characteristics of an economy that focuses on human and social services, which has limited productivity growth potential but also entails less material throughput. Finally, the authors provide a brief introduction to their Green Economy Macro-Model and Accounts framework, which integrates an understanding of the real economy, the financial economy, and the ecological and resource constraints.

"New Corporations for an Ecological Economy: A Case Study," by Richard Janda, Philip Duguay, and Richard Lehun

This chapter focuses on the role of the extraordinary presence of corporations in our current economy, arguing that the current ethical commitment of "corporate social responsibility" is subordinate to wealth maximization for the shareholders. Instead, the authors point to the recent legal innovation of the "benefit corporation" as another alternative—a business corporation that has elected to promote the "general public benefit." They review the legal structure and enabling clauses for such corporations, as well as various certification standards providers. They argue that the benefit corporation of the future can be a "virtuous hybrid" of different sets of ethical rules by which corporations can operate in an ecological economy.

"Ecological Political Economy and Liberty," by Bruce Jennings

Bruce Jennings argues that developing the ethical foundations of ecological economics should include a reconsideration of the concept of liberty and the concept of the person. Other chapters provide cross-cultural examples of ecological ways of thinking about economic activity, valuing nature, justice, and responsibility. An ecological understanding of the person and of freedom will help to complete the task of reconceptualization undertaken in this volume as a whole. Mainstream economics of the liberal and capitalist era stands in need of revision by ecological economics precisely because it has cultivated environmentally overbearing and unsustainable modes of production and consumption in the pursuit of a specific ideal of how human beings should live and how they should be free to live. Alternative modes of production and consumption need to be developed, to be sure, but that task alone will be incomplete unless the underlying ideal of human flourishing is also reconceptualized and redefined. The two components of this ideal of living involve the identity and fulfillment of the human being who lives a life (the person) and the shape of the activity of this living (the freedom or liberty of the person).

The primary purpose of this chapter is not to rehearse the ideas of personhood and liberty that have been pillars of mainstream economics and the political economy of unsustainable growth. Many critiques of libertarian individualism and negative liberty have been written. Instead, this chapter will explore a more positive vision of an ecologically informed ideal of living, supported by a relational conception of liberty and of the person. Freedom, in this view, is constituted not in spite of connections and commitments linking self to others but in and through these connections and commitments. The ecological person or self is an identity built out of an ongoing practice of building certain kinds of relationships and structured interactions that exemplify the creative and aesthetic dimensions of humanity and living nature. Such relationships involve not only social interactions with other human beings but also—and crucially—meaningful connections between the person and his or her own activity on and in the world of nature, living systems, and material objects. An ecological conception of freedom and of the person contains a countervision to notions of alienation, commodification, and the objectification of the human or natural other. Such a vision leads away from the control of natural material as a source of fulfillment and wealth and toward a notion of artistry, craftsmanship, and the accommodation of the inherent properties of natural form and the liberal tradition. In developing an account of freedom and personhood along these lines, the chapter will draw upon a critical and utopian tradition of countervisions to the economic modes of life in the modern period (e.g., in writers such as Ruskin and Morris) and in contemporary work on the concepts of liberty and autonomy and on ecological conceptions of the self.

"A New Ethos, a New Discourse, a New Economy: Change Dynamics Toward an Ecological Political Economy," by Janice E. Harvey

This chapter explores the important issue of how to realize the cultural changes in society that are needed for a transition to an ecological economics. In particular, Harvey considers the processes and interventions by which change might be realized. After discussing the relationship between discourse and cultural change, she explores how the hegemonic discourses of neoliberalism and consumerism affect institutional practice. Four barriers to a transition to ecological economics are identified: the globalization of financial markets, which makes it difficult and risky for countries to unilaterally embark on the transition; a dependence of individual well-being on the present economic growth model, which results in a similar difficulty for individuals to live according to an ecological economy; the discourse of consumer culture; and the media reinforcement of the norms of economic growth and consumption.

Recognizing that cultural change is more stable than political change and is therefore more efficacious in bringing about permanent changes in society, Harvey identifies four propositions on the characteristics of cultural change. First, cultural change usually begins at the top. Second, change is typically initiated by elites who are outside the centermost position of prestige. Third, culture change is most concentrated when the networks of elites and the institutions they lead overlap. Finally, cultural changes usually occur with some form of resistance. She concludes with considerations of how those changes may be implemented.

CHAPTER EIGHT

Toward an Ecological Macroeconomics

PETER A. VICTOR AND TIM JACKSON

1. INTRODUCTION

Three major crises are confronting the world. The first is the increasing and uneven burden of humans on the biosphere, and the observation that we have already surpassed the "safe operating space" for humanity with respect to three planetary boundaries: climate change, the nitrogen cycle, and biodiversity loss. The second is the astonishingly uneven distribution of economic output, not only among nations but increasingly within them. The third is the instability of the global financial system, which came to light in 2007–2008 and which defies an obvious solution. These crises are complex and interrelated. It is tempting to try to identify "root" causes, implying that the crises can be resolved if these root causes can be fixed. However, this is misleading. These interlocking crises are the result of systems in which causes are consequences and consequences are causes. In such circumstances, the search for root causes—causes that are not themselves consequences of other causes—can be pointless and counterproductive.

It follows that meaningful attempts to address these crises must engage not only with the dynamics of the separate systems involved—ecological, economic, and financial—but also with the interrelationships between these systems. This is a far from trivial challenge. It is a challenge in particular to the discipline of economics. Arguably, one of the factors involved in the financial crisis was the failure of economics to properly integrate the financial and the real economies. Real economic growth looked healthy even as financial balance sheets were increasingly "under water." There are now attempts to redress that failure and to understand better the linkages between these systems (e.g., Keen 2011).

Over some years, there have also been attempts—particularly within the discipline of ecological economics—to integrate an understanding of ecological limits into an understanding of the real economy. Stern's (2007) review of the economics of climate change stands as a major synthesis in this area—albeit one that found itself outdated by more recent understandings of the science of climate change and was only loosely based on the principles of ecological economics. The main conclusion of the review—that climate change can be addressed with little or no pain to the conventional model—no longer seems so viable because the conventional model has subsequently been severely tested.

Our principal argument in this chapter is that none of these attempted syntheses of real, financial, and ecological systems has yet addressed the core structural challenge presented by the combined ecological, social, and financial crises. Specifically, we seem to be missing a convincing ecological macroeconomics—that is, a conceptual framework within which macroeconomic stability is consistent with the ecological limits of a finite planet. Despite promising developments in this direction (e.g., Jackson 2009; Victor 2008), there is an urgent need for a much more fully developed ecological macroeconomics to avert immanent and massive disaster. In what follows, we discuss some of the challenges associated with this need and outline our own approach to addressing it.

The foremost challenge of building a convincing ecological macroeconomics is that the structural need for economic growth implicit in modern economics places an increasing stress on resource consumption and environmental quality. Economic stability relies on economic growth. However, ecological sustainability is already compromised by the existing levels of economic activity. We turn to this dilemma in the next section.

2. THE DILEMMA OF GROWTH

Capitalist economies place a high emphasis on the efficiency with which inputs to production (labor, capital, resources) are utilized.[1] Continuous improvements in technology mean that more output can be produced for any given input. Efficiency improvement stimulates demand by driving down costs, which contributes to a positive cycle of expansion; however, crucially, it also means that fewer people are needed to produce the same goods from one year to the next.

As long as the economy grows fast enough to offset this increase in "labor productivity," there is not a problem. If the economy does not grow quickly enough, then increased labor productivity means that there is less work to go around: someone somewhere is at risk of losing his or her job. If the economy slows for any reason—whether through a decline in consumer confidence, commodity price shocks, or a managed attempt to reduce consumption—then the systemic trend toward improved labor productivity leads to unemployment. This, in its turn, leads to diminished spending power, a loss of consumer confidence, and a further reduction in demand for consumer goods. From an environmental point of view, this outcome may be desirable because it leads to lower resource use and less polluting emissions. However, it also means that retail falters and business revenues suffer. Incomes fall. Investment is cut back. Unemployment rises further and the economy begins to fall into a spiral of recession.

Recession has a critical impact on private and public finances. Social costs rise with higher unemployment and tax revenues decline as incomes fall and fewer goods are sold. Lowering spending risks real cuts to public services. Cutting social spending inevitably hits the poorest first and represents a direct hit on a nation's well-being. To make matters worse, governments must borrow more, not just to maintain public spending but to try to restimulate demand. In doing so, they inevitably increase their own level of indebtedness. Servicing this debt in a declining economy is problematic at best. Just maintaining interest payments takes up a large proportion of the national income. The best that can be hoped for here is that demand does recover, and when it does it will be possible to begin paying off the debt. This could take decades. It took Western nations almost half a century to pay off public debts accumulated through the Second World War. It has been estimated that the United Kingdom's "debt overhang" from the financial crisis of 2008 could last into the 2030s (Chote et al. 2009).

As the recession has shown, the financial system can become very fragile very quickly when private debt runs ahead of the capacity to repay it. According to Minsky (1994) and others, this is an integral part of capitalist economies and a major cause of instability. Crucially, there is little resilience within this system. Once the economy starts to falter, feedback mechanisms that had once contributed to expansion begin to work in the opposite direction, pushing the economy further into recession. With a growing (and aging) population, these dangers are exacerbated. Higher levels of growth are required to protect the same level of average income and to provide sufficient revenues for (increased) health and social costs.

Of course, short-run fluctuations in the growth rate are an expected feature of growth-based economies. Some feedback mechanisms can bring the economy back into equilibrium. For instance, as unemployment rises, wages fall and labor becomes cheaper. This can encourage employers to employ more people and increase output again—unless the decline in wages further reduces demand, adding to the downward spiral rather than reversing it.

In short, modern economies are driven toward economic growth. For as long as the economy is growing, positive feedback mechanisms tend to push this system toward further growth. When consumption growth falters, the system is driven toward a potentially damaging collapse, with a knock on impact on human flourishing. People's jobs and livelihoods suffer. In a growth-based economy, growth is functional for stability. The capitalist model has no easy route to a steady-state position. Its natural dynamics push it toward one of two states: expansion or collapse.

A lack of growth is deeply unpalatable for all sorts of reasons. As a result, society is faced with a profound dilemma. To resist growth is to risk economic and social collapse. To pursue it relentlessly is to endanger the ecosystems on which we depend for long-term survival. This dilemma looks at first like an impossibility theorem for sustainable development, but it cannot be avoided and has to be taken seriously. The failure to do so is the single biggest threat to sustainability that we face.

3. BEYOND DECOUPLING

The conventional response to the dilemma of growth is to appeal to the concept of "decoupling" or "dematerialization." Production processes are reconfigured. Goods and services are redesigned. Economic output becomes progressively less dependent on material throughput. In this way, it is hoped that the economy can continue to grow without breaching ecological limits—or running short of resources.

It is vital here to distinguish between "relative" and "absolute" decoupling. Relative decoupling refers to a decline in the ecological intensity per unit of economic output. In this situation, resource impacts decline relative to the gross domestic product (GDP), but they do not necessarily decline in absolute terms. Impacts may still increase, but they do so at a slower pace than growth in the GDP. The situation in which resource impacts decline in absolute terms is called "absolute decoupling." Needless to say, this latter situation is essential if economic activity is to return to and remain within ecological limits. In the case of climate change, for instance, absolute reductions in global carbon emissions of 50–85 percent are required by 2050 in order to meet the 450-ppm stabilization target defined by the Intergovernmental Panel on Climate Change (IPCC).

The prevailing wisdom suggests that decoupling will allow us to increase economic activity indefinitely and at the same time stay within planetary boundaries. However, the evidence is far from convincing. Efficiency gains abound. For example, global primary energy efficiency has increased by a third since 1980. The carbon intensity of each dollar of economic output has fallen by about the same amount. However, absolute reductions in impact have been singularly elusive. Global primary energy use, carbon emissions, biodiversity loss, nutrient loadings, deforestation, and global fossil water extraction are all still increasing. Carbon dioxide emissions from fossil fuel consumption and cement production increased by 50 percent between 1990 and 2010 (Boden, Andres, and Marland 2013).

Massive investments in new technology and rapid improvements in resource productivity could, in theory, address this situation. However, the sheer scale of the challenge is daunting. In a world of nine billion people all aspiring to Western incomes, still growing at 2 percent per year, the average carbon intensity of each dollar of economic activity must decline

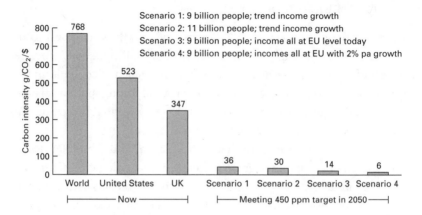

Figure 8.1. Reduction in carbon intensities needed to achieve the Intergovernmental Panel on Climate Change targets. *Source*: Adapted from Jackson (2009).

by a factor of 130 by 2050 to give us a decent chance of meeting the IPCC carbon targets (figure 8.1).

This is not simply a question of technological potential. For changes of this scale to be viable, we must ask tougher questions about the potential of our economic and social systems to be able to deliver this level of dematerialization. We must begin to build system dynamics models that link the real economy, the financial economy, and ecology. It is to this task that we now turn.

4. UNDERSTANDING SYSTEM DYNAMICS

If we are to overcome the interlocking crises identified in this chapter's introduction, we need to understand (and intervene in) the dynamics of the various systems involved. It is for this reason that we build models, be they mental or mathematical; if data are to play a major role, then computer-driven models are necessary.

Perhaps the most well-known attempt to build a computer model of the "world system" was described in *The Limits to Growth* (Meadows et al. 1972) using system dynamics. In system dynamics, stocks are interlinked through inflows and outflows that are driven by behavior, information, and feedback. The authors of *The Limits to Growth* developed various global scenarios based on assumptions about resource stocks, technology, population, pollution, and agriculture using the methodology of system dynamics. They drew attention to the "mode of behavior" of the world system as it was in 1970, which they argued is one of exponential growth followed by collapse. They also said that with appropriate and timely adjustments, expansion followed by stability was possible. With three decades of data in hand, Turner (2008) concluded that the scenarios in *The Limits to Growth* best matched the "standard run"—that is, exponential growth and collapse. Apparently, the mode of behavior of the world system has changed little and portentous signs of collapse are more evident than ever.

World 3—the model used in *The Limits to Growth*—simply (some said far too simply) described the world system in all of its complexity. However, all models are, by definition, simplifications of that which they represent, so simplification itself is hardly a relevant criticism of World 3. The question is whether the particular simplification of World 3 sacrificed structure (e.g., no price mechanism) and detail (e.g., no regions), resulting in misleading conclusions. We agree with Turner: by and large, World 3 did not result in misleading conclusions, given the scope of *The Limits to Growth* in which the primary focus was on the interplay between biophysical limits and economic growth. However, in relation to the three crises noted above, World 3 lacked the detail required for an examination of poverty and inequality; provided very little capacity for learning by governments, corporations, and individuals; and included nothing on finance. Clearly, World 3 was only a start.

Other models have gone well beyond World 3 in providing more complicated, data-rich representations of the world system; however, no model, to the best of our knowledge, has encompassed all three crises. GUMBO (Boumans et al. 2002) is a world system model in which the earth is divided into eleven biomes that encompass the entire surface area of the planet: open ocean, coastal ocean, forests, grasslands, wetlands, lakes/rivers, deserts, tundra, ice/rock, croplands, and urban. This spatial differentiation allows a much more detailed examination of the implications of economic growth for the biological systems of the planet and a more nuanced understanding of the reverse causalities. T21 is another example of a model that integrates economic and environmental systems in some detail (United Nations Environment Programme [UNEP] 2011). However, when applied to the world as a whole as in the study for UNEP on "green growth," there is no regional disaggregation, thus concealing intragenerational implications of the various growth scenarios; also, there is no financial sector, so the full implications of the assumed increases in investments cannot be assessed.[2]

5. SYSTEM LINKAGES

Disregarding the World 3 model for a moment, it is worth considering some of the ways in which the financial system, real economy, and biosphere are linked. Money and finance have a major influence on decisions about consumption, production, investment, employment, and trade. These decisions are integral to the real economy, where raw materials are processed and goods and services are produced, distributed, and consumed. In advanced economies, besides barter (which is a fringe activity), real economic activity is always matched by a financial transaction. If the financial system fails to provide the means for such transactions, then the real economy cannot function. Providing such means in the amounts required for any particular level of economic activity is not a simple task, involving careful management of interest rates and the money supply and involving the central bank, commercial banks, and a multitude of other financial institutions. The system can go badly awry as it did in 2007–2008, when asset values dropped precipitously and credit dried up, resulting in a decline in real economic output.

There are also linkages between the real economy and the financial system that go in the other direction. For example, if a country runs a balance of trade surplus, it will be matched by an outflow of financial capital affecting investment. There may also be upward pressure on the country's exchange rate, in turn affecting the balance of trade. More subtly, if participants in the real economy—companies and individuals—feel increasingly uncertain about the future, they will try to increase the proportion of their assets that are liquid, holding onto cash rather than undertaking longer-term investments and spending on consumption. These decisions in the real economy may reduce employment and profits, leading to more uncertainty. In other words, increasing uncertainty can have ramifications for the financial system and real economy that are reinforcing rather than self-correcting.

Links between the real economy and the biosphere are easy to identify once it is recognized that economies are embedded in the biosphere and are not independent of it as commonly assumed. Economies require a continual input of materials and energy that are subsequently disposed of back to the biosphere, usually in degraded form. When materials are embodied in built capital and consumer durables, they are retained within economies for extended periods of time. Eventually, they too find their way back to the biosphere through demolition and disposal. Reuse and recycling can extend the time that materials, but not energy, stay within economies; however, the second law of thermodynamics dictates that something is always lost in the process.

As we indicated previously, the principal avenue for escaping from the dilemma of growth is to appeal to decoupling. Our ability to decouple—particularly in absolute terms—depends implicitly on the dynamics of economic, social, and financial systems. In an assessment of decoupling as a way of reconciling continual economic growth with diminishing requirements for material and energy throughput, Jackson concluded the following:

> History provides little support for the plausibility of decoupling as a sufficient solution to the dilemma of growth. But neither does it rule out the possibility entirely. A massive technological shift; a significant policy effort; wholesale changes in patterns of consumer demand; a huge international drive for technology transfer to bring about substantial reductions in resource intensity right across the world: these changes are the least that will be needed to have a chance of remaining within environmental limits and avoiding an inevitable collapse in the resource base at some point in the (not too distant) future.
>
> The message here is not that decoupling is unnecessary. On the contrary, absolute reductions in throughput are essential. . . . It's far too easy to get lost in general declarations of principle: growing economies tend to become more resource efficient; efficiency allows us to decouple emissions from growth; so the best way to achieve targets is to keep growing the economy. This argument is not at all uncommon in the tangled debates about environmental quality and economic growth.
> (Jackson 2009:75–76)

The issue of decoupling is further confounded by more subtle links among financial, economic, and ecological systems. For example, it is a fundamental principle of natural resource economics that decisions about the rate at which resources are extracted from mines are influenced by the rate of interest. The same principle applies to biological resources; for example, the time period that trees are allowed to grow before harvesting for timber may depend on the rate of interest. This principle is based on the idea that owners of mines and forests have the option of extracting resources, selling them, and depositing the proceeds in a bank. An increase in the rate of interest on such deposits provides an incentive to extract resources at a faster rate in pursuit of higher profits.

The opposite happens when interest rates fall. This is not the whole story because changes in the rates of resource extraction can be expected to influence current and expected future resource prices, which complicates the connection between interest rates and extraction rates. However, the point remains that there are important links between the financial system and the burden placed by humans on the biosphere that work through the real economy. Any attempt to build a convincing ecological macroeconomics will need to understand and engage with these dynamics.

6. FOUNDATIONS FOR AN ECOLOGICAL MACROECONOMICS

Before turning to our own approach, it is worth setting out briefly some of the foundations for a systems model that addresses the challenges we have set out here. Principal amongst these is the recognition that a viable ecological macroeconomics must not rely on relentless growth in material consumption for its stability. A macroeconomy predicated on continual expansion of debt-driven, materialistic consumerism is ecologically unsustainable, socially divisive, and financially unstable.

At the same time, of course, a viable ecological macroeconomics must still elaborate the configuration of key macroeconomic variables. Consumption, government spending, investment, employment, distribution, and trade will all still matter in the new economy. However, there may well be substantial differences, such as in the balance between consumption and investment; the role of public, community, and private sectors; the nature of productivity growth; and the conditions of profitability. All of these are likely to shift as ecological and social goals come into play. New macroeconomic variables also will need to be brought explicitly into play; these will almost certainly include variables to reflect the energy and resource dependency of the economy and the limits on carbon. They will also include variables to reflect the value of ecosystem services, the stocks of critical natural capital, and wider concerns reflecting an ethical approach to nature.

The role of investment is vital. In conventional economics, investment stimulates consumption growth through the continual pursuit of productivity improvement and the expansion of consumer markets. In the new economy, investment must be focused on the long-term protection of the assets on which basic economic services depend. The new targets of investment will be low-carbon technologies and infrastructures, resource productivity improvements, the protection of ecological assets, maintaining public spaces, and building and enhancing social capital.

This new portfolio demands a different financial landscape from the one that led to the collapse of 2008. Long-term security has to be prioritized over short-term gain, and social and ecological returns must become as important as conventional financial returns. Reforming capital markets and legislating against destabilizing financial practices are not just the most obvious response to the financial crisis; they are also an essential foundation for a new ecological macroeconomy.

A new approach to investment suited to transforming the economy consistent with principles of ecology and social justice may very well entail a new approach to money itself. Contrary to popular belief, which assumes that only governments create money, in the existing monetary system virtually all money spent in the economy is created by commercial banks when they extend credit to borrowers. This circumstance has evolved out of the long history of modern money. Its implications for investment, and more broadly for how economies function in general, is still coming to light (e.g., Godley and Lavoie 2006; Minsky 1986; Ryan-Collins et al. 2012; Wray 2012). One aspect of considerable significance is the role that commercial banks play in deciding which investments to support. The financial crisis of 2007–2008 was partly attributable to the strong preference of banks to lend for purchases of financial instruments (so-called financial investment), which promised short-term returns at apparently low risk to the lender over loans to borrowers wishing to purchase real, productive assets.

Proposals for taking back control over the creation of money from the commercial banks have been around for many decades, but they are gaining impetus from publications such as Benes and Kumhof (2012) and Jackson and Dyson (2012). Under the schemes described in these publications, commercial banks would be required to maintain 100 percent reserve ratios. They would still be intermediaries in investment, but only to the extent that they managed investments on behalf of lenders explicitly prepared to place their savings at risk in investment projects. All money would be created by direct expenditure into the economy by the government working through the central bank. This transformation of the money system, for which numerous advantages are claimed, would reduce instability in the financial sector and allow the government to have a much stronger hand in steering investment toward ecologically and socially desirable projects, which commercial banks, especially under the current system, are much less inclined to support.

Above all, the new macroeconomics will need to be ecologically and socially literate, ending the folly of separating economy from society and environment. A first step in achieving this must be an urgent reform of the national accounting system so that what we measure is brought more in line with what really matters. The integration of environmental and resource variables into the national accounts and an end to the "fetishism" of GDP are essential.

However, more is required. We need to begin to build integrated systems models that incorporate the principal relationships between economic, financial, and ecological variables. These models need to be empirically grounded and calibrated as much as possible in real financial, economic, and ecological data. They must be able to address critical questions about employment, financial structure, and economic stability, as well as incorporate ecological questions about resource consumption and environmental limits.

7. TOWARD A NATIONAL GREEN ECONOMY MACRO-MODEL

Our own work on macroeconomic modeling has focused on the national rather than the global level and on developed countries. Although extreme poverty (i.e., less than $1 per day) is rare in these countries, distributive justice is a matter of growing concern, with levels of inequality not seen since the start of the Great Depression in 1929 being reached in several advanced economies (Piketty 2014). Therefore, it is important for models of developed national economies to address poverty within their own borders, along with other key aspects of their economic, financial, and environmental systems.

LowGrow, a fairly conventional macroeconomic model of the Canadian economy, was designed to answer the following question: "Is it possible to achieve full employment, achieve significant reductions in poverty and in greenhouse gas emissions, and maintain fiscal balance without relying on economic growth?" (Victor 2008; Victor and Rosenbluth 2007). LowGrow was used to simulate possible paths for the Canadian economy from 2005 to 2035, with measures introduced gradually between 2010 and 2020 to divert the economy away from "business as usual."[3]

In LowGrow, as in the economy that it represents, economic growth is driven by net investment, which adds to productive assets, growth in the labor force, increases in productivity, growth in the net trade balance, growth in government expenditures, and growth in population. Scenarios of low growth, no growth, and even degrowth can be examined by reducing the rates of increase in each of these factors singly or in combination.

Figure 8.2 shows a "business as usual" scenario for Canada derived from LowGrow. Each of the five variables are indexed to 100 in 2005 so that changes in them can more easily be appreciated. In this scenario, where the future is projected to mirror the past, real GDP per capita more than doubles between 2005 and 2035, the unemployment rate rises then falls to end above its starting value, the ratio of government debt to GDP declines by nearly 40 percent as Canadian governments continue to run budget

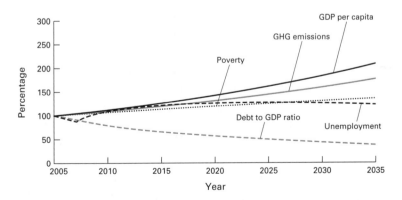

Figure 8.2. Simulation of "business as usual" in Canada using LowGrow. *Source*: Adapted from Victor (2008).

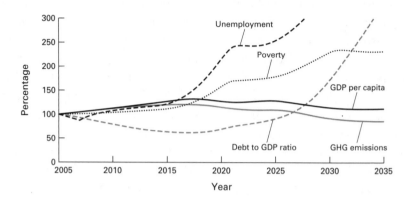

Figure 8.3. Simulation of instability in Canada using LowGrow. *Source*: Adapted from Victor (2008).

surpluses (which they were doing up to 2008), the Human Poverty Index rises due to the projected increase in the absolute number of unemployed people, and greenhouse gas emissions increase by nearly 80 percent.

Figure 8.3 illustrates what happens to the economy if all of the growth drivers (i.e., net investment, increased productivity, increased government expenditure, a positive trade balance) are eliminated over ten years starting in 2010. The economy eventually reaches a steady state in terms of GDP per capita, but the situation is clearly unstable.

A very different scenario is shown in figure 8.4. Compared with the business-as-usual scenario, GDP per capita grows more slowly, levelling off around 2028, at which time the rate of unemployment is 5.7 percent. The unemployment rate continues to decline to 4.0 percent by 2035. By 2020, the poverty index declines from 10.7 to an internationally unprecedented level of 4.9, where it remains; the debt-to-GDP ratio declines to approximately 30 percent, to be maintained at that level to 2035. Greenhouse gas emissions are 31 percent lower at the start of 2035 than 2005 and 41 percent lower than their high point in 2010. These results were obtained by projecting slower growth in government expenditure, net investment, and productivity; a small and declining net trade balance; cessation of growth in population; a reduced work week; a revenue-neutral carbon tax; and increased government expenditure on antipoverty programs, adult literacy programs, and health care.

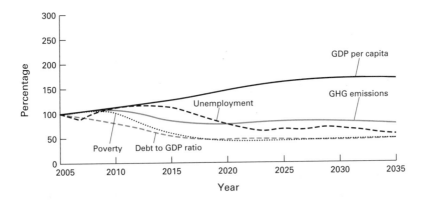

Figure 8.4. Simulation of a Canadian "steady-state economy" using LowGrow. *Source:* Adapted from Victor (2008).

In some respects, LowGrow broke new ground in terms of the question being addressed—that is, the possibility of prosperity without growth (Jackson 2009:134)—and in the modelling approach employed, which is a combination of system dynamics and Keynesian macroeconomics. In other respects, LowGrow simply pointed to the need for further, more ambitious, macroeconomic modeling.

For example, the simulations in LowGrow assume that the monetary policy of Canada's central bank will succeed in keeping inflation at or below 2 percent per year. Beyond that, the financial system is not represented in the model. The material, energy, and land use links between the economy and the environment are dealt with in a very limited fashion. Energy-related greenhouse gas emissions are assumed to be responsive to a time-varying carbon price, and the forest stocks of Canada are simulated based on the demand for timber, regeneration, road building, and random disturbances representing risks from fire and pest infestations. These are only steps toward a more complete integration of the economy and the environment through material and energy flows and land use. LowGrow is also limited in its treatment of economic behavior. Consumption, investment, production, exports, and imports are modeled in a highly aggregated manner with, for example, no differentiation among economic sectors, products, and demographic groups, and in a deterministic manner without any allowance for the role of expectations and uncertainty.

One of the conundrums addressed in LowGrow is whether full employment is possible if labor productivity continues to rise without any increase in economic output. The "simple" solution to this problem is to reduce the length of the average work year. Then, the declining total amount of work required to produce the nongrowing economic output is spread among more, increasingly productive employees. In the scenario shown in figure 8.3, the average work year is reduced by 15 percent between 2005 and 2035, sufficient to more than compensate for the still-positive increase in labor productivity and reduce unemployment to levels not seen in over half a century in Canada.

However, should we assume that labour productivity will rise, even at a reduced rate, as we think about possible and desirable futures? As Jackson (2009) argued explicitly, an economy in which—beyond the satisfaction of basic material needs—economic activity is based around the provision of human and social services has much to recommend it. Health care, education, social care, renovation and refurbishment, leisure and recreation, the protection and maintenance of green spaces, and cultural activities are sectors that contribute positively to the quality of our lives; in addition, they are also less ecologically damaging than activities predicated on the throughput of material goods. However, what is perhaps most striking about such activities is that they are inherently labor intensive and have less potential for labor productivity growth.

An aging population will require labor-intensive services that do not lend themselves to significant increases in productivity without a considerable loss in quality as human contact is replaced by machines. Likewise, many of the tasks required to reduce the burden that humans place on natural systems, such as insulating buildings and rebuilding wetlands, can be labor intensive, again defying major productivity increases. The overall trend away from goods to services in advanced economies and its implications for reduced productivity growth, known as Baumol's disease, further works against productivity growth at historical rates.

From a conventional viewpoint—in which labor productivity is seen as the foundation for a successful economy—this lack of potential for productivity growth looks distinctly unpromising. However, there is one clear and important advantage to these activities: they continue to employ people in meaningful jobs that improve people's lives. In the face of rising unemployment and declining growth rates, this is a very significant advantage.

Jackson and Victor (2011) examined this issue with a simulation model based on three production sectors: a conventional sector defined by labor productivity growth (1 percent per year), which is typical of the UK economy over the last ten years; a green infrastructure sector, characterized by the same (1 percent per year) labor productivity growth as the conventional sector but based on renewable, low-carbon technologies and infrastructures; and a "green services" sector with slightly negative (−0.3 percent per year) labor productivity growth based on the expansion of community-based, resource-light, low-carbon, service-based activities.

We developed three scenarios for the UK economy. The first involved simply expanding green technology. The second involved technological expansion with a reduction in working hours. The third included an addition to these two policies: a shift to "green services." Each scenario was constrained to provide full employment, even though growth rates in each scenario were very different. Indeed, the latter two scenarios are both essentially degrowth scenarios. Carbon emissions from each of these scenarios are shown in figure 8.5. Only the third scenario, which includes the shift to green services, is able to achieve the UK's 2050 carbon target.

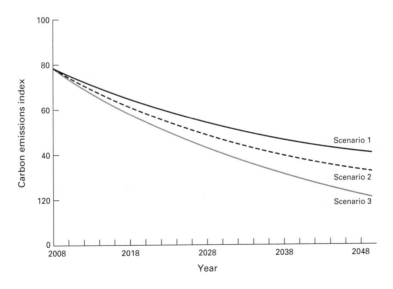

Figure 8.5. Achieving the UK 2050 carbon emission reduction targets. *Source:* Jackson and Victor (2011).

This exercise illustrates the importance of considering the expansion of human and social services as the foundation for a new economy. It also shows how a vigorous policy aimed at expanding green technologies and supporting the green services sector can achieve high employment levels and ambitious carbon reduction targets, even without relying on substantial year-on-year growth rates.

Our initial foray into this intriguing set of issues pointed to the need for ongoing work in a variety of areas:

> For instance, there is a need to understand the impacts on capital productivity, and to explore the relationship between capital productivity, resource productivity and labour productivity in more detail. The wider implications of these changes for capital markets will also need to be elucidated. More generally, this discussion raises the challenge of building a genuinely ecological macroeconomics, in which economic stability can be achieved without relentless consumption growth.
> (Jackson and Victor 2011:107–108)

8. DEVELOPING A STOCK-FLOW CONSISTENT, ECOLOGICALLY CONSTRAINED MACROECONOMIC FRAMEWORK

Model building is a matter of identifying the boundaries and specifying the structure, key components, and principal behaviors of the system that is being modeled. Our current work uses system dynamics to build a national macroeconomic model that aims to address the following questions:

1. Is growth in real economic output still required in advanced economies to simultaneously maintain high levels of employment, reduce poverty, and meet ambitious ecological and resource targets?
2. Does stability of the financial system require growth in the "real" economy?
3. Will restraints on demand and supply, such as in anticipation of or in response to ecological and resource constraints, cause instability in the real economy and or financial system?

In the remainder of this section, we outline several key features of a new "stock-flow consistent" macroeconomic model that integrates an understanding of the real economy, the financial economy, and the ecological and resource constraints under which that economy must operate. A fuller description of our Green Economy Macro-Model and Accounts (GEMMA) framework (Jackson and Victor 2013) is beyond the scope of this chapter. Here, we present some of the principal features. A simplified schematic representation of GEMMA is shown in figure 8.6.

There are several main structural components in GEMMA. Following Godley and Lavoie (2006), GEMMA incorporates a complete and consistent set of accounts that track the financial flows relating to transactions in the "real" economy. As in the standardized system of national accounts for most countries, GEMMA records assets and liabilities, and changes to them, in balance sheets for six main sectors: financial corporations, nonfinancial corporations, the central bank, government, households, and the rest of the world.

GEMMA also includes a multisector input-output model of the real economy, which links the inputs of each of twelve production sectors to

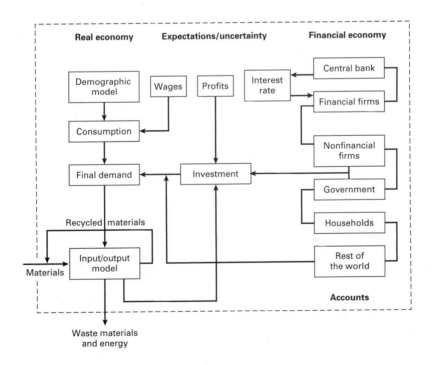

Figure 8.6. A simplified schematic diagram of GEMMA.

the outputs of each other sector so that the complete "supply chain" of each sector is represented. The input-output model includes material and energy flows between the economy and the environment so that changes in the real economy are automatically reflected positively and negatively in these flows (Victor 1972). Provision is also made for dynamic changes in the coefficients (e.g., input per unit of output, material/energy flows per unit of output) in the input-output model in response to investment.

Investment is critical to the evolution of any economy. In GEMMA, several categories of investments are distinguished, allowing for the generation of a wide variety of scenarios defined according to quantitatively and qualitatively different patterns of investment in the private sector.

GEMMA also includes a detailed demographic model of the size, age, and gender composition of the population. Demographic projections are useful for exploring the implications of an aging population for pensions, health care, and home services, which could pose a major challenge if growth is curtailed as part of a strategy to reduce the environmental burden of advanced economies. Demography is also related to consumption and investment—links that are also included in GEMMA in ways that permit consideration of a wide range of possible behaviors by actors in the economic system.

Current behavior is often influenced by expectations about the future. This point has been emphasized by Post-Keynesians, such as Paul Davidson (2011), in their interpretations of Keynes's seminal writings of the 1930s. They highlight the role of uncertainty in decisions by households and corporations regarding asset allocation—in particular, about maintaining liquidity when faced with uncertainty. They have also examined the economic and financial implications of the need for corporations to make production decisions in advance of expected sales. When credit is denied, as in the financial crisis of 2007–2008 when many financial and nonfinancial corporations and households found themselves over leveraged, problems in the financial system led to serious difficulties in the real economy, from which we have yet to recover. We are making provisions in GEMMA for key decisions to be based on endogenously formed expectations.

Finally, we note that GEMMA is a model of national economies—in particular, national economies in developed countries. It is being designed to maximize the use of statistics from the Organisation for Economic Co-operation and Development, so that it can be applied to member countries without having to be built from the ground up each time. However, we live in a world with highly integrated economies; in many instances, the global implications—economic, social, and environmental—are of greatest interest. The challenge of modelling global systems in an integrated way is much greater than at the national level; attempts to do so have generally been rather highly aggregated and restricted in scope (Costanza et al. 2007). Progress is being made in this direction, but it will require a very considerable commitment of resources and a great deal of international cooperation to make the headway called for by the urgency of the situation.

9. LOOKING AHEAD

Many societies are being faced with a low-growth or no-growth economy, despite their policy intentions. In the future, some societies may choose not to pursue economic growth as a matter of policy, having understood that prosperity in its fullest sense can best be achieved through other means. Regardless of the rationale for a low- or no-growth economy, it is essential to consider its implications in advance so that the possible adverse consequences of such a change in direction can be avoided. Developing a macroeconomic framework through which to explore these possibilities will provide policymakers, scholars, and the engaged public with greater insight into these changes.

More positively, human well-being may improve as other objectives come to the fore, as suggested by the burgeoning literatures on happiness (Layard 2005) and equity (Wilkinson and Pickett 2009). Our intention in developing this work is to help illuminate these possibilities as well.

NOTES

1. The following two sections are loosely adapted from Jackson (2011) and Jackson (2009).

2. See Victor and Jackson (2012) for more details of this critique.

3. The following description of LowGrow is adapted from Victor (2013).

REFERENCES

Benes, Jaromir, and Michael Kumhof. 2012. *The Chicago Plan Revisited, Imf Working Paper Wp12/2/02*. Accessed January 23, 2015. http://www.imf.org/external/pubs/ft/wp/2012/wp12202.pdf.

Boden, Tom, Bob Andres, and Gregg Marland. 2013. "Global CO2 Emissions from Fossil-Fuel Burning, Cement Manufacture, and Gas Flaring: 1751–2010." Accessed January 23, 2015. http://cdiac.ornl.gov/ftp/ndp030/global.1751_2010.ems.

Boumans, Roelof, Robert Costanza, Joshua Farley, Matthew A. Wilson, Rosimeiry Portela, Jan Rotmans, Ferdinando Villa, and Monica Grasso. 2002. "Modeling the Dynamics of the Integrated Earth System and the Value of Global Ecosystem Services Using the Gumbo Model." *Ecological Economics* 41 (3): 529–560. doi: 10.1016 /S0921-8009(02)00098-8.

Chote, Robert, Carl Emmerson, David Miles, and Jonathan Shaw, eds. 2009. *The Ifs Green Budget*. London: The Institute for Fiscal Studies.

Costanza, Robert, Rik Leemans, Roelof M. J. Boumans, and Erica Gaddis. 2007. "Integrated Global Models." In *Sustainability or Collapse? An Integrated History and Future of People on Earth*, edited by Robert Costanza, Lisa J. Graumlich, and Will Steffen, 417–446. Cambridge, MA: MIT Press.

Davidson, Paul. 2011. *Post Keynesian Macroeconomic Theory*. 2nd ed. Cheltenham, UK: Edward Elgar.

Godley, Wynne, and Marc Lavoie. 2006. *Monetary Economics: An Integrated Approach to Credit, Money, Income, Production and Wealth*. Basingstoke, UK: Palgrave Macmillan.

Jackson, Andrew, and Ben Dyson. 2012. *Modernising Money: Why Our Monetary System Is Broken and How It Can Be Fixed*. London: Positive Money.

Jackson, Tim. 2009. *Prosperity Without Growth: Economics for a Finite Planet*. London: Earthscan.

Jackson, Tim. 2011. "Societal Transformations for a Sustainable Economy." *Natural Resources Forum* 35 (3): 155–164. doi: 10.1111/j.1477–8947.2011.01395.x.

Jackson, Tim, and Peter Victor. 2011. "Productivity and Work in the 'Green Economy': Some Theoretical Reflections and Empirical Tests." *Environmental Innovation and Societal Transitions* 1 (1): 101–108. doi: 10.1016/j.eist.2011.04.005.

Jackson, Tim, and Peter A. Victor. 2013. *The Green Economy Macro-Model and Accounts (Gemma) Framework: A Stock-Flow Consistent Macro-Economic Model of the National Economy under Conditions of Ecological Constraint*. Guildford, UK: University of Surrey.

Keen, Steve. 2011. *Debunking Economics: The Naked Emperor Dethroned?* London: Zed Book.

Layard, Richard. 2005. *Happiness: Lessons from a New Science*. New York: Penguin Press.

Meadows, Donella H., Dennis L. Meadows, Jørgen Randers, and William W. Behrens. 1972. *The Limits to Growth: A Report for the Club of Rome's Project on the Predicament of Mankind*. New York: Universe Books.

Minsky, Hyman P. 1986. *Stabilizing an Unstable Economy*. New Haven, CT: Yale University Press.

Minsky, Hyman P. 1994. "Financial Instability Hypothesis." In *The Elgar Companion to Radical Political Economy*, edited by Philip Arestis and Malcolm Sawyer, 153–158. Aldershot, UK: Edward Elgar.

Piketty, Thomas. 2014. *Capital in the Twenty-First Century*. Translated by Arthur Goldhammer. Cambridge, MA: Belknap Press of Harvard University Press.

Ryan-Collins, Josh, Tony Greenham, Richard Werner, and Andrew Jackson. 2012. *Where Does Money Come From? A Guide to the UK Monetary and Banking System*. 2nd ed. London: New Economics Foundation.

Stern, Nicholas. 2007. *The Economics of Climate Change: The Stern Review.* Cambridge, UK: Cambridge University Press.

Turner, Graham M. 2008. "A Comparison of the Limits to Growth with 30 Years of Reality." *Global Environmental Change* 18 (3): 397–411. doi: 10.1016/j.gloenvcha.2008.05.001.

United Nations Environment Programme. 2011. "Modelling Global Green Investment Scenarios: Supporting the Transition to a Global Green Economy." In *Towards a Green Economy: Pathways to Sustainable Development and Poverty Eradication.* Accessed January 23, 2015. http://www.unep.org/greeneconomy.

Victor, Peter A. 1972. *Pollution: Economy and Environment.* Toronto: University of Toronto Press.

Victor, Peter A. 2008. *Managing Without Growth: Slower by Design, Not Disaster.* Cheltenham, UK: Edward Elgar.

Victor, Peter A. 2013. "Economic Growth: Slower by Design, Not Disaster." In *Encyclopedia of Environmental Management,* edited by Sven Erik Jørgensen. Boca Raton, FL: CRC Press.

Victor, Peter A., and Tim Jackson. 2012. "A Commentary on Unep's Green Economy Scenarios." *Ecological Economics* 77: 11–15. doi: 10.1016/j.ecolecon.2012.02.028.

Victor, Peter A., and Gideon Rosenbluth. 2007. "Managing Without Growth." *Ecological Economics* 61 (2–3): 492–504. doi: 10.1016/j.ecolecon.2006.03.022.

Wilkinson, Richard G., and Kate Pickett. 2009. *The Spirit Level: Why More Equal Societies Almost Always Do Better.* London: Allen Lane.

Wray, L. Randall. 2012. *Modern Money Theory: A Primer on Macroeconomics for Sovereign Monetary Systems.* New York: Palgrave Macmillan.

CHAPTER NINE

New Corporations for an Ecological Economy: A Case Study

RICHARD JANDA, PHILIP DUGUAY, AND RICHARD LEHUN

1. INTRODUCTION

If there is to be an ecological economy functioning sustainably in the Anthropocene and contained within planetary boundaries, it will require ecological macroeconomics and microeconomics. The contours of its macroeconomics have been traced in chapter 8. The size of the economy and its environmental footprint will have to be monitored according to a dashboard of indicators, keeping it within a safe operating space for the planet. At the same time, however, an ecological microeconomics will have to arise, ensuring that individual economic actors are no longer simply responding to the narrow profit incentives that in aggregate orient the economy as a whole toward unbridled growth. This chapter introduces the implications of ecological economics in the case of the corporation. The fiduciary principle applied to the economy as a whole—managing it within planetary boundaries in trust for future generations—must also guide economic actors and individual transactions. Corporate social responsibility standards have begun—all too modestly—to expand corporate fiduciary responsibilities. Various signals about "fair trade" and ecologically sound transactions have begun—again, all too modestly—to shift expectations about what must be accounted for in transactions. This chapter attempts to identify what it would mean to generalize and systematize these heretofore modest developments.

Ultimately, the social license to operate as an economic actor must carry with it a globally applicable test for the withdrawal of the license where externalities that transgress planetary boundaries are being generated. An account of what sorts of economic actors an ecological economy will require must begin with an account of what kinds of economic actors we now have. As will become clear, those actors—corporations—are the perfect reflection of market economy as we have constructed it.

As the chartered company evolved into the modern corporation, it shed itself of any imposed requirement to provide a public benefit. What had been a license to operate with delegated authority for purposes designated by the state became an open invitation to pursue private advantage. The hollowing out of public purpose was arguably completed with the abandonment of the *ultra vires* doctrine at the end of the nineteenth century; according to this legal doctrine, corporations could only act within the set of powers allocated to them by their articles of incorporation (Horwitz 1985). As courts accepted that there were no limits on the objects of the corporation that could be enforced in contracts with third parties, corporate law no longer sought to maintain any control—as against the world—over the purposes being pursued by the corporation. Only the shareholders in the corporation had recourse against their entity for failure by its directors and officers to pursue the purposes that had been set out by it. What the law determined to be of significance to third parties was the ability to rely on the corporation to act as a person in contractual relations without questioning whether the contract was properly motivated by the corporation's "internal" purposes. In other words, although a physical person's purposes and state of mind could be relevant in a contract to establish a "meeting of the minds," a corporation's purposes were to remain private and thus assumed to be obvious: to make a profit however it best determined for itself. Thus, the corporation pursued internally established purposes so as to produce external effects in the world. Purpose came to be policed by the shareholders and the market for shares. In this sense, the corporation became shareholder-centric and externality generating.

The corporation is now the principal actor in the economy. It is not internally constrained to produce or avoid damaging public goods, and consequently neither is the economy. Its shareholder-centric orientation has been challenged by efforts to produce corporate social responsibility (CSR)—accountability for the social and environmental impacts of corporate activity. However, CSR has largely been about reducing harm rather than about producing benefit. Essentially, CSR is about making corporations more responsive to external pressures from stakeholder groups bearing the costs of the externalities generated by corporate activity. It has been less about orienting corporations themselves, according to their own governance structures, toward producing a public benefit. Furthermore, U.S. corporations that would seek to pursue a public purpose at the same time as making a profit can find themselves facing liability under Delaware law for failure to pursue the maximization of shareholder value. To quote from the Delaware Chancery's recent eBay decision: "Delaware courts have guarded against overt risk of entrenchment and the less visible, yet more pernicious risk that [directors] acting in subjective good faith might nevertheless deprive stockholders of value-maximizing opportunities."[1]

That is, the court took the view that corporate goals other than wealth maximization (in the eBay case, community access to postings) could not be invoked to place a limit on wealth maximization. The alternative of establishing a not-for-profit corporation—or more recently, a community interest company[2] or a low-profit limited liability company[3]—that seeks to control and limit its profits presents its own difficulties and places limits on the ability to attract capital (see Tozzi 2010). Social entrepreneurs seeking to use businesses to create public benefits are thus left without a corporate form fit to a related purpose.

The "benefit corporation" is a recent legal innovation pioneered by twenty-seven jurisdictions in the United States, including Delaware, which is the main state of incorporation.[4] This example is worth exploring because it provides some insight as to how the corporate form itself might be reconfigured within an ecological economy. Benefit corporation legislation creates a special regime that works together with regular business corporations statutes. Just as nonprofits and limited liability corporations are enabled under separate legislation in most jurisdictions, benefit corporations now find themselves under a separate tier of business associations law (Honeyman 2014). The city of Philadelphia was the first municipality to offer tax incentives to benefit corporations that wish to settle within the jurisdiction, which is an indication that there could be fiscal advantages associated with this new business form (Honeyman 2014).

In its conception, the benefit corporation is designed to pursue a public benefit and to be held accountable for producing it. A benefit corporation is a business corporation that has elected to operate for "general public benefit," and it may elect to also operate in pursuit of a one or more "specific public benefits." In the model clause that has been adopted in the twenty-seven jurisdictions with benefit corporation legislation, the specific benefits include the following:[5]

1. Providing low-income or underserved individuals or communities with beneficial products or services
2. Promoting economic opportunity for individuals or communities beyond the creation of jobs in the normal course of business
3. Preserving the environment
4. Improving human health
5. Promoting the arts, sciences, or advancement of knowledge
6. Increasing the flow of capital to entities with a public benefit purpose
7. Accomplishing any other particular benefit for society or the environment

The general public benefit achieved by the benefit corporation is to have a "material positive impact on society and the environment by the operations of a benefit corporation taken as a whole, as measured by a third-party standard, through activities that promote some combination of specific public benefits."

This pursuit of public benefit is enshrined in the corporation's certificate of incorporation, in which the firm is to declare itself a benefit corporation. Perhaps surprisingly, although social and environmental performance are to be assessed by a third-party standards organization, such an assessment is not part of the certification process. The absence of state oversight of social and environmental performance helped to garner bipartisan political consensus in favor of benefit corporation legislation. The governance of the corporation toward general or specific benefit is to be assured instead by a range of internal governance requirements, notably the right of stakeholders to seek legal redress if the corporation fails to pursue the claimed public benefit.

What are we to make of the effort to bring public purposes back into the realm of the corporation? Does this not produce a fundamental contradiction between a profit-oriented entity and a publicly oriented entity? The challenges faced by creating hybrid organizations are explored in Jane Jacobs's remarkable book in dialogue form, *Systems of Survival* (1992).

2. A MONSTROUS HYBRID?

Jane Jacobs (1992) has her characters explore the claim that there are two patterns of moral behavior—they call them "syndromes"—that are applicable to all human institutions: the commercial syndrome and the guardian syndrome. The commercial syndrome, characterized by fifteen precepts (including "shun force," "compete," "be efficient," "respect contracts," "invest for productive purposes," and "be honest"), applies to businesspeople and traders—but also to scientists. The guardian syndrome, characterized by fifteen different moral precepts (including "shun trading," "be obedient and disciplined," "respect hierarchy," "be loyal," "dispense largesse," and "treasure honor"), applies to warriors, governments, religions, and charities. Jacobs has her interlocutors conclude that the two syndromes cannot be mixed and matched because, when they are combined, they produce "monstrous hybrids." The key example given is the Mafia, which mixes a guardian role with commercial dealing. However, the broader claim is made that whenever hybrids of charity and commerce arise, the result is monstrous: state-run enterprise, commercialized universities, or for-profit religion. The sets of moral precepts that provide integrity if they work separately are undermined and produce moral failure if they are intermixed.

Jacobs' interlocutors acknowledge that human behavior does not fall neatly into two entirely separate categories. Two ways have been found to keep moral zones separate—caste systems, which assigns different social roles to each, or "knowledgeable flexibility," which compartmentalizes each actor internally so as to allow different behaviors in appropriate settings (e.g., acting as a guardian to friends and family but acting commercially with strangers). According to the interlocutors, neither strategy is able to work without producing some injustice or ambivalence—there are points of friction at the boundaries.

One might observe that, even without mixing benefit and profit, the corporation already displays what Jacobs would identify as a hybrid form.

It operates in a commercial setting; however, as Coase (1937) famously explained, the frontiers of the firm management are hierarchical, and the firm becomes an internal mechanism for superseding the price mechanism of the market. That is, within the firm, market exchange does not operate and management substitutes for it. Furthermore, the governance structure of the firm underpinning management depends on loyalty—fiduciary duties exercised by directors and officers—so that internally it relies upon a guardian ethic. Indeed, what the benefit corporation idea seeks to exploit is the notion that the guardian or stewardship role of the directors and officers of the corporation could be turned toward a wider set of beneficiaries. Guarding the firm for wider social purposes, as opposed to guarding it for shareholders alone, would remain consistent with the guardian syndrome, which can be directed toward protection of society as a whole.

Nevertheless, Jacobs (1992) or her imagined interlocutors might be brought to argue that the corporation itself is already a monstrous hybrid and that any further admixture of purposes will simply produce a greater monstrosity. Can there be a virtuous rather than a monstrous hybrid lurking in the corporation? Stripped to its simplest elements, the hybrid that is being sought in the benefit corporation is one that arises between gift and exchange. The guardian or fiduciary makes a gift of care to some other. In the benefit corporation, the public benefit is a form of gift. On the other hand, the entrepreneurs who display Jacobs's "commercial syndrome" seek to profit from exchange in the market. The transformation of a gift into an exchange (or an exchange into a gift) can, in Jacobs' terms, produce a monstrous hybrid. Thus, when a gift is a form of payment in exchange for benefit, it becomes a bribe. Indeed, corporations are subject to rules against such corrupt practices. When an exchange is treated as the receipt of a gift, it becomes exploitation. Thus, for example, corporations are subject to minimum wage legislation. Bribe and exploitation are obverse ways of profiting from a gift relationship. Yet, one can pay to make the conditions of a gift possible (e.g., tuition) and give to make the conditions of exchange possible (e.g., family bequests). Can the separate but intersecting and mutually enabling circuits of gift and exchange be maintained distinct but together within the corporation?

This is where governance, accountability, and certification take on added and critical significance. They are not just about producing the positive reputation effect of a brand so as to signal to a subset of interested consumers that they should purchase and to a subset of interested investors that they should invest. They are also about maintaining integrity within the benefit corporation. What the benefit corporation gains as a return on investment is subject to its own governance, accountability, and certification standards. What the benefit corporation "gives back" to its stakeholders is subject to its own form of oversight. Separate sets of considerations apply and separate codes of conduct are elaborated. This is a version of "knowledgeable flexibility" operating within the corporation—which is a person—and premised upon the elaboration of a differentiated and accountable set of performance ethics.

In practice, must there not be trade-offs between the circuits of gift and exchange within a benefit corporation? Can we, as economic actors, become sufficiently sophisticated to differentiate between and among performance ethics? There is a certain alluring simplicity to waking up in the morning as the servant of a for-profit corporation doing what is needed to maximize profit, and then returning home in the evening to family and friends where the market is suspended. If one has to wake up in the morning as the servant of many masters (for-profit and not-for-profit) and return in the evening to a bundle of private and public obligations, there will be times when one master is sacrificed to another and the performance obligations become overburdening. This may illustrate the psychological grounding of Milton Friedman's (1970) admonition against mixing corporate profit and corporate social responsibility.

To face such overburdening, however, the benefit corporation can seek to rely upon a shared intermediation of social roles. If the entire burden of gift and exchange were to fall upon a single corporate entity, Friedman's critique would have great force and resonate with Jacobs' stern warning against monstrous hybrids. But if—as one entity that certifies benefit corporations, B Lab (2014), put it—in a "declaration of interdependence," the goal is to harness "the power of private enterprise to create public good" by acknowledging that "we are each dependent upon [one] another and thus responsible for each other and future generations" each actor is understood to bear only part of the collective burden undertaken by all hybrid benefit corporations.

Thus, B Lab interacts and intersects with a host of certification standards providers. It helps to establish the conditions for the emergence of benefit corporations and, through the parallel initiative called the Global Impact

Investing Ratings System, conditions for the emergence of an investor community focused on benefit corporations. Financing through social impact investors, or using social impact or pay-for-performance bonds, can provide impetus for the capitalization of benefit corporations (Social Finance 2014). Stakeholders such as environmental nongovernmental organizations in turn scrutinize and provide impetus for higher levels of accountability and performance. This helps to mobilize consumers to shift their own demand toward sources of products and services that participate in the gift circuit. If the signaling among all the actors in the ecosystem is transparent, coordinated, and socially networked, the burden of producing public benefit can be shared. Although the invisible hand seeks to align the supply and demand of private goods through exchange, visible joined hands can seek to give public goods.

A hybrid is a monster if it cannot reproduce and if it disrupts and cannot function within an existing ecosystem. A hybrid is a new species when it can indeed find its niche within an ecosystem so as to function and flourish. We can no longer treat the social and environmental externalities generated by corporations as absorbed by and treated within the purview of governmental guardians alone. They have lost their capacity to generate adequate resources to produce countervailing public goods and are outstripped by the scale of the externalities. An economic ecosystem full of social impact investors, a supply chain of benefit corporations, publicly monitored standard setting and accountability mechanisms, as well as socially responsive consumers might help produce some of the environmental and social goods we need.

If the benefit corporation can become a new species of hybrid corporation, could it respond to the diagnosis of the business corporation posed by Joel Bakan (2003), namely that the corporation is a psychopath—certainly one kind of monster? On Bakan's analysis, the corporation is a person displaying callous disregard for the feelings of other people, the incapacity to maintain human relationships, reckless disregard for the safety of others, deceitfulness (continual lying to deceive for profit), the incapacity to experience guilt, and the failure to conform to social norms and respect for the law—hence a psychopath (Bakan 2003). Bakan has criticized corporate social responsibility as being the equivalent of seeking to get a psychopath to agree to behave well: "The fundamental difficulty with social responsibility remains the fact that we haven't changed the nature of the corporation.

It is, and continues to be pathologically constituted, in the sense that it still must put its own interests above all others" (Bakan 2008). Bakan believes that the only solution to the pathology of the corporation is to control and punish it through government regulation. Would the benefit corporation provide a different diagnosis? Could it be a corporation with a conscience, despite being an artificial person seeking to make a return on investment "with no soul to be damned, and no body to be kicked"? (Edward, First Baron Thurlow, quoted in King 1977:1)

To have a conscience is to have knowledge with oneself or with another. It is the ability to look into oneself as if one were another and to judge one's own actions. Does the benefit corporation have the ability to look into itself as if it were another and judge its own actions? In principle, any corporation—which is constituted as a body politic—is a person governed for the sake of others (the shareholders). The fiduciary principle that underlies its governance is inherently judged by the standard of "the punctilio of an honor the most sensitive."[6] Therefore, the absence of conscience in the business corporation, rather than its presence, proves to be mysterious. How does an entity governed for the sake of producing benefit for others become one that Bakan (2003) characterized as having callous disregard for the feelings of others? This can arise only if the benefit it seeks to generate for some disregards the costs generated for others. Yet, that is precisely the result produced by a fiduciary duty which, within the business corporation, compels directors and officers to pursue only the interests of shareholders and defines those interests narrowly to maximize returns. The pathology arises therefore not from the inability to produce other-regarding behavior by the corporation, but rather from the fact that its fiduciary regard is so narrowly blinkered.[7]

3. CONCLUSION

The benefit corporation seeks to become a corporation with the blinkers taken off the fiduciary duties owed in its name. In principle, it will not sacrifice its obligation to operate as a going concern for shareholders. However, it will monitor itself and be held accountable for its impacts on others, seeking in the process to demonstrate that its existence is of net social benefit, even and especially with respect to the public goods partially consumed through its activities.[8] This is how it is meant to become a virtuous rather than monstrous hybrid.

Although the specific features of benefit corporation certification and oversight may not yet provide full and adequate accountability for operation within an ecological economy, this new form points in the direction of the kind of shift that will be required. Transforming existing corporations into benefit corporations is a significant challenge, especially when such a fundamental change requires the agreement of existing shareholders. Nevertheless, if the expectations and requirements of those who enter into transactions with corporations were such that strong incentives would exist to be able to signal the production of public benefit accompanying each exchange, the benefit corporation or its equivalents could gain more than a toehold in the economy.

NOTES

1. *eBay Domestic Holdings v. Newmark C.A.* No. 3705-CC (Del. Ch. Sept. 9, 2010)

2. The community interest company (CIC) was created in the United Kingdom under Part 2 of the Companies (Audit, Investigations and Community Enterprise) Act 2004, c. 27; available at http://origin-www.legislation.gov.uk/ukpga/2004/27. CICs are subject to an "asset lock"—they are not to distribute assets to their members except in accordance with regulatory authorization; any remaining assets on dissolution are protected for the community. A CIC must satisfy a general test, showing that the activities it conducts are for public benefit, as determined by the Regulator of Community Interest Companies. As of March 2014, there were 9177 CICs in the United Kingdom (UK Office of the Regulator of Community Interest Companies 2014). The United Kingdom has also created cooperative and community benefit societies, which act either for the benefit of a group of members (cooperative society) or for the benefit of people who are not members (community benefit society). These must demonstrate their social objectives to the Financial Conduct Authority and Prudential Regulation Authority of the Bank of England to be registered. See http://www.fca.org.uk/firms/being-regulated/meeting-your-obligations/firm-guides/cooperative-and-community-benefit-societies.

3. The low-profit limited liability company (called L3C) is a business form that has been adopted in Illinois, Louisiana, Maine, Michigan, North Carolina, Utah, Vermont, and Wyoming as an amendment to limited liability corporation (LLC) status; see, for example, Vermont 11 V.S.A. § 3001(23), available at http://www.leg.state.vt.us/docs/legdoc.cfm?URL=/docs/2008/acts/ACT106.HTM. Like charities, L3Cs must serve a charitable purpose of benefit to their community; in particular, under the Vermont legislation, "no significant purpose of the company is the production of income or the appreciation of property; provided, however, that the fact that a person produces significant income or capital appreciation shall not, in the absence of other factors, be conclusive evidence of a significant purpose involving the production of income or the appreciation of property." Unlike a nonprofit, the L3C can issue equity and seek for-profit investors (see Rosenthal 2011).

4. At the time of this writing 26 states and Washington D.C. had adopted benefit corporation legislation and 7 states had legislation pending.

5. The Model Legislation promoted by B Lab can be found at http://www.benefitcorp .net/attorneys/model-legislation.

6. *Meinhard v. Salmon* 164 N.E. 545 (N.Y. 1928) per Cardozo J. Although this characterization of the trust relationship is not, strictly speaking, currently applicable to corporate fiduciary duties, it still provides a benchmark for the fiduciary role.

7. It might be objected that, as a general matter, legally imposed fiduciary duties are blinkered in the sense that they do not involve an obligation to display other-regarding behavior in general but only with respect to the specific beneficiaries of the fiduciary duty. When that fiduciary duty flows from a relationship of dependency, such as parent–child, doctor–patient, or attorney–client, fiduciary attention and resources are indeed devoted to the vulnerable other. It is far from obvious that shareholders are in analogous situation of dependency given their comparative ease of exit from the relationship with the corporation. Also, their vulnerability and dependency are not clearly greater than that of employees, creditors, suppliers, customers, and third parties bearing externalities. Furthermore, even if the situation of shareholders were analogous, the attention of the fiduciary devoted to the vulnerable dependent beneficiary need not be exclusive—and indeed will fail if it is. Thus, a parent will be mindful of a child's general social integration, the doctor will be mindful of impacts of treatment on public health, and the lawyer will be mindful of his or her role as an officer of the court. Parental and professional responsibilities can only be fulfilled if the dependent beneficiary is enabled to take up social relationships with integrity. Thus, even a fiduciary duty undertaken toward shareholders should enable them and the entity constituted through them to take up their social relationships with integrity.

8. Note that once Berle and Means (1967:312–313) acknowledged that the modern business corporation owed fiduciary duties to stakeholders and not simply to shareholders, they came to believe that "the 'control' of the great corporations should develop into a purely neutral technocracy, balancing a variety of claims by various groups in the community and assigning to each a portion of the income stream on the basis of public policy rather than private cupidity." Perhaps their formulation anticipated the benefit corporation, although neutrality is not the key characteristic of the fiduciary's role. Rather, it is fidelity to the purposes of the corporation and capacity to make effective use of corporate resources to meet a polycentric set of claims upon it.

REFERENCES

B Lab. 2014. "The B Corp Declaration." Accessed September 24, 2014. http://www .bcorporation.net/what-are-b-corps/the-b-corp-declaration.

Bakan, Joel. 2003. *The Corporation: The Pathological Pursuit of Profit and Power.* Toronto: Penguin.

Bakan, Joel. 2008. *Interview by Wayne Visser.* Cambridge Institute for Sustainability Leadership. Accessed January 23, 2015. http://www.cisl.cam.ac.uk/Resources /Videos/Joel-Bakan.aspx.

Berle, Adolf Augustus, Jr., and Gardiner C. Means. 1967. *The Modern Corporation and Private Property*. Rev. ed. New York: Harcourt, Brace and World. First published 1932 by Macmillan, New York.

Coase, R. H. 1937. "The Nature of the Firm." *Economica* 4 (16): 386–405. doi: 10.1111/j.1468–0335.1937.tb00002.x.

Friedman, Milton. 1970. "The Corporate Social Responsibility of Business Is to Increase Its Profits." *The New York Times Magazine*, September 13.

Honeyman, Ryan. 2014. *The B Corp Handbook: How to Use Business as a Force for Good*. San Francisco: Berrett-Koehler Publishers.

Horwitz, Morton J. 1985. "Santa Clara Revisited: The Development of Corporate Theory." *West Virginia Law Review* 88: 173–224.

Jacobs, Jane. 1992. *Systems of Survival: A Dialogue on the Moral Foundations of Commerce and Politics*. New York: Random House.

King, Mervyn A. 1977. *Public Policy and the Corporation*. New York: Wiley.

Rosenthal, John. 2011. "A Hybrid Form of Non-Profit Is on the Rise in Illinois." Accessed September 24, 2014. http://www.intersectorl3c.com/blog/104163/5815/.

Social Finance. 2014. "Social Impact Bonds." Accessed September 24, 2014. http://www.socialfinance.org.uk/services/social-impact-bonds/.

Tozzi, John. 2010. "Maryland Passes 'Benefit Corp.' Law for Social Entrepreneurs." *Bloomberg Businessweek*, April 13. Accessed January 23, 2014. http://www.businessweek.com/smallbiz/running_small_business/archives/2010/04/benefit_corp_bi.html.

UK Office of the Regulator of Community Interest Companies. 2014. *Annual Report 2013/2014*. Accessed September 24, 2014. https://www.gov.uk/government/publications/cic-regulator-annual-report-2013-to-2014.

This new society . . . has only just begun to come into being. Time has not
yet shaped its definite form. The great revolution which brought it about
is still continuing, and of all that is taking place in our day, it is almost
impossible to judge what will vanish with the revolution itself and what
will survive thereafter. The world which is arising is still half buried in
the ruins of the world falling into decay and in the vast confusion of all
human affairs at present, no one can know which of the old institutions
and former mores will continue to hold up their heads and which will in
the end go under. . . . Working back through the centuries to the remotest
antiquity, I see nothing at all similar to what is taking place before our
eyes. The past throws no light on the future, and the spirit of man walks
through the night.
—Alexis de Tocqueville, *Democracy in America*

Culture is . . . a gamble played with nature, in the course of which,
wittingly or unwittingly . . . the old names that are still on everyone's lips
acquire connotations that are far removed from their original meaning.
—Marshall Sahlins, *Islands of History*

CHAPTER TEN

Ecological Political Economy and Liberty

BRUCE JENNINGS

We are coming to the end of an era. The place of humans within nature
and the biophysical impacts of human activity on natural systems will
have to change. To facilitate that change in actual behavior, our conceptual understanding and normative assessment of human activity will also
have to change. I refer to the new system of human activity as an "ecological political economy" and the new conceptual and normative order as an
"ecological public philosophy." The field of ecological economics will play
an important role in creating an ecological political economy. It can also
play an important role in shaping a new public philosophy, if the normative assumptions and implications of ecological economics can be carefully
developed and articulated.

   This book as a whole is a contribution to that enterprise. In this chapter,
I focus on one important component of that conceptual and normative
analysis by discussing the concept of liberty. My critical goal is to scrutinize

received conceptions of liberty that are too individualistic and atomistic to realistically and rationally guide human norms and self-understandings in the coming era. My constructive goal is to formulate an understanding of the liberty or freedom (I use the terms interchangeably)[1] of human agents that is consonant with the demands of the new era. In the final analysis, I argue for a theory and practice of individual liberty that fits with an ecological political economy and an ecological public philosophy. It is a "relational" theory of liberty, which is subject to certain normative standards of acceptable relationship. Justice may be viewed as providing such a standard insofar as it is a normative theory of relationality. Here, however, I propose to focus on two other considerations, which may be aspects of an overall account of justice: membership (understood as equal respect and parity of social participation and voice) and mutuality (understood as equality of concern and care and as solidarity). I explore, in short, how liberty is to be conceived and lived if it is to find a home in the coming ecological age.

1. THE ECOLOGICAL IMPERATIVE

Our entire economic system is fundamentally dependent on the functional integrity of natural and living systems that are losing patience with us. That is to say, these systems have a limited capacity to tolerate human extraction from them and excretion of waste products and byproducts into them. Human economic activity worldwide is colliding with those limits.

Why? The reasons are many, but one key factor is our deep ontological misprision. A hallmark of the modern era is that we think of the human realm as set apart from the rest of the world; we think that we can manipulate it, engineer it, as we see fit in accordance with what we find meaningful and valuable. We are still wedded to that worldview and seem blithely—and blindly—determined to pursue it to its logical extremes. Biophysical systems, even when they are scientifically well understood, are mistakenly seen as things we live off of, not as places we live within.

1.1. Ecological Political Economy

It is essential to change the way we conduct economic activity. In the past, the main argument has been between those who emphasize efficiency and those who emphasize equity; between those who stress growth and those who stress just redistribution of existing wealth; or between those who own and control capital and those who own mainly their own labor and skills. These arguments are not passé, by any means; the struggle for fairness and equality has not been won.

However, a new struggle must be, and is being, added to it. New forms of right and action—ecological public philosophy and ecological political economy—are emerging to infuse and inform our normative discourse about economics, power, and justice. Ecological political economy is "ecological" in that it places economic activity in the context of the operation of physical and biological systems. It includes the important subfield of economics known as ecological economics, but is broader in the way it brings ethical and governance issues together with economic ones— hence the return to the traditional phrase, "political economy," understood as a social system and not simply as "economics," a social scientific discipline.

Ecological political economy calls us to take into account the fact that the planetary systems that support life in its most fundamental physical, chemical, and organic manifestations have boundaries, tolerances, and thresholds (Hansen et al. 2013; Rockström et al. 2009; Schlesinger 2009). These boundaries should—and ultimately will—constrain the extractive and the excretory activity of human individuals and societies. What individuals do one at a time is important, but the social, institutional level is a more essential focus here because the effects of human action are greatly magnified by the collective capacity of institutionally structured economies and technologies.

Ecological political economy offers a radical alternative to key assumptions that have informed virtually all of Western political and social thought in the modern era, beginning roughly in the seventeenth century (Brown 2001; Brown and Garver 2009; Ophuls 1998). The modern perspective sees economic activity as drawing on the material of nature, but not as being fundamentally dependent on the possibilities and limits of natural and living systems that have a functional integrity of their own. Ecological economics is distinctive in that it seeks to place economic activity (and the social scientific study of it) in the context of the operation of physical and biological systems (Costanza 1991; Daly and Farley 2011). This embraces a thermodynamic perspective having far-reaching policy and governance implications: economics must be treated as an open system, involving transfers of both energy and matter and operating within the tolerances of the planet, which is a closed material system.[2]

Other chapters in this volume attest to the importance of respecting planetary system boundaries and thresholds. These boundaries constrain the extractive and the excretory activity of human individuals and societies. It is essential that these constraints be recognized and obeyed. Biological and ecological systems can process the material waste and excess energy produced by economic activity, but only within certain scales and tolerances. As planetary boundaries are approached (or exceeded), ecosystem functions are undermined and overwhelmed, thereby rendering them—and the social systems that depend on them—less able to support either human or natural communities that are flourishing and healthy, diverse, and resilient. No longer are only justice and dignity at stake; now, minimally decent survival is in question.

What is true for the quality—indeed, the very possibility—of life generally is true for *human* life and living as well. We fully partake of, and depend on, the systemic preconditions of life for our own biological survival. However, perhaps an equally fundamental insight of this ecological perspective is that human beings also depend on the systemic preconditions of flourishing life for the concomitant flourishing of our own *humanity*. When Hans Jonas (1985) spoke of the "ontological imperative," I think he should be taken to mean not only that the survival of life is at stake, but also that the survival of a particular form of living—our very humanity—is at stake. Our accelerating extractive assault on planetary resources and ecosystems and the unprecedented extensions of our technologic reach, especially in biotechnology (Habermas 2003; Lee 2005; Rose 2006; Sandel 2006), represent a departure from the normal human condition of the past. Jonas' colleague and contemporary, Hannah Arendt, perceived the implications of this in the mid-twentieth century:

> The human artifice of the world separates human existence from all mere animal environment, but life itself is outside this artificial world, and through life man remains related to all other living organisms. For some time now, a great many scientific endeavors have been directed toward making life also "artificial," toward cutting the last tie through which even man belongs among the children of nature.
>
> (Arendt 1958:2–3)

There can be only one conclusion. Our accelerating, global extractive assault on planetary resources and ecosystems, as well as the unprecedented extensions of our technologic reach, do not truly represent progress and the triumph of human freedom or the human destiny. Why not? For one thing, they are not sustainable or viable as a road to the future. No less important, but less often noted, is the fact that technological advance and extractive assault contain an inner contradiction. Although seeming to extend human freedom, they are laying the groundwork for its repression. Being at liberty to behave in ways that are ecological irresponsible and destructive is not to be liberated; it is to be dominated by technology and desire.[3] While seemingly representing the advanced expression of human capability, technological advance and extractive assault are actually undermining what is most precious in humanness.

How then might we find a healed relationship between humans and nature, or at least conceptual tools with which to think about such a relationship? To point the way, I again turn to Arendt. In *The Human Condition* (1958), she developed an anatomy of our humanness, using a suggestive but rather idiosyncratic terminology. According to Arendt, human beings are creatures of "labor," who are subject to the biological rhythms of their organic needs; practitioners of "work," who are subject to the creative encounter between natural materials and imaginative form; and performers of "action," especially speech acts or communicative acts, through participation in the deliberative process of shaping common meanings in the public, symbolic order.[4]

The failure to live within planetary boundaries and limits—thereby turning our back on our interdependence with the earth and our own earthly, creaturely condition—will fundamentally threaten and transform the dimensions of labor, work, and action. Labor will produce illness rather than health. Creative work will become increasingly unavailable and unavailing. Action will devolve into bargaining and positioning for strategic advantage. At the dawn of the twenty-first century, precisely those baleful transformations in the human condition, this hobbling of human possibility, seem well advanced.

Therefore, we would do well not to underestimate the task facing ecological political economy. It is both an ontological reorientation and an ethical innovation. It goes beyond the physical and life sciences in a descriptive sense and implicates the normative foundations of social order and human agency. All structures of organized human activity must give a sense of meaning to the purposive, self-conscious human agents who comprise them; thus, a new ecological political economy will need a new foundational story and a new conceptual framework of norms and ideals. This story is the narrative of a journey of discovery concerning how to imagine, construct, comprehend, and govern a new form of social order that will achieve justice and empower flourishing life and living. This new conceptual framework is a "public philosophy" suitable to the unprecedented challenges of our time.

1.2. Ecological Public Philosophy

A public philosophy provides normative guidance and a context for the legitimation of governance and public policy (Brown 1994; Jennings and Prewitt 1985; Sullivan 1986; Tully 2008–2009). It can also provide a framework for the development of democratic social intelligence and problem solving through participation and critical deliberation. Grounded in solid natural and social scientific knowledge, as it ideally should be, a public philosophy is emergent and dynamic, not dogmatic. It reflects an ethical vision of what the ends of economic agency and democratic citizenship should be. All economies, including a future ecological one, will appropriate natural matter and energy and, through work (in Arendt's sense), transform them into products for human use and exchange. For this, the coordination and organization of very large numbers of people—a vast massing of human agency—will be required in agriculture, mining and manufacturing, science and technology, transportation, construction, and the like. Such coordination requires a sense of common purpose, and a public philosophy is essential in the imagination and discovery of what that purpose should be.

A public philosophy also holds a moral mirror up to each one of us. This mirror helps us discover not only a new form of social order, but also a new self-identity and a new way to live. In the market-oriented public philosophy of neoliberalism dominant in the world today, the individual must live out the following narrative, the selfhood of *homo economicus*: To survive and flourish, as an economic self you must fulfill (biological and psychological) needs. To meet your needs, you must compete successfully to extract value from the labor of others or to secure access to positions in which your own labor can provide the necessary income. To compete, you must understand and come to dominate the natural and social systems you inhabit.

In this narrative, the desire to acquire and consume is taken to be psychologically unlimited. The individual then is compelled by inner nature and external circumstances to appropriate and strive to dominate both the social and natural environment. As a result, growth in the activities of extraction and excretion knows no bounds and perforce overcomes all other considerations. We hurtle toward barriers ahead and apply the accelerator rather than the brakes. In this we have been taught to understand ourselves as free and responsible members of society. We provide for ourselves and our families. We pull ourselves up by our bootstraps and stand on our own two feet. This story of selfhood is hollow and self-defeating. It is a cathedral with no altar.

In contrast, the kind of self called forth by an ecological public philosophy to dwell in an ecological political economy has a quite different narrative, a counternarrative to that of *homo economicus*: To survive and flourish, as an ecological self you must fulfill (biological and psychological) needs. To meet your needs, you must compete and cooperate successfully with others to link your work (again in Arendt's sense) to that of others in respectful and accommodating designs. These designs will take advantage of the value and energy produced and reproduced by cyclic geophysical resources and counterentropic living systems. Social arrangements must be such that you will receive just sustenance and provision in return for your work. In this way, your needs will be met, and you will have the wherewithal to develop and pursue multiple human capabilities and courses of personal self-realization. To coordinate your work with others in such a social arrangement (an ecological political economy), you must understand, care for, and respect the natural systems in which you reside, and you must critically and constructively participate in the cultural and social systems you inhabit.

In this counternarrative, the desire to acquire and consume is not taken to be biologically impelled, psychologically imperative, or ethically unlimited. The individual then is not compelled by inner nature and external circumstance to strive to dominate either the social or the natural environment. As a result, growth in the activities of economic extraction and excretion will be bridled. They will be much more efficiently and intelligently designed *with* nature, not *against* it. They will be kept within safe ecological operating margins and normatively reasonable bounds.

We do not quite know yet how to foster the psychosocial development of such ecological selves or write their collective biography on a large scale. However, we desperately need to learn.

2. THE LIBERTY OF LIBERALISM

A fateful hallmark of the modern era is that we have based our political economy, our law and governance, and much of our moral philosophy on a distorted and self-defeating understanding of liberty. This notion of freedom as the unimpeded individualistic pursuit of limitless need and desire, which is often called "negative liberty," is fully consonant with our blindness concerning our true ontological place in nature.

This limited understanding of liberty goes back a long way. Already in the early modern period, economic thought was arising to challenge the western medieval worldview with its purposive, hierarchical cosmology (the Great Chain of Being), its normative order of natural law doctrines constraining economic activity (e.g., the prohibition of usury or money lending at interest), and its Christian theocratic and monarchical state. Out of this challenge (essentially the Renaissance, the Reformation, and the Enlightenment), a new understanding of human social and natural relationships—a new sense of human being and doing in the world—emerged (Hirschman 1977; Lovejoy 1961; Nelson 1969; Polanyi 1944; Taylor 1989). Then in the eighteenth century, the idea of *homo economicus* arose as a psychological benchmark and an anthropological ideal, and it grew in strength with the rise of industrial capitalism in the nineteenth century (Halévy 1966).

By the mid-twentieth century, with the political ascendency of social democracy and the welfare state, negative liberty was challenged and seemed to be temporarily on the wane. However, it has experienced a striking recrudescence in the past generation. In the 1970s, throughout the developed world, the perceived failure of national fiscal policy and neo-Keynesian institutional economics to address falling rates of profit and stagflation gave new currency to negative liberty. In the aftermath of the collapse of communism in 1989, the emphasis on negative liberty in neo-liberalism took on a virtually unchallenged ideological hegemony (Harvey 2005; Jones 2012; Peck 2010). There has been a return to antiregulatory price theory and free-market orthodoxy in economics and public policy. In the social sciences, mainstream discourse legitimates the norm of the self-interested rational actor of game theory modeling (Marglin 2010; Rodgers 2011). Also, new forms of finance capital and corporate management have brought about the complex structural transformations and power shifts often referred to as globalization (Duménil and Lévy 2011).

2.1. The Economic Interpretation of Liberty

Liberty is often used to designate a condition (or a potentiality) of mind and agency that inheres in individuals as a matter of right. Specifically, liberty as negative liberty is self-directed individual agency free from interference by others in the pursuit of the subjective satisfaction of needs and desires that are ontologically and psychologically limitless and insatiable.[5] The right to freedom so understood is a moral claim that can be made by a person against others who are in a position to impede, impel, or coerce the person's behaviour in ways that conflict with the person's own purposes or interests. Private individuals may fall into this category, but the right to be free quintessentially applies against those who wield corporate institutional power or the legal police power of the state. The most potent source of the individual's security and protection can also be the most dangerous threat to his or her freedom. This is a paradox that liberalism has struggled to solve, yet it is also a paradox of liberalism's own creation.

The public philosophy of neoliberal political economy esteems the right of individual freedom very highly, often above all else. It is ambivalent, however, about how far that right to liberty extends. One interpretation holds that the freedom of the individual creates a negative obligation of forbearance (noninterference) on the part of others only. A contrasting interpretation maintains that the individual right to liberty sets up a positive obligation of assistance by others to aid the individual in obtaining the resources and capacities that will make his or her freedom productive. The libertarian face of political economy in the modern age, and neoliberal political economy today, wants to protect individuals against the exercise of power by others. The egalitarian face wants to enable individuals, with the assistance of others, to exercise power for themselves. This is the distinction (roughly) between libertarians and market liberals, on the one hand, and egalitarians and social liberals, on the other.

These nuances, although very important, are not at the root of the problem today. A more fundamental question for the ethical foundations of ecological economics concerns the moral justification for limiting or overriding the individual moral claim to noninterference by others, especially the state. One way to do this is to assign other principles or values a higher moral priority than liberty and subordinate it when it comes into conflict with those values. Another is to seek internal self-limiting conditions within the logic and meaning of the concept of liberty itself. Does the concept of individual liberty, in anything like the form that we have inherited it from the liberal political tradition, contain within itself the basis for its own moral limitation?

In fact, as was just noted, the liberal tradition does not speak with a single voice on the proper way to understand liberty. Historically, it was not liberalism, but anarchism, that held out an almost unrestricted libertarian philosophy of individual freedom (Wolff 1970). Liberalism has always been about reconciling individual subjectivity and self-determination with more constraining norms that are required for the social order and cooperation necessary for the security and well-being of each and all. Here, the modern liberal tradition splits into two streams. One stream, much influenced by the philosopher Immanuel Kant, seeks the reconciliation of liberty with the objective constraints of reason and duty. Kant solves the problem with his notion of autonomy and the acceptance of moral duty for its own sake. The other stream, influenced by John Stuart Mill, wants to reconcile individual liberty with the constraining norm of the general welfare, or the greatest good of the greatest number. Mill solves the problem by using the standard of preventing harm to others, which he viewed as justified by the principle of utility but which might also be interpreted as pitting the liberty of one person to do something against the liberty of another person not to be harmed. For him, the prevention of involuntary harm to another is the principal (even the sole) justification for curtailing the individual's unimpeded discretion in the pursuit of his or her subjective interests or desires.

Mill's solution has come to be called the "harm principle" or the "liberty principle." In *On Liberty*, he formulated it as follows:

> The only purpose for which power can be rightfully exercised over any member of a civilized community, against his will, is to prevent harm to others. His own good, either physical or moral, is not sufficient

warrant. . . . The only part of the conduct of anyone for which he is amenable to society is that which concerns others. In the part which merely
concerns himself, his independence is, of right, absolute. Over himself,
over his own body and mind, the individual is sovereign.
(Mill [1859] 1956:13)

In other words, your freedom ends where my nose begins. If social or state
power is to be used to curtail individual freedom for any purpose of communal well-being, common good, or public interest—including environmental conservation and protection—that provides the logic of a justifying
argument with which to do it. But, what if something more is at stake than
the protection of the interests of affected others, or the prevention of harm
to my nose?

To be sure, it is very tempting for ecological economics to rely on the
harm principle in order to justify controls on economic activity that threaten
the planet; surely if harm means anything morally at all, it means the
undercutting of the ecosystemic basis for biodiversity and for human and
animal life and health on regional and planetary scales. Nonetheless, it is a
mistake to settle for an ethical justification of limiting individual liberty on
the grounds of harm prevention alone. Doing so effectively shields extractive power under the protection of the right to negative liberty. Extractive
power, which I shall consider more fully below, expropriates value (useful energy) from natural and human resources up to, but not beyond, the
point of collapse. It does not "harm" nature or workers—extractive power
drains, it degrades, it wrongs. If we broaden our moral lens beyond harming to encompass wronging, then the ethical justification of constraining
individual liberty can be much more than merely defensive and preventive;
it can be transformative, enabling, and empowering. It can appeal beyond
protection and the prevention of harm to the promotion of justice, greater
equality, and the achievement of currently stifled and unrealized possibility
inherent in the evolution of our human capabilities.

Shall the state in an ecological political economy embrace some standard of natural or rational duty—such as Jonas's imperative of responsibility or a notion of right relationship or the land ethic—and permit
individual liberty to expand only to the point of wrong-doing? Or, should
it expand liberty further to the boundary of involuntary harm and indeed
direct harm to humans only, which is a standard that permits unfettered

exploitation and depletion of the biosphere? This question lives on in elite intellectual circles. In mainstream political culture, it is liberty expanded to the point of harm that has carried the day.[6] The Millian stream of liberalism (supplemented, at least in the United States, by an occasionally influential libertarian anarchist fringe) is predominant. Fatefully, it has been the economic branch of liberalism that has provided much of the impetus for this. Mill, and most liberal economists before and after him, did not turn to right reason as a check on freedom or as a solution to the dilemma of liberty. Reason for Mill, as for Hobbes and Hume before him, is a strategic faculty for determining effective and efficient conduct, not a standard for justice. It presides over the choice of means to promote individual interests and desires.

Hence, with economic liberalism one finds a version of subjectivity introduced into the notion of freedom, because each person is the most reasonable custodian and definer of his or her own interests and objectives. If the power to determine those interests is exercised by others, especially by officials of the state, a person is deprived of liberty and is hampered in the development of independent mindedness, skill, and self-reliance. Mill considered these capacities to be some of the hallmarks of human flourishing, and economic liberals have viewed them as essential for economic efficiency and growth under capitalism. (Mill [1859] 1956; Schumpeter 1950).

Moreover, for economic liberalism, there is no independent standard of reason to determine if one person's use of freedom is inherently superior to another's. There is no intrinsic right or intrinsic wrong: the free person is not subject to such elitist and arbitrary value judgments imposed from above or outside. In the modern ideology of neoliberal capitalist political economics, the individual need or desire to acquire and consume is taken to be psychologically and ethically unlimited (Macpherson 1973:18–19). Absent tangible and serious harm to others resulting from an action, individuals *should* be allowed to make their own choices and life courses. If individuals are permitted by social and political arrangements to have this liberty, Mill thought, the society as a whole will prosper, and this social arrangement of well-ordered, but not stiflingly conventional, liberty will be justified from a utilitarian point of view. For the neoclassical political economists, the thought experiment of the ideal competitive market was the mechanism whereby the self-defining morality of each person could be reconciled with welfare or the common good, at least when the good is hedonistically defined in terms of economic growth, consumption, and the satisfaction of subjective, material wants.

During the heyday of industrial capitalism at the end of the nineteenth century, and particularly with the neoliberal market revival of recent decades, this influential line of thinking has taken on a much more one-dimensional and morally impoverished character than it had in Mill's own thought. When liberty is based on self-defining or market-defined utility alone, the distinction between liberty and license withers away, and liberty ceases to be answerable to anything higher than itself (Gaylin and Jennings 2003; Taylor 1985b:211–229, 1991). Preventing harm—at least in the fairly narrow sense in which liberalism has defined it—does not constitute the entire range of ethics in individual life or in public policy and governance. The absence of harm does not entail the absence of wrong or injustice. Protection from harm may preserve and promote life, but it is not sufficient for achieving a good life.

In sum, ecological economics must morally indict neoliberal political economy on several counts besides harm to the planet. It should press con-temporary science and charge the current political economy with an erro-neous view of knowledge and a false cosmology. It should also press on the promise of a new era of human freedom inherent within a different way of structuring human productive economic agency and governing it.

2.2. A Farewell to Economic Liberty

This volume has identified three fundamental normative concepts around which an ethical foundation for ecological economics should be built: membership, householding, and entropic thrift. These tenets hold that humans are members, not masters, of the community of life; that we main-tain our own life from other living systems, which must be respected and cared for; and that natural systems contain entropic limits and scarcities that require wise use and just sharing. These norms are profoundly at odds with negative liberty and neoliberalism. They demand that we ask what liberty would look like if it were reconciled with, and served to reinforce, these norms. That is what I seek here: liberty in the context of a responsible and just human–nature relationship; individual agency and responsible self-direction with an ecological face.

There is no gainsaying the fact that the scope and degree of negative liberty currently afforded to individuals and corporations in the political economy of neoliberalism and global capitalism, and now vigilantly protected by both parliamentary (representative) government and the courts, will have to be curtailed considerably if planetary limits and ecological constraints on energy usage and economic activity are to be respected (Klein 2011, 2014). Negative liberty recognizes only voluntary or contractual obligation—grounded on the rational self-interest of each person to live in a society of mutual security and protection from involuntary harm—as a restraining norm on individual will. Therefore, it is not compatible with the principles of membership, householding, and entropic thrift, which are written in the key of natural, not contractual, duty and obligation, and from an other-regarding, not a self-regarding, moral point of view.

These principles capture well many of the fundamental moral imperatives of future conduct and modes of life in societies that can endure. We must move beyond the contractual morality of negative liberty and of much of the liberal political tradition of the past three centuries. But in so doing, we should also undertake the corrective work that needs to be done to reclaim the concept of liberty, to move toward a more adequate philosophy of freedom for an ecological political economy and its public philosophy.

However, why should we even attempt to put new ideas into an old word that has been so deeply molded by the historical era we are now leaving behind? As Marshall Sahlins (1985) puts it, why take the gamble that culture plays with nature at times such as these? There are three reasons. First, we should make this effort because a human notion that is as fundamental as liberty should not be conceded to a historically peculiar libertarian or negative interpretation without a contest. Second, we must reclaim the concept of liberty because any ecological ethic will rely on some conception of human free agency and any ecological democracy will depend on a conception of citizenship in the service of liberty, as well as in the service of life and the common good. This is because the concept of moral and legal responsibility, to which an ecological ethics and politics must appeal, is grounded on some notion of free agency and individual liberty (Pettit 2001). The contradiction between liberty and obligation or responsibility (as understood by the notions of membership, householding, and thrift) cannot be taken as given or inevitable. That seeming contradiction is an illusion of liberal market ideology and an artifact of societies that rest on cheap fossil carbon energy and have not yet filled biophysical sinks with the waste products of their activity. Only in such an age and such a world could one seriously take oneself to be free while blithely and blindly destroying the very preconditions of that freedom.

Third, and finally, if the inadequacy of a negative, individualistic concept of liberty makes it philosophically and ethically necessary to replace it with a more adequate relational conception, the cultural and political power of the concept of negative liberty make it culturally and politically essential to do so. As a normative ideal, freedom is a genie out of the bottle in the world today. For us in the advanced technological economies—who are the most ecologically destructive economic agents on the planet today—erasing liberty from our personal and social self-conceptions would be culturally catastrophic, as it has been in other societies that have suddenly lost fundamental aspects of their way of life. If many of the ways we currently use our freedom are becoming ecologically untenable—and no doubt they are— then it is far more desirable to reconceptualize freedom so that it is more in tune with natural reality than to accept the idea that freedom is no longer available in our lives at all as a meaningful concept, aspiration, or ideal.

When a society loses the concepts through which it has traditionally made sense of itself, it experiences a debilitating disorientation (Diamond 1988). Jonathan Lear's remarkable study of the experience of cultural devastation among the Crow people starting in the late nineteenth century provides a cautionary tale for any society facing a fundamental shift from one way of life to another in the span of one or two generations:

> If a people genuinely are at the historical limit of their way of life, there is precious little they can do to "peek over to the other side." Precisely because they are about to endure a historical rupture, the detailed texture of life on the other side has to be beyond their ken. In the face of such a cultural challenge, . . . there is ever more pressure to explain things in the traditional ways, yet there is an inchoate sense that the old ways of explaining are leaving something unsaid. And yet one doesn't yet have the concepts with which to say it.
> (Lear 2006:76, 78)

There is much still to be said about social arrangements that provide the space for self-directing individual agency and the flourishing of individuated human capabilities, if we can find a new concept of liberty with which to say them. This new concept will have an important place in the ethical framework that will allow ecological economics to finish its journey and contribute to a new ecological political economy and public philosophy. It will be a concept of liberty that is symbiotic with—and indeed, perhaps tacitly embedded in—the principles of membership, householding, and entropic thrift. Moreover, because any conception of liberty inherently rests on a background conception of human agency and the human condition, a new conception of liberty will go hand in hand with a new conception of the person or the self.

How can we regain contact with the dimension of the human condition that Arendt called "labor"—namely, our fundamental being as biological creatures, subject to the rhythms and requirements of life? How, in short, can we reconnect with our own animality and materiality, which we seem to have forgotten?

How can we rediscover our humanity in "work"? How may *homo economicus* (the self as a gaming, calculating maximizer of personal utility) be transformed into *homo faber* (the self as a craftsman, responsible for and respectful of his materials)?

Finally, how can we regain contact with "action" as the expression of our civic capabilities? How can privatized selves—trained to think only of a politics of who gets what, when, and how—be nurtured so as to become deliberative democratic citizens attentive to the common good and obligations of trusteeship for the natural world?

I believe that these questions are bound up inextricably with the future of liberty. The rediscovery of these dimensions of our human condition requires the transformation of the current neoliberal world of extractive liberty and possessive individualism into a world of relational liberty exercised through the practices of justice or parity of social membership and solidarity.

3. LIBERTY BEYOND LIBERALISM

The time has come to inquire more deeply into the concept of liberty. I eventually will offer a notion of relational liberty as a promising theory of freedom for our time. However, the force of that notion cannot be appreciated unless we first look closely at, and work our way through, prior conceptions of liberty from which relational liberty emerges as a corrective. I propose to do this by focusing briefly on the work of two of the most important political theorists of the twentieth century, Sir Isaiah Berlin, the liberal tradition's most humanistic revisionist, and C. B. Macpherson, one of its most powerful and humane democratic critics.

3.1. The Liberal Revisionism of Isaiah Berlin

Berlin set up an opposition between what he called negative and positive liberty (freedom from and freedom to), and he thereby reconstructed an account of what was most important about social arrangements that permit individual liberty. He exposed a crucial ambiguity in the liberal tradition (its embrace of a developmental idea, the cultivation of a more perfect self), which fatefully weakened it as a bulwark against totalitarianism. (He was writing at the height of the Cold War.)

Berlin drove a sharp wedge between the warrant for the freedom from interference in an attempt to live one's life in one's own way (negative liberty) and the warrant for the freedom to live in accordance with one's highest and best, most rational, ideals (positive liberty). In positive liberty, there is a paradox because we tend to see liberty as antithetical to duty or obligation. Yet the freedom to live in accordance with one's best self is very close to the idea of a duty to obey the rules laid down by one's best self; one is really free when you listen to the angel perched on your right shoulder and ignore the devil perched on your left. As I will discuss further, Berlin read the positive conception of liberty in light of twentieth-century experience, and viewed it as a philosophical route to a unitary or totalizing ideology. As such, it is politically dangerous and profoundly at odds with the pluralism and openness of society that he so deeply embraced. That pluralism was, for Berlin, the essence of liberalism.

As a counter to liberty defined by objective reason, however, Berlin did not wish to adopt a libertarian subjectivism or voluntarism, which both Friedrich Nietzsche and Max Weber had done in their own ways when they encountered the totalizing aspect of reason (rationalization) in history. Instead, Berlin adopted an ethical pluralism for which individual negative liberty (freedom from interference and domination by others or the state) is the keystone. He also rejected any notion that history should be viewed teleologically as the story of the growth of liberty over time and its eventual triumph—a view sometimes called the Whig interpretation of history (Gray 1996, 2000). Most important for the question at issue here about the relationship between ecological economics and liberty, Berlin held that freedom must be embraced for its own sake, not for its instrumental relation to social utility. We honor the freedom of another when we forbear interference and give space for his or her agency, even when—and especially when—we have a moral objection to the acts and the consequences that will result.

Berlin would regard ideals such as membership, householding, and entropic thrift as alternative competing values alongside (and perhaps in conflict with) liberty, and he would hold that the ethical priorities among these different, incommensurate values is a problem that moral philosophy cannot solve.[7] For pluralistic liberalism, there is no final condition of human flourishing or the good toward which any political practice or moral value will lead because the good is necessarily plural and necessarily evolving, open ended, and incomplete. Historical attempts to impose a hegemonic conception of the good lead, all too often, not to the greater realization of our humanity but to its degradation. This includes conceptions of the individual or the common good that are seemingly warranted by the science of the time and by arguments of necessity and survival made by those in authority. Thus, as Berlin wrote, employing the spatial metaphor of freedom that Mill characteristically used, "there are frontiers, not artificially drawn, within which men should be inviolable, these frontiers being defined in terms of rules so long and widely accepted that their observance has entered into the very conception of what it is to be a normal human being" (Berlin 1969:165).[8]

As Berlin used the term, *negative liberty* has to do with establishing a zone of privacy and noninterference around each person—a zone within which individuals can exercise their own faculties and pursue their own lives in their own ways. It rests on a conflict-ridden and antagonistic picture of social existence, in which each individual struggles with everyone else to control his or her own patch of ground.[9]

Positive liberty, by contrast, is a form of self-mastery.[10] One is free in the positive sense when one's reason—one's higher self—is in charge of one's conduct. Negative liberty is the absence of control by others; positive liberty is the presence of control by one's self so long as—and here is the rub—it is an ideal (i.e., rational and autonomous) self. Berlin was suspicious of positive liberty as a manifestation of a dangerous, dogmatic conception of the fully human person and the fully human life. In the name of attaining this ideal, individuals have been asked, or required, to subordinate their personal freedoms and interests to other people's causes. This is the dark side that Berlin saw not only in various romantic and irrationalist modes of thought, but also in the legacy of Kantian rationalism and other forms of objectivist, deontological ethics.

3.2. C. B. Macpherson's Liberationist Critique

An important critique of Berlin's defense of negative liberty as the bulwark of moral and political pluralism and his assimilation of positive liberty to the totalitarian temptation was offered by the Canadian political theorist C. B. Macpherson. Macpherson deconstructed Berlin's theory by showing, much as he had done in his reading of John Locke (Macpherson 1962), that embedded within a seemingly abstract and universal philosophical conception of freedom was actually an ontological and ethical legitimation of a very specific form of governance, political economy, and social power. Such a critique opens the door to an alternative, democratic practice of freedom and ideal of the self (Macpherson 1977).

Macpherson was concerned with showing that negative liberty in Berlin's interpretation is too narrow and does not offer a general philosophy of freedom. Instead, it is historically and ideologically bounded. As such, it reproduces—albeit perhaps not intentionally—a sophisticated version of the traditional capitalistic concept of freedom as the right of appropriation and of the notion of free agency as the agency of the possessively individualistic self, which at the structural level of society and personality is actually quite unfree.

Macpherson argued that Berlin's account of negative liberty is based on an overly literal understanding of the exercise of power over an individual such that liberty is violated (1973:97–104). It is also based on misleading spatial metaphors, such as zones of privacy and noninterference, within which the individual is free to choose and act. As a result, Berlin's understanding of domination or unfreedom is limited to coercion and the use of physical force, as well as to other nonconsensual courses of action, without considering adequately the phenomenon of the ideological or circumstantial prestructuring of perceived possibilities of choice and consent as such.

For Macpherson, the types of power that curtail liberty are various, indirect, and in some societies pervasive. Berlin was concerned about the situation in which a person has only one road to follow, only one door to go through, or in which no doors lead to any desirable destination. What should be of equal or greater concern, said Macpherson, is the power to compel the choice or the journey itself—to prestructure the very perception of choice in a certain direction and in the service of interests other than the individual's own. Macpherson referred to this as "extractive power," and it is actually facilitated by Berlin's concept of negative liberty, especially under conditions of a market society and a capitalist political economy. What an adequate concept of liberty requires is not noninterference in a narrow sense, but a form of life within institutional activities and power relationships that afford each person "immunity from the extractive power of others" (1973:118). Macpherson referred to individual freedom within this social condition as a state of "counterextractive liberty."

Turning to the other dimension of Berlin's theory of freedom, Macpherson distinguished three different conceptions that are conflated in Berlin's critique of positive liberty (1973:108–109). First, there is the idea that the individual should be a self-directing agent and that such agency often requires the active cooperation (and not merely the forbearance and noninterference) of others. In any complex society, positive liberty for the individual typically also requires certain forms of social arrangements, such as access to various services and resources that enable individuals to effectively implement projects to achieve their goals. Second, there is the idea that an individual may be positively forced to be free either by his or her own "higher," more rational, alter ego or by others who have attained the rationality and objectivity that the individual lacks. This is the notion that Berlin mainly feared and that Macpherson also roundly rejected. Third, positive liberty contains the democratic idea that each individual should be free to take an active part in shared decision-making and collective governance.

Having made these distinctions, Macpherson then argued that there is no inherent logic that should transform the notion of cooperation in the service of self-directed living and agency into the authoritarian notion of paternalistic direction by a superior intelligence or expertise (1973:111–114). The logic of moving from the first conception to the second is entirely a matter of social context and historical circumstance. Indeed, this is not simply a conceptual shift but also a transfer of social and economic power.

Positive liberty, the freedom to engage in self-determining forms of activity, is itself a type of power, which Macpherson called "developmental power." The move from authentic cooperative self-determination to what Berlin had called "self-mastery" (i.e., the paternalism of the higher self, particularly in the context of a market society) is a surrendering of the developmental power of the self to the extractive power of others—and that, not positive liberty or cooperative agency per se, is the root of the problem of unfreedom for Macpherson. If that is correct, and if uncontrolled and ecologically careless extractive power is an important part of the problem that ecological economics is trying to solve, then there is an important connection between the ethics of ecological economics and the ethics of liberty.

No less than Berlin, Macpherson was a philosopher of freedom, and he was centrally concerned with liberty construed as a democratic and liberationist ideal (Carens 1993). However, he regarded Berlin's theory of freedom as an impediment to the emergence of a more egalitarian mode of democratic society in the future. Unfortunately, Macpherson's work did not engage directly with the ecological aspects of the problem of how freedom and agency are distorted and unjustly shaped by the economic system of property and power. At times, like others on the left, Macpherson (1973:36–38) seemed attracted to a technological utopianism.

Although Macpherson did not take his critique in an ecological direction, there seems to be no reason in principle why we cannot do so. By placing the problem of liberty squarely in the context of the tradition of liberalism and liberal political economy, Macpherson did illuminate aspects of the task of developing a conception of liberty for an ecological political economy. I attempt to take up that task here, with a notion of relational liberty and the structuring norms of its practice. I argue that the structuring norms of relational liberty are membership (parity of participation and voice and equality of civic respect) and mutuality (equality of civic care and concern, as well as solidarity).

3.3. Extractive Power

We can now step back into the domain of political economy and the challenge of crafting the enabling acts of mind that are necessary for the emergence of a new and ecologically viable society, which fosters individual agency and freedom.

The liberty enshrined in mainstream economics and protected in neoliberal political economy is basically negative liberty that permits the exercise of extractive power. All economic activity involves the appropriation of natural matter and energy from geological sources and biological energy systems (and the disposal of waste back into those systems). Necessary to this appropriation are two elements: (1) the social organization of human productive agency—labor and work; and (2) the control and use of technology to supplement, extend, and often to replace the labor of the human body in the exercise of agency by the human person. These social conditions of appropriative economic activity can take many forms that are compatible with a society's cycle of production and reproduction over time. In the modern era, initially national capitalist, then colonial, and now global capitalist modes of organization have produced a distinctive set of human relationships (and cultural meanings) that mediate the exercise of individual agency and the practice of individual freedom. These relationships in the neoliberal political economy have several dimensions—structural (corporate management), financial (wages, liquidity, debt), and legal (property, contract, securities, rights).

It is within these structured relationships of the political economy (and its corresponding cultural significations and normative justifications) that the exercise of extractive power comes to the fore. In the social domain, Macpherson's analysis of the exercise of extractive power harkens back to Marx's account of the creation and appropriation of surplus value (Harvey 2010; Marx [1867] 1990). Those who own or control land and capital (the means of economic production) exercise extractive power over those who only own their labor and are compelled by their place in the economic system to sell it in order to live (the "relations" of economic production). In a capitalist market society, negative liberty—in the form of private property and other legal entitlements in land and capital—puts some individuals in a structural position to exercise extractive power over others. In spite of this, negative liberty is perceived as a precious benefit and is jealously guarded by all.

John Lanchester provided an insightful synopsis of surplus value in an article on Marx:

Marx's model works like this: competition pressures will always force down the cost of labour, so that workers are employed for the minimum price, always paid just enough to keep themselves going, and no more. The employer then sells the commodity not for what it cost to make, but

for the best price he can get: a price which in turn is subject to competition pressures, and therefore will always tend over time to go down. In the meantime, however, there is a gap between what the labourer sells his labour for, and the price the employer gets for the commodity, and that difference is the money which accumulates to the employer and which Marx called surplus value. In Marx's judgment surplus value is the entire basis of capitalism: all value in capitalism is the surplus value created by labour. That's what makes up the cost of the thing; as Marx put it, "price is the money-name of the labour objectified in a commodity." *And in examining that question he creates a model which allows us to see deeply into the structure of the world, and see the labour hidden in the things all around us. He makes labour legible in objects and relationships.*
(Lanchester 2012:8, emphasis added)

Macpherson (like Marx before him for the most part) kept to the social, interhuman dimension of this logic of extractive power; however, there is no good reason not to extend it to the relationship with nature. The normative, critical force of the notion of extractive power underscores the important difference between seeing nature as the intrinsically valuable physical and chemical context of life, rather than as simply an instrumental source of raw material (a stock) to extract and as a disposal site (a sink) within which to excrete economic waste.

All human work appropriates energy and value from nature or from the capabilities of other humans.[11] As noted earlier, all economies, including a future ecological one, will appropriate natural matter and energy and, through human work and technology, transform them into products for human use and exchange. However, only in certain types of economic system will this appropriation turn into insatiable extraction and excretion. Historically, this transformation has been associated with wage labor, the division of labor, and the concentration of extractive power in the hands of an elite who owns or controls the means of production. Under such conditions, the logic of extractive power pushes the application of this labor, this managed human agency, toward a scale and pace of extraction and excretion that threatens to exceed planetary tolerances, deplete nonreplenishable stocks, and overwhelm ecosystemic capacities. The remedy lies in curtailing extractive power and its exercise over those vulnerable to it by dint of class inequality of wealth and power. This remedy is blocked by the blanket protection of extractive power and extractive liberty that the liberal defense of individual negative liberty (e.g., property rights) provides.

At the dawn of the Anthropocene age, the liberal philosophy of freedom has reached an aporia—a political and economic dead end and a form of cognitive dissonance from which it seemingly cannot escape. Yet, the way out for liberalism may in fact be the same as the way out for the planet. Ecological political economy demands a regime of social control that is not compatible with the life narrative of *homo economicus* or with the wide scope of individual and corporate freedom from (state) interference in the use of extractive power, which has been widespread among the affluent of the developed world in the last century (Jennings 2010a, 2010b). What is the best way to frame this issue: As a balancing or trade-off of conflicting values? As a regrettable but necessary contraction of the sphere of individual freedom of choice made necessary by ecosystemic limits? Or perhaps, as I believe to be the case, our task is to reclaim and reconstruct the concept of liberty so that, in our moral imagination and our public philosophy, ecologically destructive behavior would not be seen as a manifestation of freedom at all; rather, it would come to be repudiated as a manifestation of ignorance, irresponsibility, and alienation that negates freedom.[12] If we frame this problem as one of balancing values, then who controls the scales? If we frame it as a devolution of freedom for the sake of survival, then what level of coercion will be used against the recalcitrantly self-destructive among us?

Reclaiming our understanding of freedom, taking it back from the destructive market ideology of neoliberalism that has colonized our political culture today, could tap into and rechannel in ecologically constructive ways the powerful motivational psychology that valorization of individual freedom has established. If we could pull this off culturally, it might minimize the need for repressive forms of behavioral control legally and politically. Self-directing ecological citizens who use their freedom and agency in the service of a sense of ecological trusteeship and responsibility are the democratic hope of the future.

4. RELATIONAL LIBERTY

I contend that the absence of outside interference (individualistic negative liberty) and the presence of active self-development (individualistic positive liberty) are insufficient for a robust ecological public philosophy and a resilient ecological political economy in the twenty-first century. They are lacking because they fail to convey adequately that free agency relies on modes of relationality that have a particular ethical structure: the experience and exercise of freedom is a social practice. A conception of relational liberty explains how such a practice works and what it involves. I believe that such a conception is necessary in order to give a full interpretation of the notions of ecological membership, householding, and entropic thrift.

Relational liberty is freedom in and through relationships of interdependence. More specifically, my notion of relational liberty can be defined as freedom through transactions and relationships with others that exemplify just membership (parity of respect, participation, and voice) and just mutuality (parity of care, concern, and solidarity). The essence of the philosophical strategy I propose is to internalize freedom into responsibility; independence into interdependence; and the common good into the individual good, to read the *We* into each *I*.[13] More specifically, my approach is to internalize the freedom and well-being of all (both human and nonhuman creatures and systems of life) into the freedom and well-being of each. Only in this way, I believe, can we break out of the cultural trap and dead end of pitting individual freedom and collective restraint and limitation in opposition to one another. Only in this way can we counter the moral license that economic liberalism has given to self-centered liberty, extractive power, and insatiable desire—market-oriented economic progress and growth. I believe this will provide an essential lynchpin for an ecological public philosophy.

Relational liberty resists the nihilistic "creative destruction" that the neoliberal political economy has institutionalized and that is so socially disruptive and environmentally rapacious. Relational liberty—freedom through interdependence—is a warrant to live one's own life in one's own way that results from embedding that way of life in a tradition, a civic life of shared purpose, and rooting that life in a sense of ecological place and in a sensibility of care for place and care for Earth's life-support systems. Just as there are certain kinds of practice or activity that by their very nature cannot be done alone, so there is a kind of freedom that subsists not in separation from others but through connection with them. Not in protections but in pacts of association; not in locked futures but in open ones; not in fences but in circles; not in extraction but in conserving; not in artificing but in accommodating.

Relational liberty rejects two constitutive features that have character-ized theories of freedom in the liberal tradition. I have alluded to these features already, but it is useful to reiterate them here. One is the privileging of individualistic values over communal ones—individual liberty trumps community solidarity (Mulhall and Swift 1996; Marglin 2010). The other is setting up a conflict or antithesis between the individual and the com-munity in the first place.

These two features make liberal theories of freedom remarkably devoid of the web of interdependencies—that is, culturally meaningful roles, styles, and self-identities; shared values, rituals, and practices. These theories, par-ticularly when applied to the realm of economic agency, tend to portray an idiotic world of atomistic individuals, each with their own self-regarding interests and life-plans. This requires, at most, a *modus vivendi* of bare tol-eration and opportunistic cooperation—a social existence of peaceful and predictable transactions for mutual advantage. A thicker sort of mutual-ity or solidarity based on care, compassion, and empathy is rarely, if ever, deemed necessary, and is readily disposable if it leads to inefficiency. Fur-thermore, it is this feature of atomistic abstraction that makes such philo-sophical accounts ideologically functional for a political economy founded on the logic of extractive power. Interestingly, and paradoxically, atomistic individualism both conceals and normatively justifies extractive power at the same time.

Again, on a relational conception of liberty, freedom is constituted not in spite of connections and commitments linking self to others, but in and through these connections and commitments. Enacting relational freedom in one's life develops a self-identity built out of ongoing practices that exem-plify the creative and aesthetic dimensions of a humanity naturally flourish-ing. Such relationships involve not only social interactions with other human beings but also—and crucially—meaningful connections between the per-son and his or her own activity on and in the world of fellow creatures, living systems, and material objects. A relational conception of freedom and of the person contains a countervision to notions of alienation, commodification, and the objectification of the human or the natural other. Such a vision leads away from the control of natural material as a source of "wealth" defined as material accumulation, relative social status, and utility maximization. It leads toward a notion of artistry, craftsmanship, and the accommodation of the inherent properties of natural form (Schor 2010; Sennett 2009).[14]

One central task of an ecological public philosophy is to reconcile individual liberty with interdependence, community, and the common good (Daly and Cobb 1994). The theory of relational liberty is designed to do that, and the task is not nearly as philosophically difficult as liberal thinkers have made it appear. The fact of the matter is that individual agency (and individual self-identity and motivation) is already thoroughly social and relational in character (Giddens 1984; Habermas 1984; Taylor 1985a). Moreover, and without contradiction, social, structural change is nothing other than a change in the ways in which individuals experience and live their own social being.[15]

Planetary thermodynamic processes may be systems that operate impersonally without a locus of intention, control, or responsibility. However, I contend that economies, societies, and political communities are not systems in that sense but rather are structures of purposive human agency. Moreover, adopting an agent-centered rather than a systems theoretic approach, as I do in this chapter, does not entail an ontological or atomistic individualism. Human acts are intentional, purposive, and meaningful both to the actors and to others who share in the rule-governed forms of life and communication within a society and culture. The ethical norms that fit into human agency therefore are not limited to self-referential states of interest or desire. To understand ethical conduct—or to engage in ethical discourses of justification and other forms of argument—one must have recourse to concepts and categories that reflect the relational nature of the human self and the contextual, socially and symbolically mediated nature of the self's interactions with others (Harré 1998; Taylor 1989).

The implication of this for ecological economics is quite important.[16] An ecological political economy will come about only through change at both the level of individual behavior and the level of social norms and institutions. In practice, this means that we must learn to articulate the values and ideals that the members of these societies would express if they thought and acted like interdependent and relational selves—ecological selves, or ecological citizens and trustees. Part of the task of ecological economics as a public philosophy is to shape this self-identity and to foster a moral imagination that can see the good and freedom in relational terms. Mainstream economics over the years has helped to build a population of possessive individualists through its doctrines and through the institutions it has legitimated (Marglin 2010). Ecological economics must be no less intentional about the task of educating a new generation of social people. The time has come for economic knowledge and discourse to show all of us—specialists and ordinary citizens alike—that our personal flourishing is inextricably linked to the flourishing of others and to the flourishing of the natural world. Ecological economics does not abandon the concept of economic self-interest but transforms it.

For this task, ecological public philosophy needs the vocabulary of solidarity, mutuality, reciprocity, community, and the common good—norms that are certainly contained, or at least alluded to, in the principles of membership, householding, and entropic thrift. The pioneers of the field have already begun to develop this vocabulary (Daly and Cobb 1994). Beyond the notion of moral obligations that are correlative to the individual rights and interests of others and the obligation to do no harm, Ecological economics also needs to appeal to a motivational structure that is informed by ecological trusteeship. Ecological trusteeship is not so much a role, or a legal status, as it is an orientation and a disposition of living that is grounded in a sense of responsibility for promoting and sustaining the common good of the community as a whole and its natural context.[17] People who have this sense are, in Aldo Leopold's words, "plain members and citizens" of the biotic community (Leopold 1989:204). Ecological economists can be the teachers and counselors of just plain citizens; an ecological public philosophy can be their creed; and an ecological political economy is their focus of study, their vision, and their goal.

The field of ecological economics thus both relies on and has the potential to instill in people the capacity to comprehend the meaning of a common danger or a common good. If the people around the world, and especially in those nations whose economic behavior has to change most drastically, have lost the capacity to comprehend these ideas, then it will not be possible either to coerce or to empower them to undertake the kind of collective institutional and behavioral change that creating an ecological political economy will require (Honneth 1996).

5. THE NORMATIVE STRUCTURE OF RELATIONAL LIBERTY

I define relational liberty as freedom through transactions and relationships with others that exemplify just membership and mutuality. In the preceding discussion, I have been using the notion of "relationship" as a term of art with a specific and deliberate normative meaning. Now, I should try to explicate the internal normative structure of the concept of relational liberty more fully and explicitly. My intention is to build standards of justice, dignity, membership, and mutuality into the concept of relational liberty as constitutive and functional aspects of it.

Not any form of human interaction or transaction constitutes a relationship through which ethically valuable human liberty is constituted. Interactions of domination, exploitation, coercion, violence, seduction, or duplicity—each of which effectively reduces human beings from the status of subjects to the status of objects—do not count as "relationships" in the requisite sense of the term. When it comes to relationships with nonhuman beings, such interactions may be ruled out ethically for the same reasons insofar as the nonhuman organisms possess agency—as clearly many animal species do (Burghardt 2006)—and thus are subjects to be respected and not simply objects to be used. Nonhuman beings that cannot be understood as purposive agents are still subject to norms and restraints on human action that follow from intelligent, experience-based, and evidence-based recognition of ontological interdependence within the web of life and existence. In short, neither in relationship with other subjects nor when using natural objects are human beings completely at liberty to do what they will. Relational liberty does not give us license to behave as if we were living in the moral equivalent of a frictionless surface—an atomized, nonrelational world of particles in motion.

5.1. Just Membership: Parity of Respect and Participation

The social philosopher Nancy Fraser argued for a notion of justice that centers on what she referred to as "participatory parity." This captures much of what I am trying to get at with my notion of membership as a component of relational liberty:

> Justice requires social arrangements that permit all (adult) members of society to interact with one another as peers. For participatory parity to be possible, I claim, at least two conditions must be satisfied. First the distribution of material resources must be such as to ensure participants' independence and "voice." . . . The second condition requires that

institutionalized patterns of cultural value express equal respect for all participants and ensure equal opportunity for achieving social esteem. (Fraser and Honneth 2003:36)

To be free is to be in certain kinds of relationships with others. The moral point of those relationships is the individual flourishing of each participant. Therefore, it follows that liberty requires parity of participation, engagement, and capacity for creative agency. The denial of parity in relational participation—disenfranchisement, exclusion, marginalization, oppression, exploitation, violence—is the denial of freedom. Relational liberty cannot exist within the context of unjust structures of "voice"—power that silences, wealth that dominates, institutions that deny social opportunity to some, and cultural meanings that efface individual self-esteem—any more than effective human economic activity can exist amid the degradation and breakdown of geophysical and bioenergetic systems. This provides a criterion for evaluating which types of relationships (transactions/interactions) are to be nurtured, facilitated, and promoted by common rules and public policy, and which are to be discouraged or prohibited.

Accustomed as we are to a negative and individualistic notion of liberty, this connection between community and liberty will seem naïve and dangerous. It is not. The reason lies in the difference among various sociological types of communal organization. Different implications for liberty exist in hegemonic and pluralistic communities—or in what the sociologist Robert Putnam (2000) referred to as "bonding" and "bridging" communities, or what the philosopher Seyla Benhabib (1992) calls "integrationist" and "participationist" orientations. The former types of communal formation impose on its members a convergence of patterns of belief and agency; the latter types are open-textured, dynamic, and fully open to cultural and personal differences. They allow community to arise from an interaction of difference and a diversity of patterns of belief and agency, as long as a second-order commitment to justice and solidarity is also present and shared by all members. This is not implausible. Cultural plurality or diversity and the expression of individual meaning are not inherently incompatible with solidarity and consensus or a shared sense of goods and purposes held in common (Walzer 2004).

Only the integrationist orientation need conflict with the concept of individual liberty inasmuch as it enjoins a closed circle of group membership and a tightly constituted self-identity. A participationist orientation is the type of community conducive to relational liberty structured by parity of membership and participation and by modes of solidarity amid plurality. (This will be discussed more fully in the next section.) Respect for difference bespeaks humility and an avoidance of the arrogance of certainty and control—a kind of moral technocracy that integrationist forms of community often espouse. Solidarity develops the moral imagination toward empathy and the embrace of individual lifeworlds that are different and yet symbiotic with one's own.

5.2. Just Mutuality: Relational Liberty and Solidarity

A student in the United States once wrote that "Abraham Lincoln became America's greatest Precedent [sic]. Lincoln's mother died in infancy, and he was born in a log cabin which he built with his own hands" (Lederer 1989:18). Solidarity is essential in order to prevent the story we tell ourselves about our own agency from becoming a fantasy of control and self-absorption—one in which we think that we can build, like Lincoln, our great precedent, the place of our own birth. Whenever it is evoked, mutuality counters that narrative of independence with one of interdependence. Mutuality recalls the structural context of individual agency and the functional integration that is necessary to that agency. It has to do with the social glue that gives stability to the creativity of action and agency. It has to do with the cultural and symbolic order that gives the originality of action and agency meaning. Also, it has to do with the historical memory and tradition that give continuity to innovative action and agency, thereby binding past, present, and future (Jennings 1981).

One of the most powerful facets of mutuality is the social stance and practice of solidarity. Given its history, its role in social movements, and its semantic resonance, solidarity is a concept that inherently leads us to view moral action in particular social, cultural, and institutional contexts (Benhabib 1987; Fraser 1986).

In considering solidarity and relational liberty, we should remember that both notions point toward the processes through which intentions are formed, possibilities are defined, and moral principles and ideals themselves are made meaningful. This occurs in the social perceptions and self-understandings of individuals precisely through the types and networks of social action in which they engage. Solidarity is a form of active engage-ment, not simply passive support. It is an intentional, engaged form of agency that is purposive and is motivated by a public commitment that exposes the agent to social visibility and potentially to risk and harmful consequences.

For our purposes here, I want to explicate the concept of solidarity in the following way. Solidarity's characteristic gesture and stance as a moral action is *standing up beside*. This stance then has three relational dimensions: *standing up for*, *standing up with*, and *standing up as*. These formulations stress the function of prepositions as markers for nuances of relationality. As in my preceding discussion of liberty as freedom from, freedom to, and freedom through, prepositions are central to the grammar of relational and positional connections, as well as to interpreting their meaning.

Standing up beside. Solidarity requires both taking a stand and standing up. When you stand up beside a person, a group, an organization, a spe-cies, a habitat, or even an idea or ideal, you make yourself visible. It is a public gesture—a communication in which saying and doing merge. This public posture also carries with it a sense of urgency and moral importance to both the agent being seen and to those who are looking. You stand up beside because you have something of importance to say, so that you can be seen and heard. The force of this comes from the fact that you are elevating your moral and social awareness and commitment: you are moving upward toward justice, such as the redress of the oppression or denigration of oth-ers or the protection of a watershed, a forest, or an endangered species. You are moving laterally from where you are (apart) to a place (beside) where you ought to be and are needed by others.

Standing up for. The first relational dimension of solidarity is standing up *for*. This suggests an intention to assist or to advocate for the "other" (oftentimes a stranger, and again not necessarily a human individual—one can stand up for other species, an ecosystem, or a cultural way of life). The other for whom you stand up in solidarity is someone whose situation presumably is vulnerable or represents a valid moral claim. One limitation of this mode of solidarity is that it falls short of establishing a relation-ship of intrinsic value to both parties. It is clearly possible to engage in a practice of solidarity as standing up for in an instrumental way, on the basis of enlightened self-interest. However, it is moral commitment, not strategic advantage, that lies at the motivational core of solidarity as such, including the solidarity of standing up for. At least as a first step, the act of standing up for begins to establish mutuality and reciprocity in a common struggle against injustice and in opposition to the abusive use of arbitrary and extractive power.

Standing up with. The second dimension is solidarity as standing up *with*. It takes another step in the direction of mutuality and recognition of shared moral standing. Moving from a mode of *relationality for* to *relationality with* in the practice of solidarity is meant to signal further entry into the lifeworld of the other. Doing so entails changes in your initial prejudgments and perspectives, and solidarity as standing with requires an openness to this possibility.

To put this point slightly differently, there is something in the imaginative dynamic of moving from *for* to *with* that is transformative of the solidarity relationship so that a (supportive) stranger-to-stranger relationship begins to develop—perhaps not all the way to a relationship that should be called friendship but at least to a stronger kind of fellowship and mutual recognition of one's self in the face of the other (*"mon semblable, mon frère,"* as Baudelaire put it). Relating to other people or groups in the specificity of their values and vocabularies of self-interpretation simultaneously develops respect for the specific standpoints of others (Dean 1995, 1996). Being with in this sense also reveals a level of commonality between the parties to this kind of solidarity which resides in the capacity for intercultural and transpersonal interpretive understanding (Forst 1992). The difference created by the specificity of lifeworlds resides within the commonality of the ability to understand lifeworlds other than one's own. Solidarity contains the possibility of being common readers of the diverse and distinct lives we each author.

Standing up as. The third dimension is solidarity as standing up *as*. Obviously, this suggests an even stronger degree of identification between the agents of solidaristic support and the recipients of such support. For agents engaged in the practice of solidarity who reach this mode of relationship, it is not a question of denying diversity or doing away with the continuing obligation to recognize and respect difference. Solidarity as *standing up as* operates at a higher level of ethical and legal abstraction, where the concepts of solidarity and justice converge (Calhoun 2002, 2005; Habermas 2005). The solidarity of standing up as involves finding a kind of covering connection that does not negate diversity at all; rather, it establishes the grounds of its respect, protection, and perpetuation. Onto-logically, it amounts to saying that I stand here both as a human person, dependent on the integrity of a complex human community, and as a human organism, a creature dependent on the integrity of a complex natu-ral ecosystem. Politically, the solidarity of standing up as is the solidarity of the civic and of citizenship, fostered by spaces of democratic delibera-tion and discourse, upon which the law and ethics of human rights and equal concern and respect depend for their vitality in the lifeworld of con-temporary society.

To move through these dimensions of solidarity is to enhance one's lib-erty in a relational sense. To move through the trajectory of solidarity is to move in the direction of greater imaginative creativity and range in the moral life. Standing up for depends on a kind of abstract moral commit-ment to support the application of general norms to the life situation of the other as a creature with a certain moral status. Standing up with involves adopting a perspective that is more internal to the lifeworld and the contex-tually meaningful agency of the other. Standing up as is the solidarity of a common ontology or, more fully expressed, the solidarity of being embod-ied, vulnerable, metabolic, and social organisms.

As the moral recognition of the other is altered by this interpretive jour-ney, so is the moral imagination of the self. Strong bonds of attachment and identification and empathy may not be the destination of this journey. However, arguably, a growth in one's capacity to project oneself imagina-tively into the perspective and viewpoint of the other person or creature, and a growth in moral awareness or the ability to see connections pre-viously unseen, are plausible outcomes of the interpretive transformation affected by the trajectory of solidarity. This is also the trajectory of liberty.

6. THICK FREEDOM, RICH LIVES

The individualism of liberal theories of freedom subtly reinforces the assumption that relationality, membership, mutuality, and solidarity are inherently confining and restrictive; that individuals require an expan-sive and unencumbered environment within which to grow and flourish, as if it is somehow more in keeping with the human good to have thin, instrumental, and strategic relationships only and to eschew thick, morally engaging ones (Bellah et al. 1985; Putnam 2000, Sandel 1998).

Today, the chasm is growing between the vision of ecological economics and the worldview of both laborers and consumers in the contemporary
West—and indeed throughout much of the developing world as well. The
everyday lifeworld of individuals is permeated by mass consumer culture.
The vocabularies that people use to define a self-identity and to comprehend
their situation are growing increasingly thin and impoverished from both an
ecological and a humanist point of view. People with a consumer's sense of
relationship and a tourist's sense of place cannot grasp that our well-being
depends on healthy natural and social systems; that we have responsibility
for preserving and restoring them; or that our freedom is threatened by those
very same institutions and practices that undermine the natural world.

Jonathan Lear's (2006) reflections on the history of the Crow people and
Alexis de Tocqueville's striking image of an old world dying and a new
world gradually being born are not to be ignored in our present situation.
We need not feel that our lifeworld of meaningful agency has disappeared
or that we are wandering in darkness, however, for there is much constructive intellectual (and political) work to do. The most urgent need, no doubt,
is to stem the tide of thermodynamic degradation to our natural world.
A second imperative task, nearly as urgent, is to stem the onslaught leading
to the degradation of our conceptual world, to prevent further conceptual
loss and the erosion of meaning. We have lived for some time now with a
conceptual vocabulary for describing our moral lives that is much sparser
and less articulate than our actual lives themselves. Despite expressing
ourselves for the most part individualistically, we nonetheless manage to
tap into an underlying moral resiliency and thereby preserve pockets—no,
actually rather expansive landscapes—of life lived relationally and caringly,
justly, and solidaristically (Bellah et al. 1985).

There are signs, however, that this fund of moral resiliency is becoming depleted, even as its natural counterpart of ecological resiliency is also
being stressed beyond its tolerances. On the surface of our lives, most especially in institutional spaces of impersonality and market mores, we behave
with instrumental rationality, self-referential interest, and the stratagems of
insatiable desire and extractive power. Can we reclaim a vital core beneath
this surface? Can we win through to the promise of a richer kind of freedom, meaning, and flourishing?

This promise is within our grasp. Beneath the surface of possessive
individualism lies a deep relationality that endures in a psychological

disposition and a symbolic cultural order of attachment, character, and care. What sociologist Robert Bellah called the "disposition to nurture" (2011:191–192)is the fundamental imperative of a creature whose evolutionary destiny is premature birth, neurological and cultural codetermination, and a prolonged period of dependency on relational others.[18] The public philosophy of the coming ecological political economy needs to tap into that deep motivational structure.

If it can do so, then the message of planetary boundaries and the end of the liberal era of cheap fossil carbon is not the bad news of lost liberty but the promise of a newfound freedom—a more humanly fulfilling kind of liberty. The ethics of relational liberty can justify and motivate the kinds of economic and social change needed nationally and globally in the next generation.[19] This is a recipe for rich lives in a socially and naturally thick and interdependent world. This sense has not been extinguished; this meaning is not lost. It remains the perennial possibility of humankind, even if thus far it has been fugitive and fleeting.

NOTES

1. For certain purposes, it is useful to differentiate the concepts of freedom and liberty. Freedom often has a more private or personal connotation, while liberty has a resonance that links it more specifically to the political realm. Moreover, the idea of freedom is often associated with the philosophical issues of determinism and freedom of the will in a metaphysical sense. Here, my focus is on moral and political liberty and the relationship between these normative ideals and the conditions of human agency and responsibility. If one is free or at liberty in this sense, one can reasonably be said to be the author of one's own actions and to bear responsibility for them. Liberty therefore has to do both with the will and motivation of an individual and with the social capacities and relationships available to an individual as they bear on the reasonable preconditions and ascription of moral responsibility for action. In this, I follow Pettit (2001). For a discussion of the differences between the two concepts, see Pitkin (1988).

2. The ecological economist Peter Victor formulated this perspective in the following terms: "One definition is that an economy is 'the system of production and distribution and consumption.' . . . A different conception of an economy . . . is as an 'open system' with biophysical dimensions. An open system is any complex arrangement that maintains itself through an inflow and outflow of energy and material from and to its environment. . . . Ecosystems are open systems. . . . Planet Earth is a closed system, or virtually so. A closed system exchanges energy with its environment but not material. . . . And here is the rub. Economies are open systems but they exist within and depend upon planet Earth which is a closed system. All of the materials used by economies come from the planet and end up as wastes disposed of back in the environment. . . .

As a result of [the physical laws of thermodynamics], open systems that depend upon their environment for material and energy must keep going back for more and must keep finding places to deposit their wastes. . . . Natural systems can be very effective in breaking down many of the wastes produced by people and machines, but often local environments are overloaded causing polluted land, water and air. . . . [T]he scale and complexity of environmental problems have increased. Now we are confronted by broad regional problems...and global problems" (Victor 2008:27–29, 32).

3. Domination is a state of such narrow and thoroughgoing closure in one's life that it negates the purposive agency of the subject, and, at an extreme, the underlying capability to exercise agency that could be deemed the person's own or to even conceive of oneself as a subject of agency. Domination must be distinguished from determination. Freedom from domination is compatible with the scientific theories and explanations that identify the determining conditions of human behavior, to which ecological economics appeals. Freedom from domination is freedom from arbitrary and contingent social and psychological determinants of behavior (and also of thought). This aspiration, which is in the modern world a moral imperative, does not require the abstract fantasy of an organically disembodied and socially disembedded individual.

4. For Arendt, action was the facet of the human condition associated with political life, the polis or public realm, while labor and work were at home in the household, the oikos or the private economic and reproductive realm. In my view, her dualism in this regard is problematic and need not be accepted in order to gain insight from her account of the three dimensions of humanness. Indeed, ecological economics suggests that labor, work, and action are dimensions of all realms of social life and that the differentiation between public and private—politics and economics, polis and oikos—is not the ethically desirable ideal that Arendt sometimes made it out to be (cf. Benhabib 1996:89–120).

5. It is worth emphasizing that the idea and ideal of negative liberty is inseparable from an atomistic conception of the unencumbered self and from a form of agency that involves the exercise of extractive power in the interactions between the self and others and between human beings and the natural world. This conception of liberty is deeply ingrained in the public philosophy of contemporary political culture, quintessentially in the United States, but increasingly throughout the global North as a whole.

My discussion of liberty could take place on the level of general social interpretation and criticism. It could be illustrated by numerous instances and indices drawn from popular culture and political discourse. Elsewhere, indeed, I have explored some of these sources (Gaylin and Jennings 2003). For the purposes of this chapter, however, it is most useful to examine particular thinkers who have constructed systematic and philosophically well-articulated conceptions of liberty. I begin with a brief orientation through a contrast between duty limited ideas of liberty understood as the manifestation of reason (Kant); and harm limited views of liberty understood as the manifestation of subjective self-assertion (Mill and liberal political economists). I then turn to a more detailed consideration of important twentieth century accounts of liberty in Isaiah Berlin and C. B. Macpherson. To be sure, it is the concepts and categories—the norms and ideals—of the broader culture, and not the works of particular philosophers per se, that concern me here. For it is political culture more broadly that influences the structure of economic and political behavior and that is what affects the planet. Still, individual thinkers can reveal ambiguities and tensions in fundamental ideas, such as liberty, that are glossed over and rendered invisible in more general modes of political and ethical discourse.

6. It should be noted that there has been a revival of interest in contractarian, deontological, and neo-Kantian approaches in moral philosophy and political theory owing to the work of Rawls (1971, 1993); Habermas (1996); and Scanlon (1998) among others. For a cogent historical and philosophical overview of the liberal tradition, see Plamenatz (1973).

7. In my view, this type of liberal value pluralism is a dead end formulation that ecological ethics must eschew. The goal is to internalize norms of right relationship among humans and for humans in nature within the concept of liberty itself, not as something to be balanced and traded off with liberty. The key to internalizing these norms within freedom itself without recourse to totalitarian elitism is to ground these norms of right relationship within the concepts of agency and self-realization of the human individual.

8. Berlin went on to flesh out at least one of the underlying conceptualizations of human agency and the self that travels with this idea of liberty: "Man differs from animals primarily neither as the possessor of reason, nor as an inventor of tools and methods, but as a being capable of choice, one who is most himself in choosing and not being chosen for; the rider and not the horse; the seeker of ends, and not merely of means, ends that he pursues, each in his own fashion: with the corollary that the more various these fashions, the richer the lives of men become; the larger the field of interplay between individuals, the greater the opportunities of the new and the unexpected; the more numerous the possibilities for altering his own character in some fresh or unexplored direction, the more paths open before each individual, and the wider will be his freedom of action and thought" (1969:178).

9. Berlin explained the concept this way: "The defence of liberty consists in the 'negative' goal of warding off interference. To threaten a man with persecution unless he submits to a life in which he exercises no choices of his goals; to block before him every door but one, no matter how noble the prospect upon which it opens, or how benevolent the motives of those who arrange this, is to sin against the truth that he is a man, a being with a life of his own to live" (1969:127).

10. Berlin explicated positive liberty this way: "The 'positive' sense of the word 'liberty' derives from the wish on the part of the individual to be his own master. I wish my life and decisions to depend on myself, not on external forces of whatever kind. I wish to be the instrument of my own, not of other men's, acts of will. I wish to be a subject, not an object; to be moved by reasons, by conscious purposes, which are my own, not by causes which affect me, as it were, from outside. I wish to be somebody, not nobody; a doer—deciding, not being decided for, self-directed and not acted upon by external nature or by other men as if I were a thing, or an animal, or a slave incapable of playing a human role, that is, of conceiving goals and policies of my own and realizing them" (1969:131).

11. Again I am using the term "work" here in Arendt's sense of creative transformation of natural matter into humanly useful or expressive form. Today, ordinary parlance does not make a distinction between "work" and "labor" as Arendt did, and she disapprovingly accused Marx of conflating the two ideas, which in her view is to equate all craftsmanship and artisanship with slavery. But for the purposes of discussion

here, registering this distinction is important for a somewhat different reason. I am not pursuing the Marxist argument that the capitalist system of wage labor is essentially a mode of slavery (the class domination of the proletariat) and that it necessarily entails the reduction of these human beings (subjects) to mindless bodies (objects). Instead, I am interested in the point that the highly inegalitarian and competitive forms that the neoliberal economic system takes reinforce and perpetuate the culture and behavior of *homo economicus*. This self and the norms characteristic of this type of political economy are blind to the integrity and limits of natural materials and to the value that they inherently possess. Such a political economy regularly and systematically fails to seek value from nature via accommodation and ecologically informed appropriation rather than via ultimately disruptive and nonresilient extraction.

12. Recall in *Utopia* Thomas More's account of the disdain shown by the Utopians toward gold and precious stones. Hythloday, the narrator, was amazed that such a radical transvaluation of values could occur (More [1516] 1975:50–52).

13. In this chapter, I am drawing on and developing a view that I have elsewhere explored in the context of related issues in the field of public health. Compare Jennings (2007, 2009, 2015). On this topic, generally see Honneth (2014).

14. My notion of relational liberty draws upon a neo-Aristotelian, civic republican tradition of political theory in a way that underscores its departure from liberalism. Historically, liberalism has been predominantly concerned to protect the self as an independent locus of private self-regarding (although not necessarily selfish) agency to a large extent. Civic republicanism has regarded human beings as "political animals" (*zōon politikon*). By this is meant that humans selves, far from being constrained or "de-natured" by political communities, must live in them if they are to flourish in accordance with their nature (Jennings 2011). Moreover, liberalism has defined politics in terms of an instrumental expedient designed to protect individuals from harm and to foster the progressive growth and just distribution of individual utility (variously and sometimes broadly defined). For its part, republicanism has defined politics in terms of creating a culture and social organization of individuals with a special kind of moral self-identity as citizens (*politēs, cives*), ruling themselves in common with equitable and just laws (*isonomia*), and seeking to achieve the human good together and the common good for all (*politia* or *res publica*). Citizenship in the republican tradition is active, not passive. It consists of ruling and being ruled in turn. Any notion of liberty or autonomy that is incompatible with being ruled, being limited and directed for the common good, is not in keeping with republican liberty. The negation of republican liberty is arbitrary power (Pettit 1997, 2012). For the individual citizen, this notion suggests a rhythm of alternating actions—autonomy and deference, doing and forbearing to do, innovation and accommodation, the assertion and denial of the will. Finding harmony and proportion in this rhythm of political and moral life was the work of maturity and judgment in individual human beings. For the polis as a whole, this conception of citizenship ran counter to Plato's political theory and critique of democracy. Shared and distributed power, mobile and not fixed within a segment of the people, is what citizenship in a politia or republic requires, not an exclusive concentration vested among the permanently wise. However, it also is not a distributed type of power like the liberal market that supposedly achieves an automatic harmony of interests via the mechanism of individual self-interested action and decisions. From the republican perspective, the

common good does not emerge automatically from the competitive work of many private minds and behind the back of individuals who never aim at it. The common good emerges intentionally from the cooperative work of the public minds of individual citizens and through the deliberative efforts of individuals who purposively aim at it.

15. The anthropologist Marshall Sahlins expressed this dialectical view of the interaction between the cultural context or scheme of meaning and the enactment of meaning in individual agency as follows: "History is culturally ordered, differently so in different societies, according to meaningful schemes of things. The converse is also true: cultural schemes are historically ordered, since to a greater or lesser extent the meanings are revalued as they are practically enacted. The synthesis of these contraries unfolds in the creative action of the historic subjects, the people concerned. For on the one hand, people organize their projects and give significance to their objects from the existing understandings of the cultural order. . . . On the other hand . . . as the contingent circumstances of action need not conform to the significance some group might assign them, people are known to creatively reconsider their conventional schemes" (1985:vii).

16. It is unfortunate that ecological economists, like virtually all economists, have either ignored this dialectical perspective on meaningful, intentional agency or have rejected it in favor of models of strategic action, rational gaming and choice that are actually not supported by historical and social scientific evidence. (cf. Green and Shapiro 1994).

17. In my view, citizenship does not have some external state of affairs called the "common good" as its instrumental objective. The common good is constituted by the proper institutionalization and functioning of citizenship and by the proper embedding of communal and ecological responsibility in the lifeworld. The common good is not a notion that sets up a test for particular policies or particular actions to meet (as does the parallel concept of "the public interest" in utilitarianism or liberal welfarism). It is not an outcome or an effect. However, the notion of the common good does provide a touchstone for judging and appraising a particular policy or decision. It appraises policy against criteria such as nondomination, nonarbitrariness, reasonable authority, mutual respect, reciprocity, and equity.

Contemporary utilitarianism tends to define interests or "utilities" abstractly across a population of individuals who have, as it were, only external or instrumental relationships to those interests. Utilitarianism also tends to ignore the distributional patterns in which these interests are fulfilled or their impact on discrete individual persons as such; it focuses instead on the net maximization of satisfaction or interest fulfillment in the aggregate (Robbins 1962; Walsh 1996). By contrast, the judgments that make up civic republican policy appraisal are judgments of fittingness, character, and appropriateness. They must take into consideration the conditions of power and meaning that constitute the identity and interests of each person as a unique individual. They are at the political boundary between moral and aesthetic judgment. As such, they cannot be the *only* means of policy appraisal, in economic policy, regulation, or law. However, neither should they be left out altogether (Günther 1993; Nussbaum 1995).

Shared purposes or problems are not the same as individual purposes or problems that happen to overlap for large numbers of people. Of course, they do affect persons as individuals and as members of smaller groups, but they also affect the constitution of a "people," a population of individuals as a structured social whole. An aggregation of individuals becomes a people, a public, a political community when it is capable of recognizing common purposes and problems in this way; when it achieves a certain kind of political imagination (Anderson 1991).

18. For an important discussion and review of the literature on this from an evolutionary theory perspective, see Bellah (2011:60–97, 175–182, 191–192). It should be said that in addition to the disposition to nurture, Bellah also found a second aspect in the deep evolutionary history of our species, which he called the "disposition to dominance"—namely, hierarchical social ordering with a resulting competitive aspect of purposive agency. However, the manifestation of this competitive ordering of status and position within a social hierarchy does not necessarily have to result in what I (following Macpherson following Marx) have called extractive power. In other words, the drive toward hierarchical differentiation in our deep evolutionary past *does not* provide an argument for the necessity of, let alone the justification for, capitalism. The alternative pathway for deep hierarchy to express itself in cultural and social order is provided by the capacity for *play*.

In Bellah's work and in the work of Burghardt (2006), play has many complex facets; suffice to say that in essence it is a relational, not an extractive, form of agency and activity. Most significant for our purposes here is the argument that play agency emerges under environmental and evolutionary conditions of a "relaxed field," meaning a state of relaxed competitive pressures from the point of view of natural selection. This suggests a certain kind of economy that meets the criteria of householding. It is a sustained interaction between human groups and their surrounding ecosystems that does not undercut the materials and services offered by those ecosystems. The relaxed conditions of householding in past cultures, and of ecological political economy in the future, make play possible. Play, in turn, satisfies the human propensity toward hierarchical differentiation (which sits alongside the propensity to nurture and to give care) in a way that staves off domination, the institutionalization of extractive power transactions, and the rest. Play is a kind of competition that sublimates domination into relationality.

Here is the kind of liberty I seek: a relational liberty that avoids domination and expresses the deep impulses of care and play. Such freedom is symbiotic with a social and natural condition of relaxation of competitive selective and hierarchical pressure. If the achievement in theory and practice—thought, word, and deed—of relational liberty has ecological preconditions, so too can it have ecological effects.

We are now on course to undermine planetary boundaries in such a way as to exacerbate circumstances antithetical to relational liberty—namely, an increasingly unveiled Hobbesian situation condition of *sauve qui peut* (every man for himself). We are doing this at least in part because we have taught ourselves to believe that the institutional (and personal) expression of extractive power is a condition of freedom. Yet, this is precisely what will bring about unfreedom, as that has been known in societies of domination in history and from an evolutionary point of view.

19. Foremost among these social changes are (1) institutional transformations in the economy and polity to achieve equitable access to power and social participation, (2) community renewal and engagement, (3) a broader base of ecological literacy, (4) changes in the direction of less hierarchy and relative social inequality that will reduce chronic stress and enhance individual dignity and self-esteem, (5) a renewed moral imagination and sensibility for the human responsibilities of trusteeship for a healthy

and resilient natural world, and (6) a renewed political imagination—"democratic will formation," to borrow a term from Habermas (1996:295–302)—to create the governance and public policies that can bring these changes about.

REFERENCES

Anderson, Benedict. 1991. *Imagined Communities: Reflections on the Origin and Spread of Nationalism*. 2nd ed. London: Verso.

Arendt, Hannah. 1958. *The Human Condition*. Chicago: University of Chicago Press.

Bellah, Robert N. 2011. *Religion in Human Evolution: From the Paleolithic to the Axial Age*. Cambridge, MA: Harvard University Press.

Bellah, Robert N., Richard Madsen, William M. Sullivan, Ann Swidler, and Steven M. Tipton. 1985. *Habits of the Heart: Individualism and Commitment in American Life*. Berkeley: University of California Press.

Benhabib, Seyla. 1987. "The Generalized and the Concrete Other." In *Feminism as Critique: On the Politics of Gender*, edited by Seyla Benhabib and Drucilla Cornell, 77–95. Minneapolis: University of Minnesota Press.

Benhabib, Seyla. 1992. *Situating the Self: Gender, Community, and Postmodernism in Contemporary Ethics*. New York: Routledge.

Benhabib, Seyla. 1996. *Democracy and Difference: Contesting the Boundaries of the Political*. Princeton, NJ: Princeton University Press.

Berlin, Isaiah. 1969. *Four Essays on Liberty*. New York: Oxford University Press.

Brown, Peter G. 1994. *Restoring the Public Trust: A Fresh Vision for Progressive Government in America*. Boston: Beacon Press.

Brown, Peter G. 2001. *The Commonwealth of Life: A Treatise on Stewardship Economics*. Montréal: Black Rose Books.

Brown, Peter G., and Geoffrey Garver. 2009. *Right Relationship: Building a Whole Earth Economy*. San Francisco: Berrett-Koehler Publishers.

Burghardt, Gordon M. 2006. *The Genesis of Animal Play: Testing the Limits*. Cambridge, MA: MIT Press.

Calhoun, Craig. 2002. "Imagining Solidarity: Cosmopolitanism, Constitutional Patriotism, and the Public Sphere." *Public Culture* 14 (1): 147–171. doi: 10.1215/08992363-14-1-147.

Calhoun, Craig. 2005. "Constitutional Patriotism and the Public Sphere: Interests, Identity, and Solidarity in the Integration of Europe." *International Journal of Politics, Culture, and Society* 18 (3–4): 257–280. doi: 10.1007/s10767-006-9002-0.

Carens, Joseph H. 1993. *Democracy and Possessive Individualism: The Intellectual Legacy of C.B. Macpherson*. Albany: State University of New York Press.

Costanza, Robert, ed. 1991. *Ecological Economics: The Science and Management of Sustainability*. New York: Columbia University Press.

Daly, Herman E., and John B. Cobb. 1994. *For the Common Good: Redirecting the Economy toward Community, the Environment, and a Sustainable Future*. 2nd ed. Boston: Beacon Press.

Daly, Herman E., and Joshua C. Farley. 2011. *Ecological Economics: Principles and Applications*. 2nd ed. Washington, DC: Island Press.

Dean, Jodi. 1995. "Reflective Solidarity." *Constellations* 2 (1): 114–140. doi: 10.1111 /j.1467–8675.1995.tb00023.x.

Dean, Jodi. 1996. *Solidarity of Strangers: Feminism after Identity Politics*. Berkeley: University of California Press.

Diamond, Cora. 1988. "Losing Your Concepts," *Ethics* 98 (1): 255-277.

Duménil, Gérard, and Dominique Lévy. 2011. *The Crisis of Neoliberalism*. Cambridge, MA: Harvard University Press.

Forst, Rainer. 1992. "How (Not) to Speak About Identity: The Concept of the Person in a Theory of Justice." *Philosophy & Social Criticism* 18 (3–4): 293–312. doi: 10.1177/019145379201800305.

Fraser, Nancy. 1986. "Toward a Discourse Ethic of Solidarity." *Praxis International* 5 (4): 425–429.

Fraser, Nancy, and Axel Honneth. 2003. *Redistribution or Recognition? A Political-Philosophical Exchange*. London: Verso.

Gaylin, Willard, and Bruce Jennings. 2003. *The Perversion of Autonomy: Coercion and Constraints in a Liberal Society*. 2nd ed. Washington, DC: Georgetown University Press.

Giddens, Anthony. 1984. *The Constitution of Society: Outline of the Theory of Structuration*. Berkeley: University of California Press.

Gray, John. 1996. *Isaiah Berlin*. Princeton, NJ: Princeton University Press.

Gray, John. 2000. *Two Faces of Liberalism*. New York: The New Press.

Green, Donald P., and Ian Shapiro. 1994. *Pathologies of Rational Choice Theory: A Critique of Applications in Political Science*. New Haven, CT: Yale University Press.

Günther, Klaus. 1993. *The Sense of Appropriateness: Application Discourses in Morality and Law*. Albany: State University of New York Press.

Habermas, Jürgen. 1984. *The Theory of Communicative Action*. Translated by Thomas McCarthy. 2 vols. Boston: Beacon Press.

Habermas, Jürgen. 1996. *Between Facts and Norms: Contributions to a Discourse Theory of Law and Democracy*. Cambridge, MA: MIT Press.

Habermas, Jürgen. 2003. *The Future of Human Nature*. Cambridge, UK: Polity.

Habermas, Jürgen. 2005. "Equal Treatment of Cultures and the Limits of Postmodern Liberalism." *Journal of Political Philosophy* 13 (1): 1–28. doi: 10.1111/j .1467–9760.2005.00211.x.

Halévy, Elie. 1966. *The Growth of Philosophical Radicalism*. Translated by Mary Morris. Boston: Beacon Press.

Hansen, James, Pushker Kharecha, Makiko Sato, Valerie Masson-Delmotte, Frank Ackerman, David J. Beerling, Paul J. Hearty, Ove Hoegh-Guldberg, Shi-Ling Hsu, Camille Parmesan, Johan Rockstrom, Eelco J. Rohling, Jeffrey Sachs, Pete Smith, Konrad Steffen, Lise Van Susteren, Karina von Schuckmann, and James C. Zachos. 2013. "Assessing Dangerous Climate Change: Required Reduction of Carbon Emissions to Protect Young People, Future Generations and Nature." *PLoS ONE* 8 (12): e81648. doi: 10.1371/journal.pone.0081648.

Harré, Rom. 1998. *The Singular Self: An Introduction to the Psychology of Personhood*. London: Sage Publications.

Harvey, David. 2005. *A Brief History of Neoliberalism*. Oxford: Oxford University Press.

Harvey, David. 2010. *A Companion to Marx's Capital*. London: Verso.

Hirschman, Albert O. 1977. *The Passions and the Interests: Political Arguments for Capitalism before Its Triumph*. Princeton, NJ: Princeton University Press.

Honneth, Axel. 1996. *The Struggle for Recognition: The Moral Grammar of Social Conflicts*. Translated by Joel Anderson. Cambridge, MA: MIT Press.

Honneth, Axel. 2014. *Freedom's Right: The Social Foundations of Democratic Life*. Translated by Joseph Ganahl. Chichester, England: Polity.

Jennings, Bruce. 1981. "Tradition and the Politics of Remembering," *The Georgia Review*, 36 (1): 167–182.

Jennings, Bruce. 2007. "Public Health and Civic Republicanism." In *Ethics, Prevention, and Public Health*, edited by Angus Dawson and Marcel F. Verweij, 30–58. Oxford: Oxford University Press.

Jennings, Bruce. 2009. "Public Health and Liberty: Beyond the Millian Paradigm." *Public Health Ethics* 2 (2): 123–134. doi: 10.1093/phe/php009.

Jennings, Bruce. 2010a. "Beyond the Social Contract of Consumption: Democratic Governance in the Post-Carbon Era." *Critical Policy Studies* 4 (3): 222–233. doi: 10.1080/19460171.2010.508919.

Jennings, Bruce. 2010b. "Toward an Ecological Political Economy: Accommodating Nature in a New Discourse of Public Philosophy and Policy Analysis." *Critical Policy Studies* 4 (1): 77–85. doi: 10.1080/19460171003715028.

Jennings, Bruce. 2011. "Nature as Absence: The Natural, the Cultural, and the Human in Social Contract Theory." In *The Ideal of Nature: Debates about Biotechnology and the Environment*, edited by Gregory E. Kaebnick, 29–48. Baltimore: Johns Hopkins University Press.

Jennings, Bruce. 2015. "Relational Liberty Revisited: Membership, Solidarity, and a Public Health Ethics of Place," *Public Health Ethics* 8 (1). doi: 10.1093/phe/phu045

Jennings, Bruce, and K. Prewitt. 1985. "The Humanities and the Social Sciences: Reconstructing a Public Philosophy." In *Applying the Humanities*, edited by Daniel Callahan, Arthur L. Caplan, and Bruce Jennings, 125–144. New York: Plenum Press.

Jonas, Hans. 1985. *The Imperative of Responsibility: In Search of an Ethics for the Technological Age*. Chicago: University of Chicago Press.

Jones, Daniel Stedman. 2012. *Masters of the Universe: Hayek, Friedman, and the Birth of Neoliberal Politics*. Princeton, NJ: Princeton University Press.

Klein, Naomi. 2011. "Capitalism vs. the Climate." *The Nation*, November 28. Accessed January 24, 2015. http://www.thenation.com/article/164497/capitalism-vs-climate.

Klein, Naomi. 2014. *This Changes Everything: Capitalism vs. the Climate*. New York: Simon & Schuster.

Lanchester, John. 2012. "Marx at 193." *London Review of Books* 34 (7): 7–10. http://www.lrb.co.uk/v34/n07/john-lanchester/marx-at-193.

Lear, Jonathan. 2006. *Radical Hope: Ethics in the Face of Cultural Devastation*. Cambridge, MA: Harvard University Press.

Lederer, Richard. 1989. *Anguished English: An Anthology of Accidental Assaults Upon Our Language*. New York: Dell.

Lee, Keekok. 2005. *Philosophy and Revolutions in Genetics: Deep Science and Deep Technology*. London: Palgrave Macmillan.

Leopold, Aldo. 1989. *A Sand County Almanac and Sketches Here and There*. Special Commemorative Edition. New York: Oxford University Press.

Lovejoy, Arthur O. 1961. *Reflections on Human Nature*. Baltimore: Johns Hopkins Press.

Macpherson, C. B. 1962. *The Political Theory of Possessive Individualism: Hobbes to Locke*. Oxford: Clarendon Press.

Macpherson, C. B. 1973. *Democratic Theory: Essays in Retrieval*. Oxford: Clarendon Press.

Macpherson, C. B. 1977. *The Life and Times of Liberal Democracy*. New York: Oxford University Press.

Marglin, Stephen A. 2010. *The Dismal Science: How Thinking Like an Economist Undermines Community*. Cambridge, MA: Harvard University Press.

Marx, Karl. (1867) 1990. *Capital: A Critique of Political Economy*. Vol. 1. Translated by Ben Fowkes. London: Penguin.

Mill, John Stuart. (1859) 1956. *On Liberty*. Indianapolis, IN: Bobbs-Merrill.

More, Thomas. (1516) 1975. *Utopia: A New Translation, Backgrounds, Criticism*. Edited and translated by Robert Martin Adams. New York: W.W. Norton.

Mulhall, Stephen, and Adam Swift. 1996. *Liberals and Communitarians*. 2nd ed. Oxford: Blackwell.

Nelson, Benjamin. 1969. *The Idea of Usury: From Tribal Brotherhood to Universal Otherhood*. 2nd ed. Princeton, NJ: Princeton University Press.

Nussbaum, Martha Craven. 1995. *Poetic Justice: The Literary Imagination and Public Life*. Boston: Beacon Press.

Ophuls, William. 1998. *Requiem for Modern Politics: The Tragedy of the Enlightenment and the Challenge of the New Millennium*. Boulder, CO: Westview Press.

Peck, Jamie. 2010. *Constructions of Neoliberal Reason*. Oxford, UK: Oxford University Press.

Pettit, Philip. 1997. *Republicanism: A Theory of Freedom and Government*. New York: Oxford University Press.

Pettit, Philip. 2001. *A Theory of Freedom: From the Psychology to the Politics of Agency*. New York: Oxford University Press.

Pettit, Philip. 2012. *On the People's Terms: A Republican Theory and Model of Democracy*. Cambridge, UK: Cambridge University Press.

Pitkin, Hanna Fenichel. 1988. "Are Freedom and Liberty Twins?" *Political Theory* 16 (4): 523–552. doi: 10.2307/191431.

Plamenatz, J. 1973. Liberalism. In *Dictionary of the History of Ideas: Studies of Selected Pivotal Ideas*, edited by Philip P. Wiener. New York: Charles Scribner's.

Polanyi, Karl. 1944. *The Great Transformation*. Boston: Beacon Press.

Putnam, Robert D. 2000. *Bowling Alone: The Collapse and Revival of American Community*. New York: Simon & Schuster.

Rawls, John. 1971. *A Theory of Justice*. Cambridge, MA: Harvard University Press.

Rawls, John. 1993. *Political Liberalism*. New York: Columbia University Press.

Robbins, Lionel. 1962. *An Essay on the Nature & Significance of Economic Science*. 2nd ed. London: Macmillan.

Rockström, Johan, Will Steffen, Kevin Noone, Asa Persson, F. Stuart Chapin, III, Eric F. Lambin, Timothy M. Lenton, Marten Scheffer, Carl Folke, Hans Joachim Schellnhuber, Bjorn Nykvist, Cynthia A. de Wit, Terry Hughes, Sander van der Leeuw, Henning Rodhe, Sverker Sorlin, Peter K. Snyder, Robert Costanza, Uno Svedin, Malin Falkenmark, Louise Karlberg, Robert W. Corell, Victoria J. Fabry, James Hansen,

Brian Walker, Diana Liverman, Katherine Richardson, Paul Crutzen, and Jonathan A. Foley. 2009. "Planetary Boundaries: Exploring the Safe Operating Space for Humanity." *Ecology and Society* 14 (2): 32. http://www.ecologyandsociety.org/vol14/iss2/art32/.

Rodgers, Daniel T. 2011. *Age of Fracture*. Cambridge, MA: Harvard University Press.

Rose, Nikolas S. 2006. *Politics of Life Itself: Biomedicine, Power, and Subjectivity in the Twenty-First Century*. Princeton, NJ: Princeton University Press.

Sahlins, Marshall. 1985. *Islands of History*. Chicago: University of Chicago Press.

Sandel, Michael J. 1998. *Liberalism and the Limits of Justice*. 2nd ed. Cambridge, UK: Cambridge University Press.

Sandel, Michael J. 2006. *Public Philosophy: Essays on Morality in Politics*. Cambridge, MA: Harvard University Press.

Scanlon, Thomas. 1998. *What We Owe to Each Other*. Cambridge, MA: Harvard University Press.

Schlesinger, William H. 2009. "Planetary Boundaries: Thresholds Risk Prolonged Degradation." *Nature Reports Climate Change* 3: 112–113. doi:10.1038/climate.2009.93.

Schor, Juliet. 2010. *Plenitude: The New Economics of True Wealth*. New York: Penguin.

Schumpeter, Joseph A. 1950. *Capitalism, Socialism, and Democracy*. 3rd ed. New York: Harper & Row.

Sennett, Richard. 2009. *The Craftsman*. New Haven, CT: Yale University Press.

Sullivan, William M. 1986. *Reconstructing Public Philosophy*. Berkeley: University of California Press.

Taylor, Charles. 1985a. *Philosophical Papers Vol. 1: Human Agency and Language*. Cambridge, UK: Cambridge University Press.

Taylor, Charles. 1985b. *Philosophical Papers Vol. 2: Philosophy and the Human Sciences*. Cambridge, UK: Cambridge University Press.

Taylor, Charles. 1989. *Sources of the Self: The Making of the Modern Identity*. Cambridge, MA: Harvard University Press.

Taylor, Charles. 1991. *The Ethics of Authenticity*. Cambridge, MA: Harvard University Press.

Tocqueville, Alexis de. (1840) 1969. *Democracy in America*. Edited by J. P. Mayer. Translated by George Lawrence. 2 vols. New York: Anchor Books.

Tully, James. 2008–2009. *Public Philosophy in a New Key*. 2 vols. Cambridge, UK: Cambridge University Press.

Victor, Peter A. 2008. *Managing Without Growth: Slower by Design, Not Disaster*. Cheltenham, UK: Edward Elgar.

Walsh, Vivian Charles. 1996. *Rationality, Allocation, and Reproduction*. Oxford: Clarendon Press.

Walzer, Michael. 2004. *Politics and Passion: Toward a More Egalitarian Liberalism*. New Haven, CT: Yale University.

Wolff, Robert Paul. 1970. *In Defence of Anarchism*. New York: Harper & Row.

I believe that . . . strong theories of fatality are abstract and wrong. Our degrees of freedom are not zero. There is a point to deliberating what ought to be our ends, and whether instrumental reason ought to have a lesser role in our lives than it does. But the truth in these analyses is that it is not just a matter of changing the outlook of individuals, it is not just a battle of 'hearts and minds,' important as this is. Change in this domain will have to be institutional as well.

—Charles Taylor, *The Malaise of Modernity* (1991), p.8

CHAPTER ELEVEN

A New Ethos, a New Discourse, a New Economy

Change Dynamics Toward an Ecological Political Economy

JANICE HARVEY

1. INTRODUCTION: FROM ETHICS TO ACTION

The authors of this book, although coming from diverse disciplines and perspectives, are united in their concern that conventional growth economics cannot survive planetary limits, nor can ecosystems survive relentless growth in economic production and consumption. Although change is inevitable, the direction and outcome of that change is not. Rather than the alternative of some form of economic and social collapse, the preferred direction is the replacement of the current growth economy with an ecological political economy. But how might this transition occur? Only recently has this question attracted much attention. While we still need to articulate and model the elements of an ecological economy, we must also begin to imagine the sociopolitical processes that might move us toward that goal.

It is one thing to argue, as this book does, that an ecological economy requires a new ethos of based on the principles of membership, house-holding, and entropic thrift (chapter 2), and thus a new ecological political economy (chapter 10). It is quite another to see how such an ethical shift might occur, and further, how it would contribute to the dynamics of system change. On the one hand, ecological economics as a field of knowledge should embed an ethics of right relationship and in this way facilitate its dispersion throughout society. On the other, until a new ethos becomes widely grounded, the prospects for an ecological economics to exert much sway beyond its natural constituency are limited. It is the proverbial chicken-and-egg dilemma. As Jennings asked in chapter 10 of this volume, "Where is our theoretical point of entry into the circle of social change to be found?"

There are insights that can be brought to bear on the complex project of ethical shifting and economic structural change. Given our rather discouraging history of environmental politics and the overwhelming influence of powerful interests vested in the status quo (Speth 2008), not to mention the privileged status of neoclassical economic dogma in our major institutions, it is naive to expect—and folly to wait for—political leadership in this crucial social project. Politics are mercurial, influenced by planned strategies, unplanned reactions, accidents, the weather, and of course, money. Gains and losses are ephemeral; governments and policies come and go at the rate of election cycles. Transforming the now-global growth-dependent political economy is a long-term prospect ill-suited to short-term political expediency and tactical maneuvering.

Nevertheless, political theory provides an essential insight around which to organize our thinking. Although systems tend toward self-perpetuation and are therefore inherently stable, they are dynamic and subject to stresses, both internal and external. Because external stresses are beyond the reach of conscious human agency, we will not deal with them here. Internal stresses, on the other hand, can create crises of legitimation at which point systemic change becomes possible (Bourdieu 1993; Habermas 1976). In this view, the institutions of the growth economy could be transformed into systems designed to function effectively without growth should the public broadly come to recognize the illegitimacy of economic growth as a public policy priority.

In this chapter, I draw on two seemingly distinct theoretical frameworks to propose how we might think about—and act upon—this proposition. From a critical perspective, discourse theory holds that the process of institutional-cultural change is discursive and dialectical (Fairclough, Mulderrig, and Wodak 2011; van Dijk 2011). World systems theory, on the other hand, reminds us that such social constructions occur within a unique historical context that either constrains or propels but ultimately shapes systemic change (Wallerstein 1999, 2004). The publication of this book is itself a discursive event. While intended to help ecological economics on its journey, it aspires to change our relationship with life and the world by grounding our self-understanding in an evolutionary worldview. In this task, it joins an ever-expanding corpus of ecological discourse with similar aspirations. The objective of this chapter is to come to terms with the change process of which the production and distribution of this book is a part.

2. ETHICS AND CULTURE

> We would do well not to underestimate the task facing ecological political economy. It is both an ontological reorientation and an ethical innovation. It goes beyond the physical and life sciences in a descriptive sense and implicates the normative foundations of social order and human agency. All structures of organized human activity must give a sense of meaning to the purposive, self-conscious human agents who comprise them; thus, a new ecological political economy will need a new foundational story and a new conceptual framework of norms and ideals.
> —Bruce Jennings, chapter 10, this volume

Broadly speaking, ethics is one of the constitutive elements of culture, along with the ideas, beliefs, values, norms, moral obligations, truth claims, and ways of knowing that situate us in and give meaning to our experience of the world. This "culture of the mind" (Wuthnow 1989) is prereflexive and subconscious—a way of being as well as a way of acting (Bourdieu 1993). Yet, it is manifest and made visible in real behavior, practices, and products and "exists within the dynamic contexts in which it is produced and disseminated" (Wuthnow 1989:16). It provides the lenses and filters through which we interpret the myriad stimuli of daily life, indicates what is and is not valuable, sets the rules and obligations for everyday living within communities, and locates us in the greater scheme of things relative to everything else (Milton 1996).

A change in ethics implies a cultural shift, but this is not a matter of will. Culture is contextual and historical, a mix of inherited tradition and new understandings created through social interaction (Hunter 2010).

As a hegemonic framework of meaning, culture is manifest in the institutions of society; thus, a change process must engage the nature, workings, and power of institutions (Wuthnow 1989). The market, the state, education systems, religions, the media, scientific and technological research, and the family are all carriers of culture while safeguarding their own logic, place, and history. Institutions embody particular forms of knowledge, expertise, credentials, and authority that are imbued with cultural meaning. This "symbolic capital" confers power, the ultimate expression of which is the power of legitimate naming, which in turn defines reality itself (Bourdieu 1989:20–23).

Hunter (2010) conceptualized the power dynamics of "naming" or meaning production by locating social positions, networks, and institutions along a center-periphery spectrum according to their endowment of "symbolic capital." Those most critically involved in the production of cultural "meaning"—what is seen as important or "true" and, conversely, what is seen as marginal and of little value—occupy or comprise a cultural center where prestige is highest; those least influential in the meaning production process are found on the periphery where status is low. This dynamic is mutually reinforcing. Powerful actors reproduce their own status through networks. The more "dense" and interactive the network, the more influential it is in the reproduction of meaning (Hunter 2010). The reverse is also the case. Culture changes when challenges to orthodox meaning production arise from outside the center and become embedded within dense networks of supportive elites and institutions (Hunter 2010; Wuthnow 1989). So while there is great inertia built into culture, it is an open system subject to change. Ethics have changed dramatically since the advent of industrial society, with attendant changes in practices and institutions, and they continue to change. But how much of this can be attributed to human agency? Is the field of cultural change amenable to active or strategic intervention?

3. THE RELATIONSHIP BETWEEN CULTURAL CHANGE AND DISCOURSE

As this theory increasingly pervades the social world, and more and more people become convinced by it—and in some cases benefit mightily from it; thus, people's behavior, and the social and cultural worlds that support it, alters accordingly. They proceed to see the world through the lens of that theory, and so change the world accordingly—as much as they are able.

—Peter Timmerman, chapter 1, this volume

An ecological political economy will come about only through change at both the level of individual behavior and the level of social norms and institutions. In practice, this means that we must learn to articulate the values and ideals that the members of these societies would express if they thought and acted like interdependent and relational selves— ecological selves, or ecological citizens and trustees. . . . For this task, ecological public philosophy needs the vocabulary of solidarity, mutuality, reciprocity, community, and the common good—norms that are certainly contained, or at least alluded to, in the principles of membership, householding, and entropic thrift.

—Bruce Jennings, chapter 10, this volume

The possibility and role of human agency becomes clearer when we consider that cultural meaning and reference points are produced, reproduced, and transformed through discourse. Even though these processes are interrelated, for analytical purposes we can distinguish between two layers or dimensions of discourse effect. The first is the intersection of the personal and the cultural. We come to understand the world around us and how we fit into it through complex combinations of individual perception and social interactions, interpretation and argument, transmitted through and by various forms of communication, including the telling of cultural narratives and myths (Hajer 1995; Milton 1996). These are the raw materials from which we construct, reproduce, and adapt, from one generation to the next, "basic frameworks of implicit meaning"—"a culture's deep structure"—through which we understand our world (Hunter and Wolfe 2006:91). Communicated symbolically through discourse over time, these subconscious frames or "schemata of interpretation" contain deeply rooted assumptions that become "second nature, well entrenched, and built into the way of doing things," and provide a broad account of social reality. Because deep cultural frames are implicit and widely shared across a culture, they are persistent over time; the fact that frames are resistant to change suggests that they are satisfying some important need (Reese 2003:14–18).

Not only do implicit frames provide the filters through which people receive and interpret information about the world around them, they also shape and legitimize societal institutions that arise within cultures.[1] For example, the dominant cultural frame that has humanity separated from and master of nature has served Western imperial/industrial expansion very well. As one of the core assumptions on which liberal institutions were built, this frame persists despite growing evidence of its destructive pathologies (McKibben 1989; Merchant 1980). An ecological economic discourse relies on a different cultural frame—a different "way of knowing"—that sees humans as members of the Earth community rather than masters of it (Leopold 1966). The transformation of growth-dependent political-economic institutions implies the transformation of the underlying ethical-cultural frame. At the most basic level, then, a change in ethics implies and requires modification or transformation of the dominant cultural frames that now underpin the pathological relationship between modern society and its planetary host.

This points us to the second dimension of discourse effect: the active "framing" of social reality through discourse. This process is at the heart of the call for a new cultural narrative. Because there are "different ways of knowing," discourses "always operate in relation to power—they are part of the way power circulates and is contested." According to Hall, the question of whether a discourse is true or false is less important than whether it is effective in practice—"in other worlds, whether it succeeds in organizing and regulating relations of power" (Hall 1996:201). In critical discourse studies, discourse is understood as a social practice in which power relations between and among social actors are expressed. In most instances, these relations operate at a subconscious level:

> Discourse is socially constitutive as well as socially shaped. . . . It is constitutive both in the sense that it helps to sustain and reproduce the social status quo, and in the sense that it contribute to transforming it. Since discourse is so socially influential, it gives rise to important issues of power. In a dialectical understanding, a particular configuration of the social world . . . is implicated in a particular linguistic conceptualization of the world; in language we do not simply name things but conceptualize things. Thus discursive practices may have major ideological effects . . . discourse may try to pass off assumptions (often falsifying

ones) about any aspect of social life as mere common sense. Both the ideological loading of particular ways of using language and the relations of power which underlie them are often unclear to people.
(Fairclough, Mulderrig, and Wodak 2011:358)

This locates discourse in the political as well as the cultural sphere, marrying the two. There are, in every society, dominant or hegemonic discourses that reflect an orthodox worldview—one that both shapes and is perpetuated by the major institutions of society, including the state and its security apparatuses, the judiciary, educational institutions, and the media (Bourdieu 1989). Discourses are hegemonic not only because they reside in elite institutions, however, but because they resonate with subconscious cultural frames. George Lakoff highlighted the special potency of cultural-moral frames (also constructs of discourse). Discourses that do not resonate with these implicit frames are rejected, often in the face of overwhelming empirical evidence of validity (Lakoff 2004:xv).[2] It is when this discursive process of cultural articulation fails that the conditions for cultural change arise.

While central institutions and their dominant groups propagate hegemonic cultural discourses that reproduce cultural frames, they are not impenetrable. Because discourses are sites of power, there are always "subaltern" or subordinated interests around which resistance or counterhegemonic discourses arise. Sociologist Pierre Bourdieu (1989) conceptualized the relationship between dominant and challenger discourses as a game played on a discursive "field" where discourses compete for legitimacy and authority. The extent to which a discourse succeeds or fails in this competition is a function of its symbolic capital, whether economic, social, or cultural. In economic discourses, financial capital and economic-academic credentials convey authority. In other discourses, other forms of capital may be more important. On environmental issues, for example, environmental organizations are routinely cited by the public as the source they most trust, while the private sector scores badly. This is a function of moral authority vested in the environmental movement compared to the profit motive of business. In a competition between economic and environmental discourses, however, the subconscious frame of economic growth as necessary to societal progress invests proponents of growth with far greater symbolic capital than those arguing for ecological protection. That said, these cultural frames themselves are constructed and reproduced through discourse and thus are subject to change. When the discursive logic and practice of central institutions fall out of step with the norms and expectations of the culture, or when the contradictions within the dominant discourse can no longer be contained, the resulting tension creates a leverage point for change. Through the continual presence of challenger or resistance discourses, which introduce new frames of meaning that better explain the world as it is actually experienced, dominant discourses can be delegitimized and their associated cultural frames can be transformed. Once widely disseminated, new discourses result in new ways of being and doing, resulting in social and institutional change (Wuthnow 1989).[3]

4. THE HEGEMONIC DISCOURSES OF NEOLIBERALISM AND CONSUMERISM

That we should have caused such damage to the entire functioning of planet Earth in all its major biosystems is obviously the consequence of a deep cultural pathology. Just as clearly, there is need for a deep cultural therapy if we are to proceed into the future with some assurance that we will not continue in this pathology or lapse into the same pathology at a later date.

—Thomas Berry, *Evening Thoughts*

Discourse is historical and contextual. Its meaning can only be derived through an analysis of the "times" within which it is produced (Hajer 1995). Arguably the most powerful barrier to transitioning to an ecological economics is the neoliberal discourse of late capitalism, which has permeated Western democracies and spread virally throughout the world via economic globalization. The cultural-moral system implicit in freeing the market is a two-hundred-year cultural legacy that has become particularly potent.[4] The neoliberal discourse holds that global wealth and ultimate freedom will be achieved through the unleashing of capital from the fetters of state regulation, economic planning, and national borders, while the role of the state should be limited to those interventions that facilitate capital expansion and minimize or manage social dissent (Harvey 2005).

The market as a precapitalist institution was shaped and controlled by society according to the dominant norms and structures of authority of the day.[5] Accordingly, the creation of the free market capitalist economy in the early nineteenth century required the creation of a market society and a cultural transformation (Polanyi 1944; Smelser 1959). While the postwar social welfare state acted as a check on the ravages of a market society, the neoliberal turn in the 1980s began a process in Western democracies of removing social constraints from the market and commoditizing formerly social spheres. Since that time, neoliberal discourse has become hegemonic, to the point where it now defines our collective identity. A new ecological discourse grounded in an ethos of membership, householding, and entropic thrift is, then, a counterhegemonic discourse. For this resistance discourse to gain a foothold in public discourse, the legitimacy of neoliberal discourse must be undermined in public consciousness.

As we have discussed, a discourse is hegemonic when it influences institutional practice. The practice of neoliberal discourse has been manifest in deregulation, privatization, and bilateral and multilateral free trade agreements in all regions of the world, and in the wake of the 2008 financial crisis, severe structural adjustments in debtor nations (a category that now includes Ireland, Greece, Italy, and Spain). Although these measures have failed to deliver the promised free-market utopia (they have starved states of revenue to support public works and the social wage, precipitated several financial crises, and hastened ecological destruction globally), they have succeeded in restoring processes of capital accumulation and wealth concentration to pre-1930s levels (Harvey 2005).[6] Nevertheless, apart from the Occupy movement's outing of the "one percent," the neoliberal discourse remains entrenched not only in economic and political circles but throughout civil society including, significantly, the academy and mainstream media.[7] British prime minister Margaret Thatcher's declaration that "there is no alternative" has lodged deeply in the psyche of capitalist society. Mass consent, implicit and explicit, has been granted to political elites to continue to pursue neoliberalism's seductive yet elusive promise of economic and individual emancipation. Meanwhile, the spectacular concentration of capital in private hands driven by a seemingly boundless reach for global resources has entrenched the power of transnational capital to dictate the trade and finance policies of most countries (Bakan 2003). Within the grip of this globalized neoliberal paradigm, politics can make only limited concessions to ecological and social sustainability without undermining the internal logic that gives global capitalism its life breath.

This pathological situation is mitigated from time to time by public out-cry and collective action within social movements.[8] In his classic account of the dynamics of the self-regulating market economy, Karl Polanyi (1944) observed that the inevitable social and environmental disruptions created by unregulated capitalism would generate an equally inevitable resistance—a phenomenon he called the "double movement." He warned, however, that this double movement can just as easily result in destructive regression as in constructive transformation, pointing to the rise of fascism following World War I as an important case in point. The direction change takes depends on the particular circumstances of time and place, including whether or not there are social institutions in place that can protect vulnerable people and nature, and whether alternative discourses and models for progressive social organization are circulating and readily invoked.

Despite the best intentions and endurance of social movements over the past decades, these points of resistance have not been sufficient to counteract the hegemony of the neoliberal discourse as the determinant of public policy. Writing in the mid-1990s, Raymond Rogers explained this failure as the absence of a counterdiscourse that sufficiently challenges the capitalist relations which shape industrial societies. According to Rogers, the only way we can meaningfully deal with environmental problems is to "create a perspective which problematizes . . . the structures and processes of capital and markets," which "have become so dominant that humans only know themselves according to this social context" (Rogers 1994:1–2). In other words, the sociocultural milieu of everyday life—where people and capital meet—must become a locus of contention. As the new millennium unfolds, this appears to be happening. Early twenty-first century political ecology discourse is evoking and invoking the 1970s radical critique of industrial society and the discourse of limits.[9] Authors are explicitly naming the growth economy as the primary driver of climate change and biodiversity loss, as well as social dislocation and cultural and spiritual malaise.[10] After a lifetime of advocating for environmental policy from within government, the United Nations system, nonprofit organizations, and the academy, noted scholar James Gustave Speth concluded that without challenging capitalism, ever-increasing environmental degradation is inevitable:

The system of modern capitalism as it operates today will generate ever-larger environmental consequences, outstripping efforts to manage them.

Indeed, the system will seek to undermine those efforts and constrain them within narrow limits. The main body of environmental action is carried out within the system as currently designed, but working within the system puts off-limits major efforts to correct many underlying drivers of deterioration. . . . Working only within the system will, in the end, not succeed when what is needed is transformative change in the system itself. (Speth 2008:85–86)

This analysis is echoed, although somewhat less passionately, by the United Nations Environment Programme (UNEP) report to the 2012 Earth Summit, Global Environmental Outlook (GEO) 5. In it, UNEP explicitly identified global population and economic growth as the drivers of ecological decline:

Population growth and economic development are seen as ubiquitous drivers of environmental change with particular facets exerting pressure: energy, transport, urbanization and globalization. While this list may not be exhaustive, it is useful. Understanding the growth in these drivers and the connections between them will go a long way to address their collective impact and find possible solutions, thereby preserving the environmental benefits on which human societies and economies depend. (United Nations Environment Programme 2012:5)

This is a marked change from its GEO-4 report of 2007, which recognized "economic activities" as a driver along with demographics and other considerations, but its view was distinctly neoliberal in advocating for market-based responses: "Economic instruments, such as property rights, market creation, bonds and deposits, can help correct market failures and internalize costs of protecting the environment. Valuation techniques can be used to understand the value of ecosystem services" (United Nations Environment Programme 2007:5). In 2012, UNEP acknowledged that such approaches fall within the conventional economic framework; although not rejecting them outright, UNEP rather diplomatically proposed another approach, taken directly from the ecological economics playbook:

An alternative physical approach, stemming from the industrial metabolism or industrial ecology tradition, seeks to identify the rates and volumes of material flows through the economy. A system such as material

flow accounting (MFA) is presumed to reveal more accurately the pressures on resources and the undesirable impacts on the environment from
any part of the life cycle of resources—from extraction through combustion or conversion into a usable commodity and consumer consumption,
to recycling, disposal or stewardship.
(United Nations Environment Programme 2012:11)

The discourse of this international institution is clearly changing, presumably in the face of environmental indicators that suggest the failure
of neoliberal political economy to address the environment-development
crisis. The eco-political discourses of accommodation and reform of the
1980s and 1990s, namely sustainable development and ecological modernization (e.g., see Carruthers 2001; Hajer 1995), are being delegitimized by
empirical evidence and are challenged by counterdiscourses that call for
structural transformation of the capitalist economy. Discourses that more
closely align with "reality" are providing new models for understanding
and responding to the ecological crisis. As with the presence of dissent-
ing social movements, this is a necessary condition for institutional transformation. Yet, it is not sufficient. Economic, political, and administrative
institutions have been constructed or adapted to operationalize the prescriptions inherent in neoliberal discourse. Institutional resistance to challenger discourses is a given, and the forces of inertia are structural. In the
spirit of encouraging an "eyes wide open" approach to the difficulties that
confront us, I will mention the most obvious barriers to change.

4.1. The Globalization Straightjacket

The formidable power of transnational capital has been so concentrated
in both national and international institutions, and the tight integration of
global finance has created such widespread vulnerability, that single nation-
states cannot adopt a significantly divergent economic policy direction
from the dominant economies without risking retaliation and potentially
disastrous capital flight. The 2008 global economic meltdown and its ongoing hangover brought two realities into sharp relief: global finance capitalism has become dysfunctional and destructive, and there is an appalling
lack of political courage within any and all wealthy nations to confront
this pathology. The orthodox solution to the crisis of neoliberalization

has been more neoliberalization. In Greece, for example, in order to pay interest to foreign lenders, the structural adjustments imposed on public services and the social wage have thrown many thousands into poverty. Because defaulting on those loans would destabilize the entire European Union banking system, member countries appear willing to sacrifice just about anything to avoid this eventuality. In Greece and Italy, where elected heads of state were for a time replaced by financial technocrats, the shift toward government by external fiat reveals the shallow roots of democratic processes when confronted by the demands of finance capitalism, even in Western democracies.

4.2. The Microeconomic Straightjacket

There is an equally pathological codependent relationship between the growth economy and those who work within it. Each needs the other for its survival. Households are immediately vulnerable financially to any slowdown in gross domestic product growth or reduced consumer spending, which might undermine the viability of the firms that employ them, the pension funds in which they have invested, or the sustainability of the social safety net. The perceived risks in abandoning "grey" economic sectors for the promise of more fulfilling livelihoods in a "green" economy remain untenable for the vast majority of citizens, for the simple reason that their immediate financial stability is vested in the growth economy. In the current context of high consumer debt, low savings, high cost of living, and job insecurity—the product of late capitalism—public support for systemic change is understandably low. Paradoxically, the material insecurity created by the system binds people to it. Thus, an internal momentum, continually primed by official discourses, perverse incentives, and other political interventions, keeps the system moving day to day along its fatal trajectory. Long-term well-being and biospheric integrity are traded off in myriad daily decisions that favor short-term individual satiation and household stability.

4.3. The Discourse of Consumer Culture

Several theorists have embedded the ecological crisis in a larger "crisis of modernity." They identify a general cultural malaise characterized by a loss of meaning associated with the release of individuals from the moral and social strictures of traditional authority, the shrinking of a collective consciousness in favor of self-referent individualism, technological and scientific determinism, increased complexity and rationalization, the reduction of all value to instrumental factors in cost-benefit analysis, and the commodification of things of intrinsic worth.[11] As traditional reference points wane, materialism emerges to fill the void.

The allure of consumerism, rooted in the conviction that personal fulfillment comes at the cash register, has bought a fierce mass loyalty to high consumption, no-limits lifestyles. This is not by accident. In 1955, marketing guru Victor LeBow proffered this solution to the problem of excess postwar industrial production: "Our enormously productive economy demands that we make consumption our way of life, that we convert our buying and use of goods into rituals, that we seek our spiritual satisfaction, our ego satisfactions, in consumption" (quoted in Assadourian 2010:15). This project—the production, constant adaptation, and dissemination of a very specific scientifically designed and verified discourse—has been wildly successful (Ewen 1976). Material wealth and consumption have become the generic cultural source of value and identity for most people in industrialized countries. An annual survey that queries first-year college students in the United States about their life priorities revealed a trend in this direction over forty years. In 1970, nearly 80 percent of students identified "to develop a meaningful philosophy of life" as their aspirational goal, compared to less than 40 percent who chose "to be well off financially." By 2008, those ratings were almost exactly reversed (Assadourian 2010:10). Such Western aspirations are now spreading as the consumer class grows in high-population countries such as China, India, and Brazil (Jackson 2008).

Exorbitant consumer debt and the fallout of the financial crisis that began in 2008 may be turning this around somewhat, as a function of necessity rather than choice. With 70 percent of the U.S. economy dependent on the retail sector, however, collapsing consumption would have severe economic consequences. Thus, governments make continual and paradoxical adjustments to monetary and fiscal policies to bolster "consumer confidence" and increase retail sales. Despite growing evidence of the futility and harmfulness of high-consumption lifestyles, including the delinking of income and happiness beyond a relatively low income threshold (about $15,000 per capita), increased stress related to work and debt, and diseases related to diet and lifestyle (Jackson 2009:42), making consumer culture problematic in the public consciousness remains a marginal activity. Furthermore, discourses that reject both the subjective and objective experience of mainstream society have a long tradition, but "dropping out" of a dominant culture is a marginal act. It does nothing to steer a culture's trajectory away from the brink. In short, the postwar creation of mass consumer culture fuelled by a global annual advertising budget of 1 percent of the gross world product—$643 billion in 2008 (Assadourian 2010:11)—has insulated the dominant growth-based political economy from meaningful social criticism. Even mainstream environmental discourse has "given in" to the dominant culture, proffering "green consumption" and recycling as viable responses to the ecological crisis.

4.4. Media Reinforcement of Norms of Economic Growth and Consumption

The final significant defense of the status quo is wielded by the mass media, which is still the dominant purveyor of public discourse in modern society (Newspaper Audience Databank Inc. 2012). According to Duane Elgin, media activist and author of the best-selling *Voluntary Simplicity*, "To control a society, you don't need to control its courts, you don't need to control its armies, all you need to do is control its stories. And it's television and Madison Avenue that are telling most of the stories most of the time to most of the people" (quoted in Assadourian 2010:13). Consider also Bernard Cohen's insight in the seminal book *The Press and Foreign Policy*:

> [The press] may not be successful much of the time in telling people what to think, but it is stunningly successful in telling its readers what to think about. . . . The editor may believe he is only printing the things that people want to read, but he is thereby putting a claim on their attention, powerfully determining what they will be thinking about, and talking about, until the next wave laps their shore.
> (Cohen 1963:13)

In this sense, what the media chooses to ignore is just as important as what it chooses to cover. Norman Fairclough described media language as "ideological work." It represents the world in particular ways, presenting its own particular constructions of social relations and social identities.

Furthermore, as an increasingly monopolized private sphere, the media "diverts attention away from political and social issues which helps to insulate existing relations of power and domination from serious challenge" (Fairclough 2005:12–13). The positioning of different actors or protagonists within the public discourse realm by the media is an exercise of micropower (to use Foucault's term) that influences public understanding and interpretation of the issue at hand. To use the example of global warming, the media positioning of the issue has contributed to an insidious public paralysis on this issue (e.g., see Bell 1994; Carvalho 2005; Grundmann 2007). This is not by accident. The very techniques of hegemonization that propelled neoliberal discourse into prominence in the West (Harvey 2005) have been used to counteract the climate change discourse of mainstream climate science and international institutions (Jacques, Dunlap, and Freeman 2008). The effect has been to give politicians a free pass in their blatant subservience to those corporate interests that stand to lose in any effective program to curb greenhouse gas emissions.

Gouldner (1990) provided essential insight into the power relations inherent in contemporary communications. He distinguished between producers of ideological products (intellectual elites) and the media through which these products are distributed. In industrial countries, there is considerable tension between the "cultural apparatus," largely influenced by the intelligentsia and academicians, and the "consciousness industry," largely run by technicians within the framework of profit maximization and now increasingly integrated with political functionaries and state apparatuses. The growth in reach and influence of the latter has created what Gouldner (1990:310, 314) called "a crisis for ideological discourse" as the "producers of 'culture' in modern society cannot communicate their work to mass audiences except by passing through a route controlled by media, and those who control the mass media, the consciousness industry." Mass culture is shaped not through argumentative ideological discourse or ideal communicative action in the public sphere (Habermas 1989) but through a mediated discourse controlled by the consciousness industry. Clearly, the dependency on the media as the primary disseminator of alternative discourses presents a substantial obstacle to a discursive shift toward an ecological economy.

The three-frontal combination of commercial advertising, programming, and news creation/reporting constructs a social reality that is difficult to ignore. Media exposure, much of which reinforces consumer culture norms and economic growth discourses (Good 2013), occupies from one-third to one-half of people's waking hours (Assadourian 2010:13). The Internet is touted as the way around corporate and ideological media messaging; to the extent that people have limitless access to alternative views through online sources, this may be true. It depends, however, on one's voluntary effort to go looking for a different representation, the capacity to sift through vast quantities of online information, and finally the ability to discern the credibility of what is on offer. Whether or not individual consciousness-raising through this medium can or will translate into sustained collective action for transformation remains to be seen. Nevertheless, any program of transformative change for the earth must confront the role of the media in reinforcing dominant discourses of economic growth and consumerism while also generating and disseminating alternative discourses however it can. Although one may argue that media effects on the general public are ambiguous, I would argue that the mutual dependency between media and decision-makers is clearer, with the media generally reinforcing dominant economic policy paradigms, making them unproblematic from a political perspective.

5. POLITICAL CHANGE OR CULTURAL CHANGE?

Overcoming these obstacles to cultural change appears daunting, even unlikely. Yet, to repeat Charles Taylor (1991), "Our degrees of freedom are not zero. There is a point to deliberating what ought to be our ends." Normally when we consider how to achieve substantive change, we understand the democratic political arena to be the locus of action. In theory, elected governments are stewards of the commons and of the common good; as such, they are accountable for their success or failure to meet this obligation. While political action is critical in holding governments to account, and while important victories are often won through advocacy and electoral involvement, there is an important distinction between political change and cultural change to be made (Hunter and Wolfe 2006). Political change is tactical and transient, easily won (relatively speaking), and just as easily lost. The gutting of a forty-year corpus of environmental law and policy by the Conservative government in Canada in two omnibus budget bills (2011 and 2012) is a case in point. Undoubtedly, at some point the tide will turn and at least some of these losses may be recouped, but the larger point is that political gains are always vulnerable.

Cultural change is a different matter. It derives from changes in belief systems, norms, ethics, and knowledge that ultimately become manifested in institutions and practices. It goes without saying that such change is difficult, conflictual, and typically happens gradually over decades, not years. Once realized, however, the changed culture is resilient and enduring—the very features that make culture change difficult in the first place (Hunter 2010). Such change has happened throughout human history. Over a period of several decades now known as the Enlightenment, the Church was sidelined as the sole arbiter of knowledge, social order, and moral authority (Wuthnow 1989). More recent examples in various countries include the abolition of slavery, apartheid, and segregation; the liberation of women from oppressive social roles; the establishment and protection of minority rights; and the collapse of certain colonial and totalitarian regimes. Such changes incubate for decades or longer, then happen with apparent lightning speed when historic conditions converge to create a moment of transformation.

In each of these cases, new discourses emerged to challenge the authority and legitimacy of existing dominant discourses. They provided a common language, including metaphors and symbols, to be employed by the networks of leaders, organizations, and institutions aligned around the particular issue. Yet, simply the existence of a challenger discourse is not enough. Its central principles or elements must be taken up and reflected in practice or action by individual and institutional actors, a self-reinforcing process that at once reflects a changing cultural milieu and drives the integration of new cultural values throughout society (Wuthnow 1989).

Antonio Gramsci (1999), writing from a political perspective, provided a key insight here. He understood hegemony not as an oppressive force but as an ideological unity within a culture. From the vantage point of a prison,[12] Gramsci's reflection on why workers would vote for a fascist regime (among other political puzzles of the interwar years) led him to conclude that political aims can only be reached if there is an "intellectual and moral unity" of the culture in support of such aims. Achieving and maintaining political-economic hegemony requires an ongoing dialectical relationship between political discourses and cultural frames, in which the former responds to cues from the latter in the ongoing effort to have a political agenda articulate with mass consciousness. This complex discursive process includes dissemination of cultural narratives and discourses embedded with a set of ideas, beliefs, and norms embedded in cultural narratives and discourses that legitimizes the ideological basis for the exercise of power. The goal is to gain the "consent of the governed" by having the masses assume the ideological elements of such discourses as their own.[13] Without ideological-cultural support, structural change—changes in the institutions of government and economy—can only be achieved through political fiat or coercion, and then only temporarily.

Cultural change, then, is a prerequisite for changing the institutional formations of society, including the economy; this process is fundamentally discursive. Gramsci (1999) identified two distinct conditions under which a challenger or counterhegemonic discourse achieves cultural acceptance. First, it must accommodate to some extent competing ideologies and expectations. While this accommodation must be genuine (Gramsci's intent was not cynical), it cannot undermine the primary ideological elements that distinguish the challenger from the hegemonic discourse. Thus ideological accommodation must be limited to secondary principles. Second, the challenger discourse must resonate to some degree or articulate with the subconscious "common sense" of popular culture.[14] By "common sense," Gramsci means "the traditional popular conception of the world"—in other words, the deep frames that comprise the worldview of the dominant culture. In short, there must be enough familiar territory within the new discourse that it can be recognized, while at the same time allowing its radically new steering ideas to become established and diffused. Once new ideas gain currency within the culture, the governing paradigms of institutions can be altered accordingly. As institutions are changed, the premises underlying the change become taken for granted; their assumptions become "second nature." Alternative conceptions become "impossible," their promotion tantamount to high treason (Gramsci 1999:134–161). Thus, political-ideological hegemony depends on achieving cultural-ideological unity.

Robert Wuthnow drew on Gramsci's insights in his empirical inquiry into three cultural shifts that changed the Western world: the Reformation, the Enlightenment, and the rise of socialism in response to nineteenth-century industrial capitalism. To understand these cultural shifts, he studied "the historical manifestations of culture . . . the concrete expressions of public discourse . . . speakers and audiences, discursive texts, the rituals in which discourse is embedded, and the social contexts in which it is produced" (Wuthnow 1989:538–539). He discovered that in each case alternative discourses to the dominant ones abounded, but the successful ones were those that both "articulated" (Gramsci's term) and "disarticulated" with the existing "discursive field."[15] The discursive field sets the limits on public discourse, particularly what is and is not legitimate knowledge, what is and is not up for debate, etc.—in other words, the dominant paradigm. Successful challenger discourses were those recognizable within common perceptions of reality, thus constituting a plausible alternative view. At the same time, they "disarticulated" with common perceptions such that they shifted perceptions into new territory, embracing new symbols of meaning and thus new understanding. Within a relatively short period of time (a few generations), new constellations of discourse or ideology (he used the terms interchangeably) replaced the old and the limits of discourse were changed (Wuthnow 1989:12–13). For Wuthnow, cultural change was manifest in a transformation of the discursive field itself.

In a more contemporary context, David Harvey (2005) used Gramsci's explication of the process of gaining "ideological unity" to analyze how neoliberal discourse moved from its place on the intellectual margins during the Keynesian years to become the dominant or hegemonic discourse of the late twentieth century, replacing the paradigm of "embedded liberalism" and resulting in the concrete institutional restructuring of politics, economics and social conditions in much of the world. He wrote:

So how, then, was sufficient popular consent generated to legitimize the neoliberal turn? The channels through which this was done were diverse. Powerful ideological influences circulated through the corporations, the media, and the numerous institutions that constitute civil society—such as the universities, schools, churches, and professional associations. The "long march" of neoliberal ideas through these institutions . . . the organization of think-tanks (with corporate backing and funding), the capture of certain segments of the media and the conversion of many intellectuals to neoliberal ways of thinking, created a climate of opinion in support of neoliberalism as the exclusive guarantor of freedom. These movements were later consolidated through the capture of political parties and ultimately, state power. . . . Furthermore, once the state apparatus made the neoliberal turn, it could use its powers of persuasion, co-optation, bribery, and threat to maintain the climate of consent necessary to perpetuate its power.

(Harvey 2005:40)

Neoliberal discourse appeals to "common" sensibility, according to Harvey, in its invocation of the idea of individual freedom. He argued that this value is held not only by conservatives but broadly by the "baby boomer" generation, which cut its teeth on individual and collective challenges to traditional authority structures, roles, and identities. While social movements combined the idea of freedom with a social justice agenda, the single idea of individual freedom, argued Harvey, taps more directly into values embedded in the cultural subconscious.[16] Neoliberal discourse would "disarticulate" with common perception in its demonization of the welfare state and its elevation of the free market as the vehicle through which individual welfare will be realized and protected. Harvey described the effects:

> [Neoliberalism] has pervasive effects on the ways of thought to the point where it has become incorporated into the common sense way many of us interpret, live in, and understand the world. The process of neoliberalization has, however, entailed much "creative destruction," not only of prior institutional frameworks and power (even challenging traditional forms of state sovereignty) but also of divisions of labour, social relations, welfare provisions, technological mixes, ways of life and thought, reproductive activities, attachments to the land and habits of the heart..
> (Harvey 2005:3)

Although this was an ideological process, it served very real material interests from which it generated material support. It was strategically and generously facilitated by a corporate sector, whose goal was not an ideal free market but restoration of conditions conducive to the capital accumulation that had flagged during the 1970s and 1980s. The point is important. Wuthnow discovered that successful challenger discourses were able to marshal resources from powerful sources to back their diffusion throughout society. Whether from the state, the nobility, or other powerful benefactors, material support and influence in high places ultimately determined which of the new contending discourses became institutionalized. Ultimately, success depended on gaining support from the state (Wuthnow 1989:577).

Returning to the project of diffusing a new ethos of membership, householding, and entropic thrift through a culture, we can distill two related ideas from these scholars. The first is that the dominant, albeit pathological, growth-dependent political economy is underpinned by deep-seated, pre-reflexive beliefs, values, and assumptions about how the world works. These are embedded in cultural narratives and discourses that legitimate existing institutions and power relations, while at the same time making other conceptions of society implausible. The dominant paradigm ossifies as it is institutionalized, creating systemic interdependencies. Furthermore, powerful beneficiaries of these structures expend considerable energy and resources maintaining their stake in the system. The combination of cultural belief systems, institutional inertia, and vigorous protection of vested interests creates a daunting barrier to change.

The second idea is that within these deep cultural frames are also the seeds of change. Belief systems represent the cultural accumulation of those values, stimulated by the experience of successive generations living within that culture. The system acts like a self-reinforcing feedback loop. The cycle is interrupted when peoples' direct lived experiences begin to contradict the common understanding of how the world works. When dominant discourses become dissonant with "common sense" and direct lived experiences, they lose authority and ultimately legitimacy. This provides an opening in the public sphere for new discourses that better explain a changing world, activate different but preexisting latent values,[17] and propose alternatives that are plausible within cultural frames of reference (Crompton 2010). This is the foundation on which a new ecological conscience can be built, thereby providing the necessary cultural impetus for the restructuring economic and political institutions. In short, the unhinging of hegemonic discourse from its cultural base is a prerequisite for institutional/systemic change.

6. OF PARADIGMS, SYSTEMS, AND HISTORICAL BIFURCATIONS

I have argued so far that cultural change is a prerequisite to the kind of structural-institutional change that is required to achieve an ecological economy. Yet, the construction and dissemination of challenger discourses that articulate with deep cultural frames is not sufficient to propel this change. Historian Richard Hofstadter wrote:

> Societies that are in . . . good working order have a kind of mute organic consistency. They do not foster ideas that are hostile to their fundamental working arrangements. Such ideas may appear, but they are slowly and persistently insulated, as an oyster deposits nacre around an irritant. They are confined to small groups of dissenters and alienated intellectuals, and except in revolutionary times they do not circulate among practical politicians.
>
> (quoted in Speth 2008:66–67)

Similarly, Stephen Reese (2003) observed that deep cultural frames are resistant to change because they satisfy some important need. As long as the need is being met, there is no compulsion to jettison one frame for another. On the other hand, when a frame no longer "fits" with direct experience of the world, a space is created within public discourse for new discourses to compete for public salience. In other words, timing is (almost) everything.

Wallerstein characterized particular socio-geopolitical-economic configurations as historical systems, all of which have a beginning and an end; the present capitalist world system is no exception. A historical system ends when it can no longer accommodate new demands within system parameters: "There comes a moment when it has or will have exhausted the ways in which it can contain its contradictions, and it therefore goes out of existence as a system" (Wallerstein 1999:124). Every system contains internal contradictions—gaps between claims and expectations and lived reality. New belief systems arise that challenge the old; tensions between internal contradictions and opposing value systems lead to systemic stresses. These secular trends act as "vectors moving the system away from its basic equilibrium." Once a system paradigm—embodied in institutions, norms, and conventions—loses broad social legitimacy, then it must submit to demands for change. The system initially responds with incremental accommodation of increasingly difficult demands through legislative and judicial means. Yet, these negative feedback mechanisms that function to return the system to equilibrium never bring the system back to its exact previous state. System parameters are constantly changing—a function of the "arrow of time." "Eventually social unrest accumulates to the point that the system becomes dysfunctional or breaks down. The system moves further and further from equilibrium, the fluctuations become ever wilder, and eventually a bifurcation occurs" (Wallerstein, 1999:124). Essentially, the rules change such that the system in its previous state cannot survive.

Although the system through its long development may be impervious to disturbances, even large ones, at the point of bifurcation small interventions have large impacts (Wallerstein 1999:129–130). In short, there comes a point in historic time when, as a result of historic secular trends and human agency exercised in myriad historically contextual decisions, system paradigms are changed. This process is by definition nonlinear—an indeterminate evolution toward unknown radical transformations. The actual trajectory can only be discovered and explained after the fact through historical analysis from the perspective of the "longue durée" (Wallerstein 1999:1–3, 130).

Wuthnow's historical analysis of the Reformation, the Enlightenment, and the rise of socialism affirmed Wallerstein's theory. In each case, specific historical conjunctures made cultural innovation possible. Unique institutional contexts influenced what ideologies could develop and thrive. As products of history (Wallerstein's "arrow of time"), they were not planned or controlled. At a certain point in time, institutions became vulnerable to deliberate interventions. It began with "oppositional movement discourses defining new boundaries within which discourse can be framed . . . [thereby constructing] an alternative source of authority with which to challenge the authority of prevailing ideas." It ended with selected ideologies becoming institutionalized. Society was changed.

> In each historical episode the leading contributors to the new cultural motifs recognized the extent to which the institutional conditions of their day were flawed, constraining, oppressive, arbitrary. Their criticism of these conditions was often extreme and unrelenting. It was sharpened by an alternative vision . . . constructed discursively, a vision that was pitted authoritatively against the established order, not as its replacement but as a conceptual space in which new modes of behaviour could be considered.
> (Wuthnow 1989:582–583)

Notably, it was not simply a matter of exchanging one ideology for another. The producers and disseminators of successful new discourses were tightly networked across sectors and they gained the support of certain influential elites, thereby penetrating the sphere of power. Furthermore, the successor ideology translated into practices that appeared to resolve the problems articulated by it. Wuthnow (1989:583) wrote, "The strength of their discourse lay in going beyond negative criticism and beyond idealism to identify working models of individual and social action for the future."

Thus, we see that systemic change emerges from a conjuncture of social action within a particular "historic moment," which itself is the product of the arrow of time. The process is dynamic, nonlinear, and to a great extent opportunistic. Intelligent interventions are those that anticipate, shape, recognize, and seize opportunities as they arise. The outcome, however, is unpredictable because the system has its own internal dynamic over which there is no control. There will always be unintended consequences of actions. Although the universe may not be unfolding "as it should," it is unfolding along a historic trajectory that social agents can try to influence, but they cannot control it.

For the purposes of this book, we can envision the production and dissemination of ethical-ecological discourses combined with ready-at-hand alternative economic models around which a new cultural consensus can coalesce. Most importantly, these elements must coincide with a historical conjuncture—a period of system bifurcation—that ploughs the ground in which they can root and grow. Wallerstein asserted that the capitalist world-system has "entered into a terminal crisis and is unlikely to exist in fifty years." The ecological-climate crisis is but one symptom of the inability of capitalism to contain and mitigate its inherent contradictions. We may in fact be in that historical moment when small interventions can trigger very large changes.

7. WHAT NEXT?

The whole history of the progress of human liberty shows that all concessions yet made to her august claims have been born of earnest struggle. . . . If there is no struggle, there is no progress. Those who profess to favor freedom and yet deprecate agitation, are men who want crops without plowing up the ground. They want rain without thunder and lightning. They want the ocean without the awful roar of its many waters.

This struggle may be a moral one; or it may be a physical one; or it may be both moral and physical; but it must be a struggle. Power concedes nothing without a demand. It never did and it never will. Find out just what any people will submit to, and you have found out the exact amount of injustice and wrong which will be imposed upon them; and these will continue till they are resisted with either words or blows, or with both. The limits of tyrants are prescribed by the endurance of those whom they oppress.
—Frederick Douglass, "An Address on West India Emancipation,"
August 3, 1857

Lest we are tempted to oversimplify, James Davison Hunter (2010) interrogated cultural change theories strategically, from the perspective of crafting an agenda for proactive intervention for cultural change. In this section, I present and examine four of his propositions about human agency and structural change.

Proposition 1: Culture Changes from the Top Down, Rarely If Ever from the Bottom Up

While social unrest in response to unacceptable conditions can topple authoritarian dynasties, defeat sitting governments, and force major policy reforms, these are political, not cultural, changes. To change culture, an idea must penetrate "the structure of our imagination, frameworks of knowledge and discussion, the perception of everyday reality" (Hunter 2010:41). Such penetration is rarely if ever achieved through grassroots mobilization; more likely, such mobilization is the manifestation of a deeper transformation underway. Hunter explained:

> The work of world-making and world-changing are, by and large, the work of elites: gatekeepers who provide creative direction and management within spheres of social life. Even when the impetus for change draws from popular agitation, it does not gain traction until it is embraced and propagated by elites. The reason for this . . . is that culture is about how societies define reality—what is good, bad, right, wrong, real, unreal, important, unimportant, and so on. This capacity is not evenly distributed in a society, but is concentrated in certain institutions

and among certain leadership groups who have a lopsided access to the means of cultural production. These elites operate in well-developed networks and powerful institutions.
(Hunter 2010:41)

The cultural innovation process, according to Hunter, unfolds something like this. Creative thinkers and theorists generate ideas; researchers explore, revise, expand, and validate ideas and create knowledge; educators pass ideas and knowledge on to others; popularizers turn these into simpler forms for public consumption; and practitioners turn ideas into concrete applications (Hunter 2010:42). Wuthnow described a similar process in Western Europe: ideas were formulated into discourse and related cultural artefacts; they went through a process of selection based on their perceived legitimacy and means of diffusion (Bourdieu's symbolic capital); and the ones that best "articulated" with the social circumstances finally became "a relatively stable feature of the institutional structure of a given society" (Wuthnow 1989:9–10). In the case of the Enlightenment, for example, he found that in those countries where the state provided support, particularly ensuring freedom for intellectual and artistic innovation, this cultural revolution flourished. In countries where the state did not support innovation, it languished. I would add this caveat from a contemporary perspective. When the public is mobilized behind a new idea, the state is more likely to follow. The extent to which the public can be mobilized and provide sustained pressure on the state depends on the resources or symbolic capital that can be attracted to the project from civil society and nonstate actors. Either way, without significant resources and intellectual leadership, human agency is unlikely to drive systemic change.

Proposition 2: Change Is Typically Initiated by Elites Who Are Outside the Centermost Positions of Prestige

This proposition follows from the first. Hunter conceptualized spheres of influence within society as stratified gradations of prestige from a nucleus in which power is concentrated out to the periphery from where power is least exercised. Systemic change rarely originates within the nucleus of power. Culture-changing elites[18] are situated in the zone between the nucleus and the periphery. It is from here that challenger discourses are produced and disseminated outward. As new discourses are broadly diffused, the influence and prestige of the initiating elites is increased, while that of the nucleus is diminished (Hunter 2010:42–43).

There are two insights here. First, we should not look to those located in the nucleus of cultural power and prestige to take the lead in overturning their world. What may appear to be a shortcut to social transformation—convincing those with their hands on the steering wheel to change direction—usually ends up wasting scarce time and resources. New demands can be accommodated by the center only to the extent that accommodation does not require getting off the main highway and onto a path less taken. Such route changes will be vigorously resisted.

Second, elites outside the nucleus must be willing to deploy their symbolic capital, however it is embodied, to help achieve transformational change. John Muir, Aldo Leopold, Rachel Carson, Paul and Ann Ehrlich, Barry Commoner, Dennis and Donella Meadows, Fritz Schumacher, and Amory and Hunter Lovins wrote the canon on which the modern environmental movement has been built. However, they were also activists participating in one way or another in the diffusion of their ideas into the public sphere and doing battle there. Commoner went so far as to run for president of the United States, a strategy Ralph Nader has continued. It is not enough to research, write, and publish for an audience of peers. The counterdiscourse producers and disseminators need to be actively and publicly engaged in the change process, making sure counterhegemonic cultural products articulate with the lived experience and perception of possibilities within the mass public, and then moving them out into the public sphere. This will require all the resources we can possibly muster.

Proposition 3: Culture Changing Is Most Concentrated When the Networks of Elites and the Institutions They Lead Overlap

Wuthnow's study of the Enlightenment revealed a cultural revolution generated by "an alternative network of leaders, providing an alternative base of resources, oriented toward the development of an alternative cultural vision (a new anthropology, epistemology, ethics, sociality, and politics), established in part through alternative institutions, all operating at the elite centres of cultural formation" (Hunter 2010:75). The new idea here is the network. Isolated individuals or institutions do not change cultures.

Instead, the outward flow of ideas through the various stages of refinement, transfer, and application can only diffuse broadly through the culture if the elites producing alternative discourses are networked with elites in myriad dissemination centers. The key is to build dense networks in overlapping social and cultural spheres that work together for a sustained period measured in decades, not years. "When cultural and symbolic capital overlap with social capital and economic capital and, in time, political capital, and these various resources are directed towards shared ends, the world, indeed, changes" (Hunter 2010:43). It is precisely such dense networks—deliberately constructed—that propelled the "neoliberal turn" in Western nation-states (Harvey 2005).

Many factors mitigate against the development of dense overlapping networks that can propel counterhegemonic discourses forward. Institutions of all types tend toward competition rather than cooperation. Social movements and their "organic intellectuals" are fragmented, sometimes even antagonistic. The most significant barrier is the failure of progressive elites across civil society to articulate a common problem definition. Many environmental organizations, which enjoy a certain status in civil society, not to mention social justice and labor groups, are still avoiding the hard questions about economic growth. Only recently has economic growth (or more accurately, uneconomic growth) been targeted as a driver of ecological decline by a fairly broad spectrum of thinkers and analysts—many of whom have already been cited. Fewer are ready to openly declare capitalism to be an anachronism that cannot survive in a source- and sink-constrained world (or vice versa).

Problem definition is the first step in discourse construction; the form and content of a discourse depends on this initial work (Hajer 1995). Without widespread agreement on the problem, it will be difficult to generate the critical mass of idea generation and diffusion throughout cultural and social networks. This situation makes the need for a new integrative counterdiscourse, around which all progressive change agents can coalesce, all the more urgent.[19]

Proposition 4: Cultures Change but Rarely, If Ever, Without a Fight

Here we invoke Frederick Douglass, a key protagonist in one of the greatest cultural-economic transformations of Western history. By definition, culture is a contested domain. It consists of competing values, beliefs, and ideas, albeit of unequal status. Vested interests will vigorously defend their ideological and institutional terrain against challengers, so conflict is ever-present in the change process (Hunter 2010). This is difficult to accept; thus, the appeal of discourses that suggest win-win solutions to the environmental crisis can be found. At this stage of planetary overshoot, however, to achieve the elusive "balance" of interests, the heavy end of the scale onto which the world's most powerful vested interests have piled will have to be significantly lightened. They will not jettison their cargo willingly.

Because the stakes are so high, resistance is responding in kind, in a potentially significant "historical moment." In August 2011 and again in the winter of 2013, among those arrested for acts of civil disobedience outside the White House in opposition to a pipeline from the Alberta tar sands to Texas refineries and export terminals, were several prominent scientists and academics associated with prestigious institutions, including Speth, NASA's climate scientist James Hanson, and a coeditor of this book, Peter Brown. In Canada, Mark Jaccard—an academic and consultant to federal, provincial, and state governments; member of the Intergovernmental Panel on Climate Change; and Royal Society of Canada fellow—was arrested blocking a coal train en route to the Port of Vancouver (Jaccard 2013). Although there were no arrests, Canadians also took notice when hundreds of lab-coated scientists converged on Parliament Hill and at various points across Canada in protest against their muzzling and cuts to environmental research (Semeniuk 2013). These examples are notable in that they involve local elites outside their typical roles and comfort zones, in concert with tens of thousands of others who are engaged in active resistance against expansion of the fossil fuel economy.

Equally notable and perhaps with much greater consequence, new resistance discourses are arising from outside the dominant culture. While the Idle No More movement of aboriginal peoples coalesced around the dismantling of environmental protections by the Canadian government, it has empowered many First Nations struggles for sovereignty and in defense of their lands and waters (Idle No More 2014). In September 2013, in protest against shale gas exploration and "fracking," the people of Elsipogtog, a Mi'kmaq community in eastern New Brunswick, issued a "notice of eviction" to Houston-based SWN Resources, declared stewardship over their unceded territory, and with their Acadian and Anglophone neighbors, launched a several-week-long blockage of seismic trucks owned by the company. The aggressive police response sparked solidarity actions across the country (Martens 2013), including the erection of a traditional long-house, the seat of government, across from the New Brunswick Legislature by the Wolastoq (Maliseet) nation as a declaration of their sovereign nation status in opposing harmful resource exploitation (CTV Atlantic 2013).

These actions are illustrative of new alignments and modes of intervention grounded in new discourses of resistance that suggest two possibilities: First, overlapping networks are beginning to coalesce around a common discourse of an alternative to a fossil-fuel-dependent growth economy. Second, we may be witnessing on a global scale an intensification of forces pushing the capitalist world system "far from equilibrium." The ground is being ploughed; the tempest is being engaged.

Reflecting on the forty-year history of trying to achieve reform within the parameters of capitalist institutions, Speth considered the effort of the environmental movement (of which he counts himself a part) to date to have been ultimately futile. He wrote, "My generation is a generation of great talkers, overly fond of conferences. We have analyzed, debated, discussed and negotiated these global issues almost endlessly. But on action we have fallen far short" (Speth, 2008:18–19). This cohort vested their talents and faith in the administrative state to cope with the environmental fallout of rampant consumer-capitalism. Except in ameliorating localized environmental irritants, the experiment has failed. The only way forward now, Speth concluded, is to directly challenge the institutions of capital which require and are the engines of growth.

If we see capitalism through Wallerstein's lens as a historical system with a beginning and end, then we realize that whether or not capitalism continues indefinitely is not a matter of choice. The "limits to growth" thesis explicates the most potent and intractable force that will bring this growth-dependent system to an end. To paraphrase *Limits to Growth* co-author Dennis Meadows, it is not a question of being for or against growth. On a finite planet, growth in material consumption will end (Stone 2010). Former chief economist of CIBC World Markets Jeff Rubin concurred. Our world, said Rubin, is about to get a whole lot smaller. Triple-digit oil prices due to increasing scarcity will end economic growth (Rubin 2012). The question is, how will the transition from the old system to a new one occur, and what will that successor be? Because system change is nonlinear and beyond human control, the outcome cannot be predicted. But the earth's chances (and humanity's) are enhanced if discourses that embody an ecological ethic, coupled with plausible alternative models of economic life, are widely diffused throughout the public sphere.

The theoretical work underpinning a new ethos and economy has been done. Research to refine, clarify, and validate theories in concrete terms is amassing daily across the globe, and new discourses and narratives are being constructed and tested. Although many have been conceptualized, we are less advanced in constructing and test-running new institutional models (or at least less successful in replicating or scaling them up). Penetration of educational institutions remains low, however. While the No Child Left Inside movement is boosting public school systems out of a forty-year environmental education rut (Louv 2008), universities remain bastions of defense of the dominant growth paradigm, even while they incubate challenges to it. Furthermore, environmental studies programs are failing to prepare their students for the structural challenges we face (Maniates 2013). Popularization and dissemination of alternative discourses are the bread and butter of environmental organizations; however, their influence waxes and wanes with economic fluxes and political currents. Globally, the social and aboriginal movements for economic, social, and ecological justice are expanding and well documented, constructing new discourses that reflect the cultural diversity of a global movement (Hawken 2007). There are signs of the institutionalization of these discourses, sometimes directly facilitated by states (e.g., Germany's renewable energy strategy, the entrenchment of the rights of nature in national law in Ecuador and Bolivia) and other times through civil society (e.g., the Transition Town and Ecovillage movements in the North, Kenya's Green Belt Movement, Vandana Shiva's Navdanya, other common property movements in the South).

Notwithstanding, the process of cultural transformation is largely invisible and the outcome cannot be determined. Hunter explained:

> The most profound changes in culture can be seen first as they penetrate into the linguistic and mythic fabric of a social order. In doing so, it then penetrates the hierarchy of rewards and privileges and deprivations and punishments that organize social life. It also reorganizes the structures of consciousness and character, reordering the organization of impulse and inhibition. One cannot see change taking place in these ways. It is not

perceptible as an event or set of events currently unfolding. Rather, cultural change of this depth can only be seen and described in retrospect, after the transformation has been incorporated into a new configuration of moral controls.

(Hunter 2010:47)

It is tempting to abandon this project for something that delivers more tangible, short-term results. Yet, it is the conclusion of Charles Taylor, Thomas Berry, the authors of this book, and many others that unless we arrive at a "new configuration of moral controls," we consign many species to extinction and human life on Earth to a miserable future. Most of the tools and some of the conditions needed are at hand. Creating the dense networks across institutions and throughout civil society through which the challenger discourses are linked and diffused is the next critical step.

8. CONCLUSION

In this chapter, I have made a preliminary attempt to understand the dynamics that might lead to this new moral code, and how such cultural transformation intersects with the renewal of societal institutions toward an ecological political economy. I have engaged discourse theory as well as world systems theory to account for the social construction of belief systems, values, and norms on the one hand, and the structural-historical constraints on human agency on the other. Other theories or configurations of theories may well provide more cogent insights into change processes. Conversely, it may be argued that an ecological-political economy can be realized without cultural-structural transformation of capitalist society. In either case, there is a gap to be filled between analysis of the problem—the ecological crisis—and the solution, an ecological political economy. This gap is a systematic consideration of the processes by which the solution may be realized—and most importantly, the levers and interventions available to change agents and how they may be used to best effect.

NOTES

1. Peter Timmerman in this volume provides rich examples of economic practices that follow very different cultural discourses than those which shape industrial society.

2. We can think of climate change denial as well as the "tea party" phenomenon in the United States as examples of this.

3. Wuthnow (1989) provided historical evidence of this process in his study of three major cultural shifts in Western Europe: the Reformation, the Enlightenment, and the nineteenth-century rise of socialism.

4. This was achieved through a four-decade-long struggle to displace the hegemonic social welfare discourse of Western democracies, which David Harvey documented in his book, *A Brief History of Neoliberalism* (2005).

5. Polanyi in *The Great Transformation* explained that a self-regulating market economy cannot exist outside a market society because it depends on society to create and maintain the conditions for its operation.

6. David Harvey argued that the restoration of conditions for capital accumulation was the goal all along, not some ideal free market. He pointed to the rapidity with which the principles of free-market competition are abandoned when they threaten monopolistic corporate strategies of profit maximization.

7. While the academy is the source of much critical analysis and counterdiscourses, as institutions universities are either silent on or supportive of orthodox discourse, especially economic discourse. Likewise, while the media is an instrument for the dissemination of counterdiscourses, corporate media does not challenge economic orthodoxy.

8. Until recently, the antiglobalization movement, which reached its pinnacle with the disruption of 1999 World Trade Organization talks in Seattle, Washington, was the most obvious example of this. Public discontent reemerged in the wake of the 2008 global finance crisis with mass civil disobedience led by the Occupy movement, as well as mass protests against economic structural adjustments in southern Europe. Since 2012, civil disobedience actions have been deployed against pipelines associated with the expansion of Canada's Athabasca tar sands, including the Keystone XL pipeline which would transport the bitumen from Alberta to Texas, and the Kinder Morgan pipeline from Alberta to Vancouver, British Columbia. The Global Frackdown movement against hydraulic fracturing of methane and oil trapped in shale deposits is growing in size and scope (Food & Water Watch 2014), manifested most powerfully in Canada in aboriginal sovereignty claims in New Brunswick (Schwartz and Gollom 2013).

9. See Hajer (1995) and Carruthers (2001), for insightful overviews of 1970s radical environmental discourse and its evolution into discourses of sustainable development and ecological modernization.

10. See, for instance, McFague (2001); Nadeau (2003); Meadows, Randers, and Meadows (2004); Homer-Dixon (2006); Victor (2008); Jackson (2009); Foster (2009); Brown and Garver (2009); McKibben (2010); Speth (2011); Korten (2009); Heinberg (2011); Gilding (2011); Rubin (2012); and Czech (2013). This is just a sampling of books that are circulating in the public arena that either call for or predict an end to economic growth. There is a whole other vast peer-reviewed literature that is generally circulated only within academia, particularly in the field of ecological economics but also in other social sciences.

11. See, for instance, Marcuse (1964); Habermas (1976); Catton (1982); Lyotard (1984); Winner (1986); Alexander and Sztompka (1990); Lasch (1991); Rifkin (1991); Taylor (1991); Rogers (1994); and Ophuls (1998), among others.

12. He was imprisoned by the Mussolini regime as a leader of the Communist Party in Italy.

13. The prestige or status of the discourse-sponsor group and their spokespersons is important in gaining this consent. I take this matter up later in the chapter.

14. Gramsci's use of "common sense" is similar to the anthropological understanding of culture as the mental milieu that gives meaning to the world. He distinguished this from "good sense," which is a conscious understanding based on careful consideration of relevant factors. Bourdieu's concepts of doxa and habitus appear similar to Gramsci's "common sense." In both cases, because they are subconscious, they are quite inaccessible to deliberate examination through discourse.

15. Wuthnow employed Bourdieu's concept here.

16. Harvey suggested the conscious values related to social justice were not activated by neoliberal discourse and therefore lost ground in the political-cultural milieu as the new discourse "marched through the institutions."

17. Studies in social cognition tell us that individuals possess the values needed to support an ecological ethic. Intrinsic values (concern for bigger-then-self issues, sense of community, high empathy levels, internal validation of self-worth, low concern for status and wealth) coexist with extrinsic values (concern for wealth and power, envy of high social strata, strong focus on financial success, low empathy levels, external validation of self-worth). The relative weight one gives each class of values is determined by a range of factors, such as upbringing, social norms, role models, circumstances, and life experience. Values are exercised (or moral identity enhanced) through repeated exposure to those values through peers, media, education, or experience with societal institutions. Consistent exposure to external stimuli that trigger extrinsic values will strengthen those values relative to intrinsic ones. Likewise, the more individuals are exposed to bigger-than-self intrinsic values, the stronger those values will become relative to the opposing extrinsic values (Crompton 2010). Narratives and discourses are framed to trigger particular values, whether inadvertently or deliberately, with the concrete result that particular conceptions of society become more or less plausible depending on their alignment with preexisting, subconscious belief systems (and not with evidence-based reasoning or "good sense") (Lakoff 2004).

18. These are individuals who possess high levels of symbolic capital by virtue of their wealth, education, personal achievement, or position of authority or leadership in their respective fields or institutions who are engaged in "cultural" or "ideological" production (Gramsci's organic intellectuals and Wuthnow's well-placed backers of resistance discourses).

19. The emerging degrowth movement, which began in Europe and has recently crossed the Atlantic, is an example of intellectuals joining with social movements to deepen the critique of the hegemonic economic growth discourse and to articulate new models of ecological-economic relations (Martínez-Alier et al. 2010; Schneider, Kallis, and Martinez-Alier 2010). Another hopeful development is the collaboration of environmental leader Bill McKibbon and social change activist and author Naomi Klein on the climate change issue, beginning with their joint "Do the Math" tour in 2012–2013 to promote a fossil fuel divestment campaign on university and college campuses in North America (350.org 2013). In her speech at the September 1, 2013, founding of Unifor, a new mega-union created by the Canadian Autoworkers and the Canadian Energy and Paper Workers Union, Klein argued that climate change transcends all other interests and should be the common focus of all social and environmental justice organizations (Klein 2013).

REFERENCES

350.org. 2013. "Do the Math Tour." Accessed September 10, 2014. http://math.350.org/.

Alexander, Jeffrey C., and Piotr Sztompka. 1990. *Rethinking Progress: Movements, Forces, and Ideas at the End of the 20th Century*. Boston: Unwin Hyman.

Assadourian, Erik. 2010. "The Rise and Fall of Consumer Cultures." In *State of the World 2010: Transforming Cultures; from Consumerism to Sustainability*, edited by Linda Starke and Lisa Mastny, 3–20. Washington, DC: Worldwatch Institute. http://blogs.worldwatch.org/transformingcultures/contents/.

Bakan, Joel. 2003. *The Corporation: The Pathological Pursuit of Profit and Power*. Toronto: Penguin.

Bell, Allan. 1994. "Climate of Opinion: Public and Media Discourse on the Global Environment." *Discourse & Society* 5 (1): 33–64. doi: 10.1177/0957926594005001003.

Berry, Thomas. 2006. *Evening Thoughts: Reflecting on Earth as Sacred Community*. Edited by Mary Evelyn Tucker. San Francisco: Sierra Club Books.

Bourdieu, Pierre. 1989. "Social Space and Symbolic Power." *Sociological Theory* 7 (1): 14–25. doi: 10.2307/202060.

Bourdieu, Pierre. 1993. *Sociology in Question*. Translated by Richard Nice. London: Sage Publications.

Brown, Peter G., and Geoffrey Garver. 2009. *Right Relationship: Building a Whole Earth Economy*. San Francisco: Berrett-Koehler Publishers.

Carruthers, David. 2001. "From Opposition to Orthodoxy: The Remaking of Sustainable Development." *Journal of Third World Studies* 18 (2): 93–112.

Carvalho, Anabela. 2005. "Representing the Politics of the Greenhouse Effect." *Critical Discourse Studies* 2 (1): 1–29. doi: 10.1080/17405900500052143.

Catton, William Robert, Jr. 1982. *Overshoot: The Ecological Basis of Revolutionary Change*. Urbana: University of Illinois Press.

Cohen, Bernard Cecil. 1963. *The Press and Foreign Policy*. Princeton, NJ: Princeton University Press.

Crompton, Tom. 2010. *Common Cause: The Case for Working with Our Cultural Values*. Woking, Surrey, UK: WWF-UK. http://www.wwf.org.uk/wwf_articles.cfm?unewsid=4224.

CTV Atlantic. 2013. "Anti-Fracking Demonstrators Erect Traditional Longhouse near Legislature." *CTV News*, October 28. Accessed September 10, 2014. http://atlantic.ctvnews.ca/anti-fracking-demonstrators-erect-traditional-longhouse-near-legislature-1.1516713.

Czech, Brian. 2013. *Supply Shock: Economic Growth at the Crossroads and the Steady State Solution*. Gabriola Island, BC: New Society Publishers.

Douglass, Frederick. 1999. *Frederick Douglass: Selected Speeches and Writings*. Edited by Philip Sheldon Foner. Chicago: Lawrence Hill Books.

Ewen, Stuart. 1976. *Captains of Consciousness: Advertising and the Social Roots of the Consumer Culture*. New York: McGraw-Hill.

Fairclough, N. 2005. *Media Discourse*. London: Edward Arnold.

Fairclough, N., H. Mulderrig, and R. Wodak. 2011. "Critical Discourse Analysis." In *Discourse Studies: A Multidisciplinary Introduction*, edited by Teun Adrianus van Dijk, 357–378. London: Sage Publications.

Food & Water Watch. 2014. "#Global Frackdown." Accessed September 10, 2014. http://www.globalfrackdown.org/.

Foster, John Bellamy. 2009. *The Ecological Revolution: Making Peace with the Planet.* New York: Monthly Review Press.

Gilding, Paul. 2011. *The Great Disruption: Why the Climate Crisis Will Bring on the End of Shopping and the Birth of a New World.* New York: Bloomsbury Press.

Good, Jennifer Ellen. 2013. *Television and the Earth: Not a Love Story.* Halifax, NS: Fernwood Publishing.

Gouldner, Alvin. 1990. "Ideology, the Cultural Apparatus, and the New Consciousness Industry." In *Culture and Society: Contemporary Debates,* edited by Jeffrey C. Alexander and Steven Seidman, 306–316. Cambridge, UK: Cambridge University Press.

Gramsci, Antonio. 1999. *Selections from the Prison Notebooks of Antonio Gramsci.* Translated by Quintin Hoare and Geoffrey Nowell-Smith. 11th ed. New York: International Publishers.

Grundmann, Reiner. 2007. "Climate Change and Knowledge Politics." *Environmental Politics* 16 (3): 414–432. doi: 10.1080/09644010701251656.

Habermas, Jürgen. 1976. *Legitimation Crisis.* Translated by Thomas McCarthy. London: Heinemann.

Habermas, Jürgen. 1989. *The Structural Transformation of the Public Sphere: An Inquiry into a Category of Bourgeois Society.* Translated by Thomas Burger. Cambridge, MA: MIT Press.

Hajer, Maarten A. 1995. *The Politics of Environmental Discourse: Ecological Modernization and the Policy Process.* Oxford: Clarendon Press.

Hall, Stuart. 1996. "The West and the Rest: Discourse and Power." In *Modernity: An Introduction to Modern Societies,* edited by Stuart Hall, David Held, Don Hubert and Kenneth Thompson, 184–228. Malden, MA: Blackwell.

Harvey, David. 2005. *A Brief History of Neoliberalism.* Oxford: Oxford University Press.

Hawken, P. 2007. *Blessed Unrest.* New York: Viking Press.

Heinberg, Richard. 2011. *The End of Growth: Adapting to Our New Economic Reality.* Gabriola Island, BC, Canada: New Society Publishers.

Homer-Dixon, Thomas. 2006. *The Upside of Down: Catastrophe, Creativity and the Renewal of Civilization.* Toronto: A.A. Knopf.

Hunter, James Davison. 2010. *To Change the World: The Irony, Tragedy, and Possibility of Christianity in the Late Modern World.* New York: Oxford University Press.

Hunter, James Davison, and Alan Wolfe. 2006. *Is There a Culture War? A Dialogue on Values and American Public Life.* Washington, DC: Pew Research Center and Brookings Institution Press.

Idle No More. 2014. "Idle No More." Accessed September 10, 2014. http://www.idlenomore.ca/.

Jaccard, Mark. 2013. "The Accidental Activist." *The Walrus,* March 2013. Accessed September 10, 2014. http://thewalrus.ca/the-accidental-activist/.

Jackson, Tim. 2008. "The Challenge of Sustainable Lifestyles." In *State of the World 2008: Innovations for a Sustainable Economy,* edited by Linda Starke, 45–60. Washington, DC: Worldwatch Institute. Available at http://www.worldwatch.org/node/5561.

Jackson, Tim. 2009. *Prosperity Without Growth: Economics for a Finite Planet.* London: Earthscan.

Jacques, Peter J., Riley E. Dunlap, and Mark Freeman. 2008. "The Organisation of Denial: Conservative Think Tanks and Environmental Scepticism." *Environmental Politics* 17 (3): 349–385. doi: 10.1080/09644010802055576.

Klein, Naomi. 2013. "Overcoming 'Overburden': The Climate Crisis and a Unified Left Agenda." *Common Dreams.* Accessed September 30, 2014. https://www.commondreams.org/view/2013/09/04.

Korten, David C. 2009. *Agenda for a New Economy: From Phantom Wealth to Real Wealth.* San Francisco: Berrett-Koehler Publishers.

Lakoff, George. 2004. *Don't Think of an Elephant! Know Your Values and Frame the Debate.* White River Junction, VT: Chelsea Green.

Lasch, Christopher. 1991. *The True and Only Heaven: Progress and Its Critics.* New York: W.W. Norton.

Leopold, Aldo. 1966. *A Sand County Almanac: With Other Essays on Conservation from Round River.* New York: Random House.

Louv, R. 2008. *Last Child in the Woods.* 2nd ed. Chapel Hill, NC: Algonquin Books.

Lyotard, Jean-François. 1984. *The Postmodern Condition: A Report on Knowledge.* Minneapolis: University of Minnesota Press.

Maniates, Michael. 2013. "Teaching for Turbulence." In *State of the World 2013: Is Sustainability Still Possible?*, edited by Linda Starke, 255–268. Washington, DC: Worldwatch Institute.

Marcuse, Herbert. 1964. *One-Dimensional Man: Studies in the Ideology of Advanced Industrial Society.* Boston: Beacon Press.

Martens, Kathleen. 2013. "Elsipogtog Solidarity Is Spreading across Canada." *APTN National News,* October 17. Accessed September 10, 2014. http://aptn.ca/news/2013/10/17/elsipogtog-solidarity-is-spreading-across-canada/.

Martínez-Alier, Joan, Unai Pascual, Franck-Dominique Vivien, and Edwin Zaccai. 2010. "Sustainable De-Growth: Mapping the Context, Criticisms and Future Prospects of an Emergent Paradigm." *Ecological Economics* 69 (9): 1741–1747. doi: 10.1016/j.ecolecon.2010.04.017.

McFague, Sallie. 2001. *Life Abundant: Rethinking Theology and Economy for a Planet in Peril.* Minneapolis, MN: Fortress Press.

McKibben, Bill. 1989. *The End of Nature.* New York: Random House.

McKibben, Bill. 2010. *Earth: Making a Life on a Tough New Planet.* New York: Times Books.

Meadows, Donella H., Jørgen Randers, and Dennis L. Meadows. 2004. *The Limits to Growth: The 30-Year Update.* White River Junction, VT: Chelsea Green.

Merchant, Carolyn. 1980. *The Death of Nature: Women, Ecology, and the Scientific Revolution.* San Francisco: Harper & Row.

Milton, Kay. 1996. *Environmentalism and Cultural Theory: Exploring the Role of Anthropology in Environmental Discourse.* London: Routledge.

Nadeau, Robert. 2003. *The Wealth of Nature: How Mainstream Economics Has Failed the Environment.* New York: Columbia University Press.

Newspaper Audience Databank Inc. 2012. "Latest Study News." Accessed September 10, 2014. http://nadbank.com/en/nadbank-news/latest-study-news.

Ophuls, William. 1998. *Requiem for Modern Politics: The Tragedy of the Enlightenment and the Challenge of the New Millennium.* Boulder, CO: Westview Press.

Polanyi, Karl. 1944. *The Great Transformation.* Boston: Beacon Press.

Reese, S. D. 2003. "Framing Public Life: A Bridging Model for Media Research." In *Framing Public Life: Perspectives on Media and Our Understanding of the Social World*, edited by S. D. Reese, J. O. Gandy and A. E. Grant, 7–31. Mahwah, NJ: Lawrence Erlbaum Associates.

Rifkin, Jeremy. 1991. *Biosphere Politics: A New Consciousness for a New Century*. New York: Crown Publishers.

Rogers, Raymond Albert. 1994. *Nature and the Crisis of Modernity: A Critique of Contemporary Discourse on Managing the Earth*. Montréal: Black Rose Books.

Rubin, Jeff. 2012. *The End of Growth*. Toronto: Random House.

Schneider, François, Giorgos Kallis, and Joan Martinez-Alier. 2010. "Crisis or Opportunity? Economic Degrowth for Social Equity and Ecological Sustainability." *Journal of Cleaner Production* 18 (6): 511–518. doi: 10.1016/j.jclepro.2010.01.014.

Schwartz, Daniel, and Mark Gollom. 2013. "N.B. Fracking Protests and the Fight for Aboriginal Rights." *CBC News*, October 19. http://www.cbc.ca/news/n-b-fracking-protests-and-the-fight-for-aboriginal-rights-1.2126515.

Semeniuk, Ivan. 2013. "Scientists Push Campaign for Evidence-Based Decision Making from Government." *The Globe and Mail*, September 16. http://www.theglobeandmail.com/news/national/scientists-aim-to-put-state-of-canadian-research-in-the-public-spotlight-with-demonstrations/article14332546/.

Smelser, Neil J. 1959. *Social Change in the Industrial Revolution: An Application of Theory to the Lancashire Cotton Industry, 1770–1840*, International Library of Sociology and Social Reconstruction Series. London: Routledge & Kegan Paul.

Speth, James Gustave. 2008. *The Bridge at the Edge of the World*. New Haven, CT: Yale University Press.

Speth, James Gustave. 2011. "Off the Pedestal: Creating a New Vision of Economic Growth." *Yale Environment 360*. Accessed September 10, 2014. http://e360.yale.edu/content/print.msp?id=2409.

Stone, Robert (director). 2010. *Earth Days* (documentary film). Arlington, VA: PBS Distribution. DVD, 102 min.

Taylor, Charles. 1991. *The Malaise of Modernity*. Cambridge, MA: Harvard University Press.

United Nations Environment Programme. 2007. "Global Environment Outlook 4 (Geo-4)." Accessed September 10, 2014. http://www.unep.org/geo/geo4.asp.

United Nations Environment Programme. 2012. "Global Environment Outlook 5 (Geo-5)." Accessed September 10, 2014. http://www.unep.org/geo/geo5.asp.

van Dijk, Teun Adrianus, ed. 2011. *Discourse Studies: A Multidisciplinary Introduction*. 2nd ed. London: Sage Publications.

Victor, Peter A. 2008. *Managing Without Growth: Slower by Design, Not Disaster*. Cheltenham, UK: Edward Elgar.

Wallerstein, Immanuel. 1999. *The End of the World as We Know It: Social Science for the Twenty-First Century*. Minneapolis: University of Minnesota Press.

Wallerstein, Immanuel. 2004. *World-Systems Analysis: An Introduction*. Durham, NC: Duke University Press.

Winner, Langdon. 1986. *The Whale and the Reactor: A Search for Limits in an Age of High Technology*. Chicago: University of Chicago Press.

Wuthnow, Robert. 1989. *Communities of Discourse: Ideology and Social Structure in the Reformation, the Enlightenment, and European Socialism*. Cambridge, MA: Harvard University Press.

CONCLUSION

Continuing the Journey of Ecological Economics

Reorientation and Research

Ecological economics is grounded in the unavoidable insight that the economy is fully embedded in the energy and material flows of Earth and subject to the laws of the universe. It is a fundamental insight that must ground our thinking and action if we are to have any chance of successfully navigating the stormy Anthropocene. However, it will fall short of providing both compass and momentum if it does not take on two essential, but related, challenges: (1) development of itself—an internal agenda; and (2) furthering the extended agenda (discussed in the introduction to this book) of applying the reorientation generated by its insight into other domains, other disciplines, modes of thought and action, and more broadly how humanity relates to life and the world. Just as standard economics now is the direct or indirect touchstone for many of these domains, so we assume that a full acceptance of the embeddedness of our activities in the physical realm of the Earth—its conditions and limits—will precipitate an equivalent shift in the basic assumptions of these domains. We fail to make these changes at our peril.

This concluding chapter sets out, based on each section of this volume, how the reorientation we argue for can be furthered by future research opportunities and a fresh vision of the human prospect.

1. PROPOSED ETHICAL FOUNDATIONS

1.1. Rethinking Ethics

Ethics and political philosophy in the twentieth century, at least in the Anglophone countries, have largely focused on whether to emphasize human rights or utility. Yet, neither tradition has been firmly grounded in more sophisticated contemporary understandings of human subjectivity and social and ecological interdependencies. In the early twentieth century, Bergson (1911), Schweitzer (1987), Whitehead (1978), and others began an "organic" countermovement that began to articulate an integrated understanding of the relationship between the human self and the world. This inspiring counterperspective has come back to life as thinkers such as Leopold (1949), Callicott (1994), Jamieson (2002), Berry (1999), Elliott (2005), and Brown (2012) have explicitly sought to connect ethics to science. Indeed, contemporary science supports and situates understandings of the self and world that are contained in many of the world's ethical and religious belief systems, and in the work of previous philosophers such as Spinoza. The insights of neuroscience help us to understand how humanity can access and experience fundamental connections to a creative universe and undergird the construction of new narratives of humanity's place in it (Lakoff and Johnson 1999; Nadeau 2013). New frontiers of investigation and understanding are emerging, such as Hauser (2006) and Wilson (2002). We must advance the dialogue between moral philosophy and the insights from contemporary neuroscience, evolutionary biology and cosmology (Chaisson 2006), and complexity theory (Kauffman 1995).

1.2. Rethinking Agency

Behavioral economics in particular and evolutionary science in general now paint a much more complex picture of human motivations than is found in neoclassical economics. One clear research agenda for ecological behavioral economics becomes to identify how the human person, and in particular the brain, as a product of evolution shapes our actions. However, this is just the beginning of a rich and far-reaching agenda: human choice and conduct must be understood as embedded in energy and material flows, as well as in social and institutional matrices that shape and direct behavior. How can this understanding be used to help meet the challenges of the Anthropocene?

1.3. Rethinking Rationality

It is essential to further develop and deploy a rationality that respects and enhances the regenerative capacity of natural systems; takes a multigenerational perspective on preserving, enhancing, and restoring the underpinnings of life; and understands that decisions necessarily involve both human and natural systems (Princen 2005). How is "rationality" to be understood in a revised conception of agency? An unavoidable question for ecological economics is whether agency is confined to people. What is the place for the insights from many traditional cultures that agency is widespread in the natural world? Is agency confined to intentional actions? Do viruses and hurricanes have agency?

1.4. Rethinking the Human

As noted in various chapters, the model of the human in standard economics is a very strange, deracinated, being. This model was designed to mimic simple physical processes so as to make the mathematics work out; its danger, as in all simple models of humans to which humans are susceptible, is its appealing social, cultural, and political power. The deepest danger is that if humans do not fit the models, they need to be made to fit the models or be disregarded, killed, or left to die. In addition, the pervasiveness of the economistic perspective has strongly influenced our fundamental senses of selves, objects, and nature (Horkheimer and Adorno 2002). The "rethinking of agency" research agenda is only one reflection of the current trend toward rethinking the human through the application of behavioral science, etc. One specific research role of ecological economics is to enrich the rethinking of the human in its ecological matrices and settings. Human preferences are dependent on social context, individual histories, and conscious preference development (Albert and Hahnel 1991). An ecological economics model incorporates a sense of fairness and a socially contingent decision-making rooted in the biology of moral reasoning (Fehr and Gächter 2002). Models of decision-making, such as "prospect theory" (Tversky and Kahneman 1981) and "biased cultural transmission" (Henrich 2004), have proved to be better predictors of economic behavior than the axiomatic rational actor model.

1.5. Changing Our Relationship to the Community of Beings

Neoclassical economics conveniently relegates all other forms of life and the processes that support it to purely instrumental standing. This could be a stepchild of the doctrine of special creation—that humanity is made in the image of God—found in the book of Genesis. On this view, other animals and Earth itself have a different moral standing than humans. This idea finds no support within an evolutionary worldview, thus opening wide the door to reimagining our place in the community of beings and the universe as a whole. Is the universe our community, as Stuart Kauffman suggested in *At Home in the Universe* (Kauffman 1995)?

1.6. Reconsidering Justice and Distribution

At its most fundamental level, ecological economics is about justice. On a planet open to energy and closed to matter, what are fair shares of Earth's life-support capacity? In an evolutionary paradigm, there is no reason to think that humans are the only ones with moral standing. Justice pertains to all creatures great and small, now and evermore.

At the human level, ecological economics has had mostly a marginal interest in the marginal people of the world, with some exceptions (e.g., Martínez-Alier 2002); much more needs to be done, drawing on more recent detailed research (e.g., Banerjee and Duflo 2011), the discussions toward the next phase of the implementation of the Millennium Development Goals (post-2015), and multiple case studies on unsustainable development (e.g., Suyanto, Khususiyah, and Leimona 2007). Marginal people are often at the "cutting edge" of forest destruction, water scarcity and degradation, and overexploited ecosystems—how does ecological economics respond to this at more than the theoretical level?

1.7. Recasting Education

Multidisciplinary and interdisciplinary initiatives have made some progress in educating students on the multifaceted nature of the current human crisis, but they rarely question whether individual disciplines are authentic and scientifically sound. Yet, many of these disciplines can be likened to "orphans." They are alive in pedagogy and practice, but their intellectual parents are often incompatible with what we know about the universe and our place in it at the dawn of the new millennium—for this reason, they should be called "orphans."

These frameworks and/or disciplines can be divided into at least two groups. First, there are several normative structures that shape and mediate humanity's relationship with life and the world. These are directly prescriptive—they tell us what we should do. Economists are happy to share their ideas about how to stimulate growth; finance to prescribe how to get rich; law to adjudicate between persons in conflict, and among various justice claims; governance for the powers of the state over the individual and how to secure ourselves from foreign enemies. Ethics offers guidelines for being a good person and living a good life. Second, there is a much broader set of disciplines that also suffer from deep ontological errors. Engineering takes it for granted that the world belongs to humanity and may be modified, often at will, to suit our purposes no matter how trivial. Conventional agriculture assumes that "food security" applies only to people. It can be argued that many of the social sciences are orphan disciplines in that they rely on the assumption of a sharp nature–culture divide (Kohn 2013).

2. MEASUREMENTS

2.1. Rethinking Price

The economy is now largely oriented around a single signal: the price signal. Price is such a remarkable, successful, and ubiquitous invention that we have come to believe it is inevitable and inherent to things. Price conveys a signal about the relationship between resources and desires: how much will I give up or gain for what I want? When the price signal is aggregated across all transactions, it produces a signal about the overall movement of goods and services toward the satisfaction of desires in the economy. That movement is currently captured poorly by the gross domestic product (GDP). Orienting ourselves toward that number orients us toward growth, leading to further declines in life's prospects.

Ecological economics has made substantial contributions to adjusting GDP to better account for income distribution, unpriced costs and benefits of economic activity, and broader tradeoffs between paid and unpaid work (Bagstad, Berik, and Gaddis 2014; Cobb, Halstead, and Rowe 1995; Daly and Cobb 1994; Lawn 2005). However, these efforts still accept price as a unit of comparability, and thus implicit substitutability between economic and non-economic human wants and needs. Vatn and Bromley (1994) challenged us to imagine "choices without prices without apologies."

An ecological economy would require other signals, relating us to other dimensions of our choices, including their ecological impacts and social transactions. Further research will be required to determine how these signals could be constructed out of the range of indicators we have providing us with information about our relationship to planetary boundaries. Can we aggregate other signals into an impact score of all our decisions? Will it relate to an ecological currency or currencies that function differently from money? Will new aggregates relate to an economy that focusses more on public goods rather than a market economy that focusses more on private goods? These questions are beginning to be explored in various fora, such as social network platforms, and require a slate of empirical investigations.

2.2. Embracing a Plurality of Values

A significant literature critiques the monetary foundation of cost-benefit analysis in economics, including the implicit assumption of substitutability between human-made and natural capital (Gowdy 1997); the phenomena of "crowding-out" moral behavior with the introduction of monetary values (Frey and Oberholzer-Gee 1997); the importance of lexicographical preferences in evaluating tradeoffs between economic, social, and environmental goods or services (Spash and Hanley 1995); and the existence of hyperbolic discounting in evaluating medium-to-distant future outcomes (Laibson 1997). Ecological economics should continue to explore "values-plural" approaches, such as multicriteria decision analyses that explicitly address the biological, social, and complex nature of human decision-making (Kosoy et al. 2010; Martinez-Alier, Munda, and O'Neill 1998).

2.3. Measuring Economic Production as a Biophysical Process

Neoclassical economics treats production as an allocation of given resources without regard to biophysical processes. Ecological economics is instead grounded in material and energy throughput analysis, including measurement and models of ecological footprints (Wackernagel and Rees 1996), human appropriation of net primary production (Haberl et al. 2007), and spatially explicit generation, delivery, and demand of ecosystem services (Kareiva et al. 2011). Ecological macroeconomic systems modeling is now emerging to investigate strategies for growth-neutral or degrowth economies (Jackson 2009; Kallis 2011; Victor 2008). Significant research remains on measuring production in economic systems as a physical transformation of low-entropy matter and energy to high-entropy waste—a return to early goals of ecological economics reflected in the work of Georgesu-Roegen, Daly, Christensen, and others.

3. IMPLICATIONS

3.1. Modeling and Growth

What is a fair distribution of income between capital and labor in the context of a slow or nongrowing economy? Was Piketty (2014) correct in arguing that slower growth favors capital over labor (given certain assumptions about the rate of return to capital and the savings rate)? Or, would slower growth result from financially "unproductive" investment, which in effect reduces the return to capital overall? What is the ease or difficulty with which capital can be substituted for labor? How can these issues be addressed within ecological macroeconomic models?

3.2. Fitting the Economy to the Earth

It is essential to find a way to tie science and its insights into the production of economic norms and the legal frameworks that underpin them. We are now in the midst of producing new ways to generate and respond to legal and economic norms. Just as the emergence of writing and then publication transformed how norms were generated and communicated, so too are computerized network communications producing a revolution in the transmission of norms, which might be called "real-time law." Real-time law allows us to produce norms reflexively: to have them adjust and respond in real time to the impacts they generate. This is because the latency between the expression of the norm and the impacts of behavior according to it can be reduced to almost zero as we gather data on networks. In the past, we would promulgate a norm, state its meaning and requirements ex ante, try to enforce it ex post, and then amend it periodically in light of shifting social consensus. All of this can now be sped up—and indeed, it needs to be if we are going to have any prospect of averting the catastrophic collective impacts of our choices.

3.3. Developing the Field of Ecological Finance

From an ecological economics point of view, wealth is the ability to support desirable far-from-equilibrium systems, such as human beings, trees, hospitals, universities, etc. What is only now beginning to be explored is the relationship between these systems and financial instruments of all kinds. The life-support capacity of Earth is arguably about the same (or less) than it was in, say, 1980; however, the total amount of paper and digital wealth is many times what it was then. Is this a problem, and if so, of what kind, magnitude, urgency, and severity? How should we respond? Is there a way to systematically relate money in all its forms to Earth's life-support systems? How should we deal with the fact that those who hold the money form of wealth control the institutions that would need to undergo radical change that are likely needed? Frederick Soddy's *The Role of Money* (1935) is a promising place to begin. Is there a path through the thicket; and, if so, who will venture along it?

3.4. Trading in the Language of "Free" Trade

The current and regrettably expanding trade regime rests on ideas such as "comparative advantage," the need for growth, "the third world," and other constructs that find little footing in an ecological framework. Indeed, these ideas serve to facilitate and legitimate the liquidation and appropriation of the low-entropy materials found in forest, farm, and fossil fuels alike. How should trade be understood in an ecological economics context and in the full world of the Anthropocene? Does the social imaginary of "development," "the third world," and "poverty" crumble and collapse in the face of a fresh and deeper understanding of wealth? How does ecological economics relate to the need for a "postdevelopment" discourse, such as that called for by Arturo Escobar (2012)?

3.5. Rethinking Transactions

Ever since Adam Smith's division of economic inquiry into a theory of moral sentiments on the one hand and an inquiry into the wealth of nations on the other, there has tended to be split between caricatures of human behavior from *homo economicus* to *homo reciprocans*. We need some further research on the role of a gift economy in an ecological economy. We tend to assume that the operation of an exchange economy will be extended, albeit under different parameters, to an ecological economy. However, this is too simple. The question for the further development of ecological economics becomes: what is the appropriate balance between gift and exchange economies within an ecological economy, and how is it to be achieved?

3.6. Rebuilding Governance and Law

A key postulate of the liberal politics that still underpins most Western democracies is that people may live how and where they wish, pursuing what John Stuart Mill called purely self-regarding actions (Mill [1859] 2011). Yet, ecology and thermodynamics, plus the overwhelming evidence concerning anthropogenic causes of climate change and the sixth mass extinction, clearly reveal that pure self-regard is a misnomer at best. Recognizing this, our project builds on work that pursues the suggestion in Rockström et al. (2009) of the need for novel and adaptive governance approaches from the local to the global level, based on planetary boundaries that define limits (Galaz et al. 2012).

A foundation of law in many liberal cultures is the strict protection of private property. However, the main thrust of current scientific understanding, particularly ecology, directly challenges the idea of severability on which a liberal understanding of property depends and underscores the interconnection between public goods and private property. In addition, the assumption that humans are the sole rightful owners of Earth is difficult to ground in an evolutionary worldview. We will need to advance a more holistic approach to the law that recognizes ecological limits and interdependencies, and that humans coevolved with the rest of life, building on such work as Rose (1997), Solan (2002), Hornstein (2005), Ruhl (2012), Garver (2013), Cullinan (2011), Bosselmann (2008), Ostrom (1990), and Ost (2003).

As a final coda to this ambitious list, it is worth pointing out that ecological economics in itself provides an ideal example of how to build a study of the human economy that is (1) viewed both as a complex social system and as one embedded in the biophysical universe; (2) grounded in the evidentiary standard of the physical and biological sciences; and (3) framed in a problem-solving approach built on methodological pluralism that borrows broadly from many fields (Gowdy and Erickson 2005).

Extending the vision of ecological economics is an opportunity that must not be missed. We are just at the beginning of a great journey of discovery of our place in the cosmos—and the implications of that place for ourselves and the rest of life with which we share heritage and destiny. The findings will be revolutionary, frightening, unsettling, and full of opportunities.

REFERENCES

Albert, Michael, and Robin Hahnel. 1991. *The Political Economy of Participatory Economics*. Princeton, NJ: Princeton University Press.

Bagstad, Kenneth J., Günseli Berik, and Erica J. Brown Gaddis. 2014. "Methodological Developments in US State-Level Genuine Progress Indicators: Toward GPI 2.0." *Ecological Indicators* 45: 474–485. doi: 10.1016/j.ecolind.2014.05.005.

Banerjee, Abhijit V., and Esther Duflo. 2011. *Poor Economics: A Radical Rethinking of the Way to Fight Global Poverty*. New York: PublicAffairs.

Bergson, Henri. 1911. *Creative Evolution*. Translated by Arthur Mitchell. New York: H. Holt and Company. http://www.gutenberg.org/ebooks/26163.

Berry, Thomas. 1999. *The Great Work: Our Way into the Future*. New York: Three Rivers Press.

Bosselmann, Klaus. 2008. *The Principle of Sustainability: Transforming Law and Governance*. Burlington, VT: Ashgate.

Brown, Peter G. 2012. Ethics for Economics in the Anthropocene. In *Teilhard Series No. 64*. Woodbridge, CT: American Teilhard Association.

Callicott, J. Baird. 1994. *Earth's Insights: A Survey of Ecological Ethics from the Mediterranean Basin to the Australian Outback*. Berkeley: University of California Press.

Chaisson, Eric. 2006. *Epic of Evolution: Seven Ages of the Cosmos*. New York: Columbia University Press.

Cobb, Clifford W., Ted Halstead, and Jonathan Rowe. 1995. *The Genuine Progress Indicator: Summary of Data and Methodology*. San Francisco: Redefining Progress.

Cullinan, Cormac. 2011. *Wild Law: A Manifesto for Earth Justice.* 2nd ed. White River Junction, VT: Chelsea Green Publishing.

Daly, Herman E., and John B. Cobb. 1994. *For the Common Good: Redirecting the Economy toward Community, the Environment, and a Sustainable Future.* 2nd ed. Boston: Beacon Press.

Elliott, Herschel. 2005. *Ethics for a Finite World: An Essay Concerning a Sustainable Future.* Golden, CO: Fulcrum Publishing.

Escobar, Arturo. 2012. *Encountering Development: The Making and Unmaking of the Third World.* Princeton, NJ: Princeton University Press.

Fehr, Ernst, and Simon Gächter. 2002. "Altruistic Punishment in Humans." *Nature* 415 (6868): 137–140. doi: 10.1038/415137a.

Frey, Bruno S., and Felix Oberholzer-Gee. 1997. "The Cost of Price Incentives: An Empirical Analysis of Motivation Crowding-Out." *The American Economic Review* 87 (4): 746–755. doi: 10.2307/2951373.

Galaz, Victor, Frank Biermann, Carl Folke, Måns Nilsson, and Per Olsson. 2012. "Global Environmental Governance and Planetary Boundaries: An Introduction." *Ecological Economics* 81: 1–3. doi: 10.1016/j.ecolecon.2012.02.023.

Garver, Geoffrey. 2013. "The Rule of Ecological Law: The Legal Complement to Degrowth Economics." *Sustainability* 5 (1): 316–337. doi: 10.3390/su5010316.

Gowdy, John, and Jon D. Erickson. 2005. "The Approach of Ecological Economics." *Cambridge Journal of Economics* 29 (2): 207–222. doi: 10.1093/cje/bei033.

Gowdy, John M. 1997. "The Value of Biodiversity: Markets, Society, and Ecosystems." *Land Economics* 73 (1): 25–41. doi: 10.2307/3147075.

Haberl, Helmut, K. Heinz Erb, Fridolin Krausmann, Veronika Gaube, Alberte Bondeau, Christoph Plutzar, Simone Gingrich, Wolfgang Lucht, and Marina Fischer-Kowalski. 2007. "Quantifying and Mapping the Human Appropriation of Net Primary Production in Earth's Terrestrial Ecosystems." *Proceedings of the National Academy of Sciences of the United States of America* 104 (31): 12942–12947. http://www.jstor.org/stable/25436409.

Hauser, Marc D. 2006. *Moral Minds: How Nature Designed Our Universal Sense of Right and Wrong.* New York: Ecco.

Henrich, Joseph. 2004. "Cultural Group Selection, Coevolutionary Processes and Large-Scale Cooperation." *Journal of Economic Behavior & Organization* 53 (1): 3–35. doi: 10.1016/S0167-2681(03)00094-5.

Horkheimer, Max, and Theodor W. Adorno. 2002. *Dialectic of Enlightenment: Philosophical Fragments.* Edited by Gunzelin Schmid Noerr. Translated by Edmund Jephcott. Stanford, Calif.: Stanford University Press.

Hornstein, Donald T. 2005. "Complexity Theory, Adaptation, and Administrative Law." *Duke Law Journal* 54 (4): 913–960. doi: 10.2307/40040504.

Jackson, Tim. 2009. *Prosperity without Growth: Economics for a Finite Planet.* London: Earthscan.

Jamieson, Dale. 2002. *Morality's Progress: Essays on Humans, Other Animals, and the Rest of Nature.* Oxford: Clarendon Press.

Kallis, Giorgos. 2011. "In Defence of Degrowth." *Ecological Economics* 70 (5): 873–880. doi: 10.1016/j.ecolecon.2010.12.007.

Kareiva, Peter, Heather Tallis, Taylor H. Ricketts, Gretchen C. Daily, and Stephen Polasky, eds. 2011. *Natural Capital: Theory and Practice of Mapping Ecosystem Services.* New York: Oxford University Press.

Kauffman, Stuart A. 1995. *At Home in the Universe: The Search for Laws of Self-Organization and Complexity.* Oxford: Oxford University Press.

Kohn, Eduardo. 2013. *How Forests Think: toward an Anthropology Beyond the Human.* Berkeley: University of California Press.

Kosoy, Nicolas, Makiko Yashiro, Carlota Molinero, and Anantha Duraiappah. 2010. "Valuation of Ecosystem Services: Methods, Opportunities and Policy Implications." In *Valuation of Regulating Services of Ecosystems: Methodology and Applications,* edited by Pushpam Kumar and Michael D. Wood, 222–234. Abingdon, Oxon, UK: Routledge.

Laibson, David. 1997. "Golden Eggs and Hyperbolic Discounting." *The Quarterly Journal of Economics* 112 (2): 443–477. doi: 10.2307/2951242.

Lakoff, George, and Mark Johnson. 1999. *Philosophy in the Flesh: The Embodied Mind and Its Challenge to Western Thought.* New York: Basic Books.

Lawn, Philip A. 2005. "An Assessment of the Valuation Methods Used to Calculate the Index of Sustainable Economic Welfare (ISEW), Genuine Progress Indicator (GPI), and Sustainable Net Benefit Index (SNBI)." *Environment, Development and Sustainability* 7 (2): 185–208. doi: 10.1007/s10668-005-7312-4.

Leopold, Aldo. 1949. *A Sand County Almanac and Sketches Here and There.* New York: Oxford University Press.

Martínez-Alier, Joan. 2002. *The Environmentalism of the Poor: A Study of Ecological Conflicts and Valuation.* Cheltenham, UK: Edward Elgar.

Martinez-Alier, Joan, Giuseppe Munda, and John O'Neill. 1998. "Weak Comparability of Values as a Foundation for Ecological Economics." *Ecological Economics* 26 (3): 277–286. doi: 10.1016/S0921-8009(97)00120-1.

Mill, John Stuart. (1859) 2011. *On Liberty.* Project Gutenberg. http://www.gutenberg.org/ebooks/34901.

Nadeau, Robert. 2013. *Rebirth of the Sacred: Science, Religion and the New Environmental Ethos.* Oxford: Oxford University Press.

Ost, François. 2003. *La Nature Hors La Loi: L'écologie À L'épreuve Du Droit.* Paris: La Découverte.

Ostrom, Elinor. 1990. *Governing the Commons: The Evolution of Institutions for Collective Action.* Cambridge, UK: Cambridge University Press.

Piketty, Thomas. 2014. *Capital in the Twenty-First Century.* Translated by Arthur Goldhammer. Cambridge, MA: Belknap Press of Harvard University Press.

Princen, Thomas. 2005. *The Logic of Sufficiency.* Cambridge, MA: MIT Press.

Rockström, Johan, Will Steffen, Kevin Noone, Asa Persson, F. Stuart Chapin, III, Eric F. Lambin, Timothy M. Lenton, Marten Scheffer, Carl Folke, Hans Joachim Schellnhuber, Bjorn Nykvist, Cynthia A. de Wit, Terry Hughes, Sander van der Leeuw, Henning Rodhe, Sverker Sorlin, Peter K. Snyder, Robert Costanza, Uno Svedin, Malin Falkenmark, Louise Karlberg, Robert W. Corell, Victoria J. Fabry, James Hansen, Brian Walker, Diana Liverman, Katherine Richardson, Paul Crutzen, and Jonathan A. Foley. 2009. "Planetary Boundaries: Exploring the Safe Operating Space for Humanity." *Ecology and Society* 14 (2): 32. http://www.ecologyandsociety.org/vol14/iss2/art32/.

Rose, Carol M. 1997. "Demystifying Ecosystem Management." *Ecology Law Quarterly* 24: 865–869.

Ruhl, J. B. 2012. "Panarchy and the Law." *Ecology and Society* 17 (3): 31. doi:10.5751 /ES-05109–170331. http://www.ecologyandsociety.org/vol17/iss3/art31/.

Schweitzer, Albert. 1987. *The Philosophy of Civilization*. Translated by C. T. Campion. Amherst, NY: Prometheus Books. First published 1949.

Soddy, Frederick. 1935. *The Role of Money: What It Should Be, Contrasted with What It Has Become*. New York: Harcourt, Brace.

Solan, Lawrence M. 2002. "Cognitive Legal Studies: Categorization and Imagination in the Mind of Law. A Conference in Celebration of the Publication of Steven L. Winter's Book, *A Clearing in the Forest: Law, Life, and Mind*—Introduction." *Brooklyn Law Review* 67 (4): 941–948.

Spash, Clive L., and Nick Hanley. 1995. "Preferences, Information and Biodiversity Preservation." *Ecological Economics* 12 (3): 191–208. doi: 10.1016/0921–8009(94)00056–2.

Suyanto, S., Noviana Khususiyah, and Beria Leimona. 2007. "Poverty and Environmental Services: Case Study in Way Besai Watershed, Lampung Province, Indonesia." *Ecology and Society* 12 (2): 13. http://www.ecologyandsociety.org/vol12/iss2/art13/.

Tversky, A., and D. Kahneman. 1981. "The Framing of Decisions and the Psychology of Choice." *Science* 211: 453–458. doi: 10.1126/science.7455683.

Vatn, Arild, and Daniel W. Bromley. 1994. "Choices without Prices without Apologies." *Journal of Environmental Economics and Management* 26 (2): 129–148. doi: 10.1006 /jeem.1994.1008.

Victor, Peter A. 2008. *Managing without Growth: Slower by Design, Not Disaster*. Cheltenham, UK: Edward Elgar.

Wackernagel, Mathis, and William E. Rees. 1996. *Our Ecological Footprint: Reducing Human Impact on the Earth*. Gabriola Island, BC: New Society Publishers.

Whitehead, Alfred North. 1978. *Process and Reality*. Edited by David Ray Griffin and Donald Wynne Sherburne. Corrected ed. New York: The Free Press.

Wilson, David Sloan. 2002. *Darwin's Cathedral: Evolution, Religion, and the Nature of Society*. Chicago: University of Chicago Press.

CONTRIBUTORS

PETER G. BROWN, PHD, professor, School of Environment and the Departments of Geography and Natural Resource Sciences, McGill University. Peter Brown's career has concentrated on the practical uses of philosophy to think critically about the goals of society. Since the 1980s, this work has centered on the deterioration of Earth's life-support capacity and the thought systems that facilitate and legitimate this decline. He is the author of *Restoring the Public Trust: A Fresh Vision for Progressive Government in America* and *The Commonwealth of Life: Economics for a Flourishing Earth.* He is also a coauthor of a book on macroeconomics and global governance entitled *Right Relationship: Building a Whole Earth Economy;* and coedited *Water Ethics: Foundational Readings for Students and Professionals.* He is currently the principal investigator for Economics for the Anthropocene: Re-Grounding the Human/Earth Relationship, a partnership among McGill, the University of Vermont, and York University in Toronto. He is a member of the Religious Society of Friends (Quakers), and the Club of Rome.

PHILIP DUGUAY works as a public-affairs adviser to the government of Québec in Boston, Massachusetts, focusing mostly on energy, environmental, and transportation policy issues in the six New England states. He holds a BCL-LLB from the McGill University Faculty of Law and a BA with honors from Dalhousie University.

JON D. ERICKSON, PHD, professor, Rubenstein School of Environment and Natural Resources, and fellow, Gund Institute for Ecological Economics University of Vermont (UVM). Jon Erickson has published widely on ecological economics, climate-change policy, renewable-energy economics, and environmental management; led international research and education programs as a Fulbright Scholar in Tanzania and visiting professor in the Dominican Republic, Iceland, and Slovakia; produced Emmy award–winning documentary films on water-, energy-, and food-system transitions; and founded and led numerous nonprofit organizations, including the U.S. Society for Ecological Economics, Adirondack Research Consortium, Deportes para la Vida, and Bright Blue EcoMedia. He currently leads UVM's participation in the Economics for the Anthropocene research and doctoral training partnership with McGill and York Universities.

JAMES W. FYLES, PHD, professor, Department of Natural Resource Sciences, McGill University. Jim Fyles is an ecosystem ecologist with broad interest and expertise in the ecology of forests, agro-ecosystems, and devastated lands. Raised in Victoria, British Columbia, he obtained his BSc and MSc in ecology from the University of Victoria, and his PhD jointly in soil science and botany at the University of Alberta in 1986. He holds the Tomlinson Chair in Forest Ecology at McGill. He has been the chair of the department since 2011. He is the director of the Molson Nature Reserve and the Morgan Arboretum, peri-urban conservation and research areas near the Macdonald Campus. Between 2004 and 2010, he was the scientific director of the Sustainable Forest Management Network Centre of Excellence (SFM-NCE), a national research network involving partners from industry, governments, aboriginal groups, and non-governmental organizations. Dr. Fyles' research interests focus on the interrelationships among human activity, organisms, soil disturbance, and climate that structure patterns of ecosystem function across multiple scales. Through his multidisciplinary work on food security and with the SFM-NCE, he has become increasingly interested in complex social-ecological systems and the relationships among scientific knowledge, policy, and management of natural landscapes. Dr. Fyles has published more than eighty articles in scientific journals and coauthored many knowledge-exchange documents. He has received the Macdonald College Award for Teaching Excellence.

GEOFFREY GARVER, PHD candidate, Department of Geography, McGill University. Geoff Garver is a doctoral candidate in geography at McGill University and project coordinator of the Economics for the Anthropocene partnership, a research, teaching, and outreach project among McGill University, York University, and the University of Vermont along with twenty-two other academic, civil-society, and governmental organizations. His degrees include a BS (chemical engineering) from Cornell University (1982), a JD cum laude from the University of Michigan Law School (1987), and an LLM from McGill University (2011). His PhD dissertation will build on his LLM thesis, "The Rule of Ecological Law: A Transformative Legal and Institutional Framework for the Human–Earth Relationship." Previously, he worked for twenty years in public service, most recently from 2000–2007 as director of submissions on enforcement matters at the Commission for Environmental Cooperation, and earlier for the Environment and Natural Resources Division of the U.S. Department of Justice, the U.S. Environmental Protection Agency, and U.S. District Court Judge Conrad Cyr. Mr. Garver is coauthor of *Right Relationship: Building a Whole Earth Economy* (2009), as well as several published articles and book chapters. From 2010 to 2013, he was a member of the Joint Public Advisory Committee of North America's Commission for Environmental Cooperation.

MARK S. GOLDBERG, PHD, professor, McGill University, Division of Clinical Epidemiology, Department of Medicine, McGill University Health Centre (MUHC)–Research Institute. Dr. Mark Goldberg holds a PhD from McGill University (1991) in epidemiology and biostatistics. He is full professor in the Department of Medicine, McGill University; a member of the Division of Clinical Epidemiology, Royal Victoria Hospital, MUHC; and an associate member in the Department of Epidemiology and Biostatistics, the McGill School of Environment, and the Department of Oncology. From 2009 until 2013, he was co–editor in chief of the scholarly journal *Environmental Research*. His main interests in research have been in occupational and environmental epidemiology, focusing especially on environmental and occupational causes of cancer as well as the short- and long-term effects of air pollution on health. His research bridges clinical and environmental epidemiology and he has published more than 135 papers in peer-reviewed journals, covering myriad topics, including the health effects arising from exposures to ambient biogas produced in municipal solid-waste sites, environmental and occupational causes of disease, and occupational and environmental investigations of breast cancer, ovarian cancer, colon cancer, and lung cancer. Notable among these are his recent findings that chronic exposure to ambient air pollution may cause breast cancer and prostate cancer. He was a member of Health Canada's Science Advisory Board and he has sat on a number of expert committees of the U.S. National Academies.

JANICE HARVEY, MPhil in policy studies, IDST PhD candidate (ABD), University of New Brunswick, Lecturer, St. Thomas University. Janice Harvey is a PhD candidate in the Interdisciplinary PhD program at the University of New Brunswick and lecturer in the Environment and Society Programme at St. Thomas University in Fredericton. She has worked for three decades as a policy analyst and advocate in the Canadian environmental movement, primarily through her work with the Conservation Council of New Brunswick. She participated in several international nongovernmental organization (NGO) preparatory conferences for the 1992 Earth Summit and in solidarity exchange projects with NGOs in Latin America. She has held directorships and positions on several NGO and federal-government boards, steering committees, and advisory committees and served on the Premier's Round Table on Environment and Economy from 1992 to 2002. From 1995 to 2011, she wrote a weekly public-affairs column for the *New Brunswick Telegraph Journal*, a provincial daily paper, under the banner "A Civil Society." For several years, she was also a member of New Brunswick CBC Radio's weekly political panel and, for two elections, CBC's election-night political panel. In 2010, she received a Social Sciences and Humanities Research Council doctoral fellowship to examine in more depth how the dominant economic discourse, as well as the cultural context of consumerism, has resulted in a postecologist discourse of "sustaining the unsustainable."

TIM JACKSON, PHD, is professor of sustainable development at the University of Surrey. He currently holds a professorial fellowship on Prosperity and Sustainability in the Green Economy funded by the Economic and Social Research Council. Tim has been at the forefront of international debates about sustainable development for over two decades and has worked closely with the UK government, the United Nations, and numerous private companies and NGOs to bring social-science research into our understanding of sustainability. His research interests focus on the economic and social dimensions of the relationship between sustainability and prosperity. Between 2004 and 2011, Tim was economics commissioner on the UK Sustainable Development Commission, where his work culminated in the publication of his controversial and groundbreaking book *Prosperity Without Growth: Economics for a Finite Planet* (Routledge 2009), which was subsequently translated into sixteen languages. Since 2010, Tim has been engaged in an ambitious collaborative project to build a new ecological macroeconomics. He and Peter Victor (York University, Canada) are developing the conceptual and empirical basis for an economy in which stability no longer depends on relentless consumption growth, and they are illustrating these possibilities with system-dynamic models of the national economy. In addition to his academic work, Tim is an award-winning playwright with numerous radio-writing credits for the BBC.

RICHARD JANDA, PHD, is an associate professor at the Faculty of Law of McGill University and an associate member of the McGill School of Environment. He has taught environmental law and sustainable development and is currently leading the McGill Social Score Project, which is seeking to produce a signaling mechanism distinct from the price signal, relating to us in real time the environmental and social impacts of our choices.

BRUCE JENNINGS, MA, is director of bioethics at the Center for Humans and Nature, Dobbs Ferry, New York, a research institute that studies ethical and policy questions in conservation, ecological economics and politics, and environmental ethics. He is the editor of the center's electronic journal, *Minding Nature*. He holds faculty appointments at Yale University and Vanderbilt University. He also is senior advisor and an elected fellow at the Hastings Center, where he served as executive director from 1991–1999. He is editor in chief of *Bioethics*, 4th ed. (formerly the *Encyclopedia of Bioethics*), a six-volume reference work in the field of bioethics published in 2014. A political scientist by training, he has written and edited numerous books and articles on ethical issues in public policy, particularly in the areas of health and environmental policy. Among his recent books are *Public Health Ethics: Theory, Policy, and Practice* and *The Perversion of Autonomy: Coercion and Constraints in a Liberal Society*.

RICHARD LEHUN, DCL, is an attorney and teaching fellow at McGill University in the areas of philosophy of law and justice theory. Richard Lehun studied under Jürgen Habermas at the Johann Wolfgang Goethe University in Frankfurt on a German Academic Exchange Service (DAAD) scholarship and completed his Magister Artium on nonconceptual truth claims in T. W. Adorno's *Negative Dialectics* and *Aesthetic Theory*. After completing his Magister Artium, Lehun pursued graduate studies in Canada, first completing a BCL/LLB and then the DCL program at McGill University. Lehun's current research is focused on the concept of the fiduciary as applied to social transformation.

QI FENG LIN is a doctoral candidate in environmental ethics at McGill University. His research interests lie in Aldo Leopold, Daoism, and alternative worldviews and economic systems. The working title of his doctoral dissertation is "Rethinking the Concept of the Self in U.S. Forestry." He majored in statistics as an undergraduate at the National University of Singapore and graduated with a master's degree from the Yale School of Forestry and Environmental Studies.

NANCY E. MAYO, PHD, is a professor in the Department of Medicine and in the School of Physical and Occupational Therapy, McGill University. At McGill, Dr. Mayo leads a Health Outcomes Research Unit. She also heads the Health Outcomes Axis and the McGill University Health Center Research Institute. Trained originally as a physical therapist, Dr. Mayo holds a PhD in epidemiology and biostatistics. Her research focus has been in measuring the health of populations and contributing evidence toward ways of improving health outcomes of vulnerable populations. She has a long-term interest in models of health and in understanding drivers of health and health change over time. She has published more than 200 research papers and presented her work around the world. Dr. Mayo is also a committed educator, teaching core research-methods courses, and she has supervised more than eighty MSc and PhD students in rehabilitation science and epidemiology.

PETER TIMMERMAN, MA, associate professor, Faculty of Environmental Studies, York University. Peter Timmerman has been working on environmental issues for many years, beginning with emergency and risk research, early work on climate change, coastal zone management, and nuclear waste management. He has been the coordinator for the joint Faculty of Environmental Studies/Schulich Business and the Environment graduate diploma for a number of years. He now works on environmental philosophy and ethics, including religion and ecology, with a special research focus on Buddhism and ecology in South and Southeast Asia. In the area of ecological economics, he is currently working on the rise of the metaphors of progress, growth, and development in the eighteenth century.

PETER A. VICTOR, PHD, is a professor in the York University Faculty of Environmental Studies. Dr. Victor is an economist who has worked on environmental issues for more than forty years as an academic, consultant, and public servant. By extending input-output models to include material flows to and from the environment, in the 1960s, Dr. Victor provided ecological economics with a practical, quantitative method for linking the economy to the environment. His more recent work on alternatives to economic growth was recognized by the award of the Molson Prize from the Canada Council for the Arts (2011) and the Boulding Memorial Prize (2014) from the International Society for Ecological Economics.

INDEX

absolute decoupling, 241

abstract reasoning, 72

abundance: ethics of, 31–35; scarcity and, 33–34

agency, 358–59

aggregate environmental impact, 153

aggregate indicators, 131–32; scale relating to, 162–63

agriculture: Leopold on, 212; transition to, 26

allied health fields, 194–96

alternative Western tradition, 27–31

anarchism, 281

animals, 41, 48

Anthropocene, 184n3, 295, 357

antiglobalization movement, 351n8

appropriative economic activity, 293

Arendt, Hannah, 275–76, 287

Aristotle: on economic growth, 30; ethics relating to, 27–31; on household management, 29; on justice, 99–100; on usury, 29–30; on value and exchange, 28–31

atmospheric boundaries, 158–59

atmospheric nitrogen loading, 176–79

atonement, 82–83

attachment, 35

authoritarianism, 116nn2–3; ethics and, 91–93

awakening, 43–44

back pain, 144–45

Balinese economics, 57–59

Balinese ethics, 57–59

banking: investments relating to, 247–48; Islamic, 48–50

Bateson, Gregory, 57

behavior, 23; moral, 264

Bellah, Robert, 307, 312n18

benchmarking, 131, 201

benefit corporations, 262–66; B Lab, 266–67; conclusions on, 268–69; as hybrid corporation, 267–68

Berlin, Isaiah, 288–90, 309nn8–10

Berry, Thomas, 8, 325

Bhagavad Gita, 39–40

biodiversity: boundary, 179–82; climate change on, 179–80; governance on, 179–82; HANPP relating to, 181–82; land use relating to, 180, 181–82; Rockström team on, 179
biogeochemical systems, 150
biophilia, 7
biophysical systems, 79; production as, 363
biosphere: human embeddedness and, 120; measurements for, 120; real economy and, 244–46
biotic community, 218, 227n4
birth ritual, 53
B Lab, 266–67
Bourdieu, Pierre, 324
brain, 70–71
Brody, Hugh, 8
Brown, Peter, 10, 17–18, 154–55; Leopold compared to, 213
Buddhist economics, 45–47
Buddhist ethics: householding and, 43; interdependent impermanence and, 42–47; mindfulness and, 43–44, 45; process theorizing and, 44–45; *Small Is Beautiful* relating to, 43, 45–46
butterfly health model, 198

Canada, 249–54
capitalism: climate change relating to, 327–28; cultural change and, 325–29; discourse theory on, 325–28; double movement from, 327; with ecological economics, 111; efficiency in, 239; evolution of, 325–26; globalization and, 329–30; growth dilemma in, 239–40, 245, 327, 330; Habermas on, 111; justice under, 107–8; labor and, 293–94; liberty and, 293–94; Marxism and, 293–94; neoliberal discourse of, 325–27; Polanyi on, 327; production and, 239; Speth on, 327–28
carbon dioxide, 152
carbon emissions: global emissions pathways, 172, 172–73; governance on, 172–73, 176; indicators on, 141–42; in UK, 253, 253

carbon intensity, 241–42, 242
carrying capacity, 153–54
case studies: alternative Western tradition, 27–31; Balinese ethics and sustainable community, 57–59; Buddhist ethics and interdependent impermanence, 42–47; conclusion on, 59–60; Gandhian ethics, 38–42; Hopi, 51–57; introduction to, 26–27; Islamic economics and ethics, 47–51; Nayaka and ethics of gift, 35–38; St. Francis and ethics of abundance, 31–35
catallactics, 39
CFCs. *See* chlorofluorocarbons
change. *See* climate change; cultural change
chlorofluorocarbons (CFCs), 112
Chomsky, Noam, 72
CICs. *See* community interest companies
citizenship, 310n14; common good and, 311n17
civic republicanism, 310n14
Clapham, Arthur Roy, 216–17
climate change: on biodiversity, 179–80; boundary, 170–76, 174–75; capitalism relating to, 327–28; cost of, 113–14; debt relating to, 81, 107; democracy and, 116nn2–3; global economy and, 96; global emissions pathways and, 172, 172–73; governance of, 170–76; IPCC on, 171–73, 176; justice relating to, 80; population relating to, 328; UNFCCC on, 113; wealth relating to, 79
climate models, 140–41
closed systems, 307n2
Cohen, Bernard, 332
commensurability, 130–35
commercial banks, 247–48
common good: citizenship and, 311n17; humanity and, 289
common sense, 336, 339, 352n14
commonwealth, 76–77, 78
communal organization, 301
communism, 108
community: biotic, 218, 227n4; changing our relationship with, 360; land, 218,

219–20; relational liberty and, 301; solidarity, 297; sustainable, 57–59

community interest companies (CICs), 269n2

complex interacting systems, 141–42

composite indicators: description of, 131–32; limitations of, 132–35

consciousness industry, 333

consumer culture, 330–32

consumption, 9; cultural change relating to, 330–34; dematerialization, 241–42; ethics, 212; hegemonic discourses of consumerism and, 325–34; materialism relating to, 73; media reinforcement on, 332–34; status relating to, 74

contaminants, 193–94

contemporary science, 9

Copernican Revolution: on freedom, 104; of justice, 102–7; knowledge relating to, 102–3; second, 104–5

corporate law, 261–62

corporate social responsibility (CSR), 262, 265, 267–68

corporations, 234; benefit, 262–69; CICs, 269n2; conclusions on, 268–69; evolution of, 261; fair trade and, 261; fiduciary duty relating to, 270nn7–8; Global Impact Investing Ratings Systems relating to, 266–67; governance of, 265–66, 268; hybrid, 264–68; introduction to, 260–64; Jacobs on, 264–65; L3C, 269n3; LLCs, 269n3

Costanza, Robert, 200–201

courage, 81–82

creation: Islam on, 51; life and, 67–68, 360; of money, 247–48; religion relating to, 8

crises, 237–38, 256; consumer culture relating to, 330–31; self relating to, 330–31

CSR. *See* corporate social responsibility

cultural change, 236, 286; capitalism and, 325–29; conclusions on, 350; consumer culture and, 330–32; consumption relating to, 330–34;

discourse theory relating to, 321–25; Douglass on, 342–43, 346–47; elites relating to, 343–46; Enlightenment and, 344, 345–46; entropic thrift, householding, membership, and, 339; ethics and, 320–21; fight and protest against, 346–50; in future, 342–50; globalization and, 329–30; Gramsci on, 335–36; by grassroots mobilization, 343; Harvey, D., on, 337–38; historical systems and, 340–41; through history, 335, 336–37, 341–42; Hunter on, 343–46, 349–50; network overlap relating to, 345–46; political change and, 334–39; power relating to, 344–45; propositions on, 343–50; social reality relating to, 323–25; Wuthnow on, 336–37, 338–39, 341–42, 345–46

cultural frames, 322–25, 339–40

cultural schemes, 311n15

culture: consumer, 330–32; popular, 336. *See also specific cultures*

Curry, John Steuart, 227n7

data: handling, 119–20; on land, 221; overabundance of, 221

debt: climate change relating to, 81, 107; justice and, 107; private, 240; recession relating to, 239–40

decoupling, 241–42; issues of, 245–46

degrowth movement, 352n19

dematerialization, 241–42

democracy, 93, 116nn2–3

democratic dysfunction, 116n2

demographics, 256

Derrida, Jacques, 36–37

determination, 308n3

de Tocqueville, Alexis, 272, 306

discourse competition, 324–25

discourse theory, 11; on capitalism, 325–28; on consumer culture, 330–32; on cultural change and political change, 334–39; cultural change relating to, 321–25; cultural frames and, 322–25; of neoliberalism and consumerism, 325–34; political ecology discourse, 327

discursive field, 337
distribution, 4–5; fair, 79–80; indicators relating to, 163–64; justice of, 99–100, 248; planetary boundaries relating to, 161; reconsidering, 360; right relationship relating to, 161
diversity-stability hypothesis, 227n6
divine mandate, 8
domination, 308n3, 312n18
double movement, 327
Douglass, Frederick, 342–43, 346–47
Duguay, Philip, 234

Earth Story, 53–55
ecological economics: agendas of, 3–9; capitalism with, 111; ecological political economy, 273–77; ethos, 85; explicit agenda of, 4–5; freedom relating to, 234–35; free trade and, 364–65; future implications of, 363–65; implicit agenda of, 5–7; indicators in, 10–11; justice in, 108–11; MFA and, 328–29; normative criteria for, 284; original premise of, 2–3, 15–16; physical systems relating to, 23–24; pivotal significance of, 9–12; planetary boundaries in, 115; premises of, 75–78; principles of, 78–81; reconstruction agenda of, 7–9, 80–81, 83; transition to, 107, 111–16, 318–20, 358–65; virtues of, 81–84
ecological finance, 364
ecological macroeconomics, 11, 233–34; decoupling, 241, 245–46; beyond decoupling, 241–42; foundations for, 246–48; future of, 257; GEMMA framework, 255, 255–57; growth dilemma in, 239–40, 245, 327, 330; interest rates in, 245–46; introduction to, 237–38; investment relating to, 247–48; national accounting reform, 247–48; national green economy macro-model, 248–54; stock-flow consistent model of, 254–57; system dynamics, 242–44; system linkages, 244–46; variables in, 246–47

ecological political economy: liberty in, 273–77; on planetary boundaries, 274–75
ecological politics, 10
ecological public philosophy, 277–79
ecology, 211, 327
economic growth: Aristotle on, 30; decoupling and, 241–42; degrowth movement, 352n19; dilemma of, 239–40, 245, 327, 330; *The Limits to Growth* on, 242–43, 348; in LowGrow, 249–51, *250, 251*; media reinforcement of, 332–34; microeconomics and, 330; modeling and, 363; in recession, 239–40; slowing of, 239; stationary state relating to, 30
economic liberty: defining, 280; interpretation of, 280–84; leaving behind, 284–87; Mill on, 281–84; power relating to, 292–95
economics. *See specific topics*
economic theory, 21–26
economy: gift, 35–37, 55–56; global, 96; real, 244–46, 255–56; Victor on, 307n2; work, labor, and, 276, 287, 293–95
ecosystem: heterogeneity of, 126; humans and, 217–22; services, 102; terminology background, 216–17; Victor on, 307n2
ecosystem health, 127, 146n1; benchmarking, 201; Costanza on, 200–201; criticisms of, 201–2; definitions of, 199–203; discussion and conclusions on, 203–4; human health metaphor and, 190–99, 202–3, 222; humans relating to, 220–22; indicators of, 190–91, 202; introduction to, 190–92; land health, 209–10, 213–16, 220, 222; Leopold relating to, 209–10, 213–16, 216–25
education, 360–61
efficiency, 4–5, 81; in capitalism, 239; freedom and, 99; indicators relating to, 163–64; justice and, 99

Elgin, Duane, 332
elites, 343–46
emancipation project, 67–68, 75–76,
    85; Enlightenment and, 103;
    exemptionalism relating to, 103–4;
    intelligence and, 73; justice relating
    to, 103–4; truisms, 103
emancipatory claims, 97–98
embeddedness: human, 104–5, 120, 214,
    218–19, 226; of human health, 191; of
    real economy and biosphere, 244–46;
    of traditions, 24–26
embedded permeable person, 69–70
embedded self, 69–70
employment, 252
endangered species, 180–81
energy, 77
Enlightenment: cultural change and, 344,
    345–46; emancipation project and,
    103; freedom during, 105; knowledge
    relating to, 102–6
entropic thrift, 16, 77–78, 79; cultural
    change and, 339; Leopold and, 213;
    money and, 81
environmental economics, 101
environmental justice, 10
epistemological humility, 82
eradicated nature, 103–4
ethical foundations: entropic thrift as,
    16; householding as, 16; introduction
    and summary of, 15–19, 66–68;
    membership as, 16; proposed,
    358–61
ethics, 9–10; of abundance, 31–35;
    arguments in, 90–91; Aristotle
    relating to, 27–31; authoritarianism
    and, 91–93; Balinese, 57–59;
    Buddhist, 42–47; consumption, 212;
    CSR and, 262, 265, 267–68; cultural
    change and, 320–21; ecological
    politics, 10; ecology relating to,
    211; economic theory relating
    to, 21–26; environmental justice,
    10; freedom relating to, 234–35;
    Gandhian, 38–42; of gift, 35–38;
    Hopi and, 51–57; intrinsic right
    and wrong, 283; Islamic economics

and, 47–51; justice and, 90–92; land,
    209–13, 214, 224; moral behavior,
    264; morality, liberty, and, 285–87;
    Nayaka and, 35–38; person and,
    68–71; relational liberty and, 298–99;
    religion and, 25; rethinking, 358;
    science, knowledge, and, 71–75;
    social concern, 10; St. Francis and,
    31–35; waste and, 77
exchange: Aristotle on, 28–31; justice
    of, 99
exemptionalism, 75–76, 103–4
explicit agenda, 4–5
extinction rates, 180–81
extraction rates, 245–46
extractive power, 282, 291, 292–95
extrinsic values, 352n17

Fairclough, Norman, 332–33
fair distribution, 79–80
fair share, 83–84
fair trade, 261
fear, 106
Feng Lin, Qi, 11, 123–24
fertilizers, 191–92
fiduciary duty, 270nn7–8
fiduciary principle, 109–11
fiduciary stewardship, 115–16
financial crisis, 238, 256
fossil fuels, 24–25
Franciscan movement, 32–35
Fraser, Nancy, 300–301
freedom: Berlin on, 288–89;
    conclusions on, 305–7; Copernican
    Revolution on, 104; defining, 307n1;
    ecological economics relating to,
    234–35; efficiency and, 99; during
    Enlightenment, 105; ethics relating
    to, 234–35; justice and, 99, 103, 106;
    land ethic relating to, 224; liberty
    compared to, 280, 307n1; Locke on,
    74; Mill on, 74; relational liberty and,
    295–97, 300–301; standard economics
    and, 22, 24
free trade, 364–65
Friedman, Milton, 22
Fyles, James, 11, 123–24

Game Management (Leopold), 216
Gandhian ethics, 38–42
Garver, Geoffrey, 10–11, 121–23, 154–55
GDP. *See* gross domestic product
Gellner, Ernst, 24–25
GEMMA. *See* Green Economy Macro-
Model and Accounts
Georgescu-Roegen, Nicolas, 4
Giddens, Anthony, 24
gift economy, 35–36; of Hopi, 55–56;
nonreciprocal gift and, 36–38
gift ethics, 35–38
global carrying capacity, 153–54
global economy, 96
global emissions pathways, *172*, 172–73
global hegemony, 75–76
Global Impact Investing Ratings Systems,
266–67
global indicators, 129; uncertainty and,
157–58
globalization, 329–30; antiglobalization
movement, 351*n*8
global warming: average rate of, 171–72;
measurements of, 136–39, *137*, *138*
Goldberg, Mark, 10–11, 121–23
Good Work (Schumacher), 46
governance: on biodiversity, 179–82;
on carbon emissions, 172–73, 176;
challenges of, 164–66; of climate
change, 170–76; of corporations,
265–66, 268; criteria for, 168, 170;
filter, 166–67, *167*; framework for,
174–75; indicators relating to, 133–35,
164–70; land use restrictions, 180;
on nitrogen loading, 176–79; of
planetary boundaries, 122, 155–56,
164–70, *174*–75; rebuilding, 365–66;
scarcity relating to, 166
Gramsci, Antonio, 335–36, 352*n*14
grassroots mobilization, 343
Greek culture, 8
Green Economy Macro-Model and
Accounts (GEMMA), *255*, 255–57
gridbox mean, 137
gross domestic product (GDP), 154,
156–57; decoupling relating to, 241;

in LowGrow, 249–51, *250*, *251*;
rethinking, 361–62
A Guide for the Perplexed (Schumacher),
46–47
GUMBO model, 243

Habermas, Jürgen, 111
habits, 70–71
HANPP. *See* human appropriation of net
primary production
hard-line objection, 92–93
harm principle, 281–82, 284
Harvey, David, 337–38
Harvey, Janice, 11, 236
HCFCs. *See* hydrochlorofluorocarbons
HDI. *See* Human Development Index
health, 120–21, 122–23; allied health
fields, 194–96; butterfly model of, 198;
connotative concept of, 220, 221–22;
contaminants relating to, 193–94;
contextual factors of, *198*; definitions
of, 196–99, 220; ecosystem, 127,
146*n*1, 190–204; human health
definitions, 196–97; human health
metaphor, 190–99, 202–3, 222;
indicators and measurements of,
127, 192, 194–96; International
Classification of Functioning,
Disability, and Health, *198*, 199; land,
209–10, 213–16, 220, 222; Larson on,
197–98; physician roles, 191, 192–94,
196–97; population, 194, 195; Saracci
on, 197; Smith on, 196–97; WHO on,
197–99
hedonism, 69
hegemonic discourse. *See* discourse
theory
heterogeneity, 126
Hinduism, 39–40, 41
historical cultural change, 335, 336–37,
341–42
historical systems, 340–41
Hofstadter, Richard, 340
homo economicus, 277–79, 287
Hopi: background on, 51–53; birth ritual
for, 53; conclusion on, 56–57; Earth

Story, 53–55; ethics, economics, and, 51–57; gift economy of, 55–56
householding, 16, 76–77; Buddhist ethics and, 43; cultural change and, 339; Leopold on, 213
household management, 29
human appropriation of net primary production (HANPP), 151, 156; biodiversity relating to, 181–82
human-associated microbiota, 218, 221
The Human Condition (Arendt), 276
Human Development Index (HDI), 156–57
human-Earth relationship, 150–51, 217–22
human embeddedness, 104–5; biosphere and, 120; in land, 214, 218–19, 226
human health. *See* health
humanity: common good and, 289; liberty and, 275, 287
humanness, 308n4, 359–60
humility, 82
Hunter, James Davison, 321; on cultural change, 343–46, 349–50; on elites, 343–46
hunter-gatherers, 26, 35–38
hybrid corporations, 264–68
hydrochlorofluorocarbons (HCFCs), 112

impermanence, 42–47
implicit agenda, 5–7
incommensurability, 130–31
indicators: aggregate, 131–32, 162–63; application of, 155–56, 182–83; on carbon emissions, 141–42; combining information, 132–33; commensurability relating to, 130–35; complex interacting systems and, 141–42; composite, 131–35; conclusion on, 142–43; considerations, of measurement process, 135–39; contextual considerations, 127–28, 154; distribution relating to, 163–64; in ecological economics, 10–11; of ecosystem health, 190–91, 202;

efficiency relating to, 163–64; global, 129, 157–58; governance relating to, 133–35, 164–70; of health, 127, 192, 194–96; integrative approach to, 182–83; interpreting changes in, 133; introduction to, 125–26, 154–57; IPAT formulation and, 163; issues related to, 143–46; mixing disparate quantities, 132; models and prediction, 139–41; normative criteria for, 161–62; objectives for, 143; paradigm of accuracy, 139; on phosphorus cycle, 142; planetary boundaries relating to, 150–55, 157–58, 182–83; in policy, 133–35, 168; for production, 129–30; right relationship and, 157–58, 160; Rockström team on, 150–51, 153, 157–61, 164, 167–68, 184n4; safe operating space relating to, 155, 157–62; scale relating to, 128–30, 150, 162–63; scope relating to, 128; time relating to, 129; uncertainty and, 141, 157–58; validity of, 143–46, 156–57; variables of, 131, 135–36
individual labor, 297
individual welfare, 338
Ingold, Tim, 209–10
intelligence: adaptive advantage of, 72–74; Chomsky on, 72; emancipation project and, 73; swarm, 94
interest rates, 245–46
intergenerational fairness, 79–80
Intergovernmental Panel on Climate Change (IPCC), 171–73, 176
International Classification of Functioning, Disability, and Health, 198, 199
intrinsic right and wrong, 283
intrinsic values, 352n17
investment: ecological macroeconomics relating to, 247–48; in GEMMA, 256; Global Impact Investing Ratings Systems, 266–67
IPAT formulation, 153–54, 163

IPCC. *See* Intergovernmental Panel on Climate Change

Islam: on animals, 48; on creation, 51; ethics, economics, and, 47–51; foundational belief, 47–48; *Koran* and, 48; on usury, 48–49; on zakat, 48

Islamic banking, 48–50

Jackson, Tim, 11, 233–34
Jacobs, Jane, 264–65
Janda, Richard, 10, 11, 18–19, 234
Jennings, Bruce, 11, 234–35, 320, 322
justice, 75; Aristotle on, 99–100; under capitalism, 107–8; climate change relating to, 80; after communism, 108; considerations, 98–102; Copernican Revolution of, 102–7; debt and, 107; dimensions of, 96–97, 97; of distribution, 99–100, 248; in ecological economics, 108–11; efficiency and, 99; emancipation project relating to, 103–4; emancipatory claims relating to, 97–98; environmental, 10; ethics and, 90–92; of exchange, 99; fear relating to, 106; Fraser on, 300–301; freedom and, 99, 103, 106; hard-line objection and, 92–93; introduction to, 89–95; liberal justice model, 103, 107–8; matrix, 97, 98, 100–101, 109; measurements for, 106; metaphysical, 100–101; metatheoretical, 94–98, 107–11; participatory parity relating to, 300–302; planetary boundaries and, 102–7, 163–64; reconsidering, 360; self-interest and, 99; social justice model, 107–8; soft-line objection and, 93–94; in standard economics, 98–102; theories, 95–96, 103; transitional, 111–16; truisms, 103

Kant, Immanuel, 281
Kantians, 73
Keynesians, 279
knowledge: adaptive advantage of intelligence and, 72–74; as

approximate and provisional, 2; Copernican Revolution relating to, 102–3; Enlightenment relating to, 102–6; ethics, science, and, 71–75

Koran, 48

L3C. *See* low-profit limited liability company

labor, 287; capitalism and, 293–94; economy, work, and, 276, 287, 293–95; planetary boundaries relating to, 276; productivity, 239, 252

Lady Poverty, 31

Lanchester, John, 293–94

land: aesthetic, 212–13; community, 218, 219–20; data on, 221; health, 209–10, 213–16, 220, 222; human embeddedness in, 214, 218–19, 226; pyramid, 214, 227n8

land ethic, 209–13, 214; freedom relating to, 224

The Land Ethic (Leopold), 211

"The Land-Health Concept and Conservation" (Leopold), 216

"Land Pathology" (Leopold), 216

land use: biodiversity relating to, 180, 181–82; Leopold on, 211–13; restrictions, 180

Lansing, John Stephen, 58–59

Larson, James, 197–98

law: corporate, 261–62; rebuilding, 365–66

Lear, Jonathan, 286, 306

LeBow, Victor, 331

legal norms, 109–10

legitimation, 111

Lehun, Richard, 10, 18–19, 234

Leopold, Aldo, 123–24; on agriculture, 212; on beauty, 212–13; Brown compared to, 213; on consumption ethic, 212; Curry relating to, 227n7; ecosystem health relating, 209–10, 213–16, 216–25; entropic thrift and, 213; final thoughts on, 225–26; *Game Management*, 216; on householding, 213; on humans, in ecosystem, 217–20; introduction and background on, 208–10; "The Land Ethic," 211; land ethic of, 209–13, 214, 224; on land health, 209–10, 213–16, 220, 222; "The Land-Health Concept and Conservation," 216; "Land Pathology," 216; on land pyramid, 214, 227n8; on land use, 211–13; on membership, 213; "A Mighty Fortress," 223–25; personal history of, 209–10, 227nn1–2; "The Role of Wildlife in a Liberal Education," 215; on science, 214

liberalism, 90–91; liberty beyond, 287–95; liberty of, 279–87; relational liberty compared to, 310n14. *See also* neoliberalism

liberal justice model, 103, 107–8

liberal revisionism, 288–90

liberationist critique, 290–92

liberty, 11; Berlin on, 288–90, 309nn8–10; capitalism and, 293–94; defining, 307n1; in ecological political economy, 273–77; ecological public philosophy and, 277–79; economic, 280–87, 292–95; extractive power relating to, 282, 291, 292–95; freedom compared to, 280, 307n1; humanity and, 275, 287; individual, 297; introduction to, 272–73; beyond liberalism, 287–95; of liberalism, 279–87; Macpherson on, 290–92; Mill on, 281–84; morality, ethics, and, 285–87; negative, 279–80, 284–86, 289–91, 308n5; positive, 289–92; relational, 295–305, 310n14; technological advance relating to, 276, 286

life, 67–68, 360

limited liability corporations (LLCs), 269n3

The Limits to Growth (Meadows et al.), 242–43, 348

Lindeman, Raymond, 217

LLCs. *See* limited liability corporations

Locke, John, 74, 77

Lovelock, James, 116n2

LowGrow: description of, 249; economic growth in, 249–51, *250, 251*; employment under, 252; GDP in, 249–51, *250, 251*; labor productivity in, 252; as national green economy macro-model, 249–54; production in, 253–54

low-profit limited liability company (L3C), 269n3

Macpherson, C. B., 290–92

Malthus, Thomas, 5

Marglin, Steven, 70

Marx, Karl, 5; Schumpeter on, 27; on value, 28

Marxism, 293–94

material flow accounting (MFA), 328–29

materialism, 73

Mauss, Marcel, 36

Mawdidi, Sayyid Abu A'la, 51

Mayo, Nancy, 10–11, 122–23

Meadows, Dennis, 348

Meadows, Donella H., 242–43

measurements, 11; for biosphere, 120; challenges in, 126; errors in, 145; of global warming, 136–39, *137, 138*; of health, 127, 192, 194–96; introduction and summary of, 119–26; for justice, 106; metric, 119; models compared to, 139–40; of monetary valuation, 130; of price, 361–62; rethinking, 361–63; scope relating to, 128; of temperature, 135–39, *137, 138*; validity relating to, 143–46. *See also* indicators

media, 332–34

membership, 16, 75–76; cultural change and, 339; Leopold on, 213

Merchant, Carolyn, 8

metaphysical justice, 100–101

metatheoretical justice: approach to, 94–98; fiduciary principle within, 109–11; shift to, 107–9

metatheory, 94

metric, 119

MFA. *See* material flow accounting

microbiota, 218, 221

microeconomics, 11, 330

"A Mighty Fortress" (Leopold), 223–25

Mill, John Stuart: on freedom, 74; on liberty, 281–84
mindfulness, 43–44, 45
modeling, 363
monetary valuation, 130
money: carbon intensity, per dollar, 241–42, 242; creation of, 247–48; entropic thrift and, 81; real economy and, 244
Montreal Protocol on Substances that Deplete the Ozone Layer (Montreal Protocol), 112–13
moral behavior, 264
morality, 285–87
More, Thomas, 310n12
Muir, John, 8
Musgrave, Richard, 5
mutuality, 302–5

Nadeau, Robert, 6–7, 69
national accounting reform, 247–48
national green economy macro-model: importance of, 248–49; LowGrow as, 249–54
naturalism, 8
nature-deficit disorder, 7
Nayaka, 35–38
negative liberty, 279–80, 284–86, 308n5; Berlin on, 289; Macpherson on, 290–91
neoliberalism, 293–95; appeal of, 338; globalization and, 329–30; hegemonic discourses of, 325–34; shift to, 337–38
network overlap, 345–46, 348
New Story, 8
nitrogen flux, 177, 178
nitrogen loading, 176–79
nonreciprocal gift, 36–38
nonviolence, 40–42
normative boundaries, 152
normative criteria: for ecological economics, 284; for indicators, 161–62; for planetary boundaries, 161–62, 164–65
normative pluralism, 95–96

Odum, Eugene, 217

oil spills, 157
ontological dualism, 76
open systems, 307n2
otherness, 7
ownership. See property and ownership

paradise, 8
Pareto, Vilfredo, 28
participatory parity, 300–302
person: embedded permeable, 69–70; essential questions relating to, 68; ethics and, 68–71; market manufacturing, 70–71; Nadeau on, 69; nature of, 68–69; quantum physics on, 69; rational, 68–69; systems theory on, 69–70
philosophy, 277–79
phosphorus, 142
physical systems, 23–24
physician roles, 191, 192–94, 196–97
planetary boundaries: atmospheric boundaries, 158–59; for atmospheric carbon dioxide, 152; biodiversity boundary, 179–82; categories of, 151; climate change boundary, 170–76, 174–75; common features of, 151–52, 160; critique of, 160; distribution relating to, 161; in ecological economics, 115; ecological political economy on, 274–75; exemptionalism relating to, 103–4; governance of, 122, 155–56, 164–70, 174–75; HANPP, 151; human-Earth relationship guided by, 150–51; indicators relating to, 150–55, 157–58, 182–83; interrelatedness of, 152–53; introduction to, 150–54; IPAT formulation and, 153–54, 163; justice and, 102–7, 163–64; labor relating to, 276; nitrogen loading boundary, 176–79; normative boundaries, 152; normative criteria for, 161–62, 164–65; precautionary principle on, 159–60, 167–68; Rockström team on, 150–51,

153, 157–61, 164, 167–68, 184*n*4;
thresholds and, 159, 160–61; tipping
points, 157–58; trends in, *169*;
uncertainty relating to, 157–60, *169*
play, 312*n*18
pluralism, 95–96
plurality, of values, 362
Polanyi, Karl, 327
policy: criteria, 168, 170; indicators in,
133–35, 168
political change, 334–39
political ecology discourse, 327
popular culture, 336
population: carrying capacity relating to,
153–54; climate change relating to,
328; expansion of, 9; health, 194, 195;
impact of, 153–54; projections, 153;
rural, 211–12
positive liberty, 289–92
possessiveness, 35
post-Keynesians, 256
poverty: addressing, 248–49; Lady
Poverty, 31; St. Francis and, 31–33
power: cultural change relating to,
344–45; economic liberty relating to,
292–95; extractive, 282, 291, 292–95;
relations, 333
precautionary principle, 159–60,
167–68
prestige, 344–45
price, 361–62
primitive religion, 8
private debt, 240
process theorizing, 44–45
production: as biophysical process, 363;
capitalism and, 239; HANPP, 151,
156, 181–82; indicators for, 129–30;
in LowGrow, 253–54; in neoliberal
political economy, 293
productivity, 239, 252
progress, 85
property and ownership, 6; attachment
to, 35; Locke on, 77; for Nayaka, 38;
scarcity relating to, 34–35
protest, 346–50
public philosophy, 277–79

quantification, 119–20
quantum physics, 69
quasi-science, 2

rationality, 359
rational person, 68–69
Rawls, John, 77
real economy, 244–46, 255–56
reality: social, 323–25, 333–34; validity
compared to, 145, *146*
reasoning, 72
recession, 239–40
reciprocity: Derrida on, 36–37; in gift
ethics, 35–38; nonreciprocal gift,
36–38; Sahlins on, 36
reconstruction agenda, 7–9, 80–81, 83
Reese, Stephen, 340
relational liberty: community and, 301;
defining, 295–96, 300–301; ethics
and, 298–99; freedom and, 295–97,
300–301; on individual liberty and
community solidarity, 297; liberalism
compared to, 310*n*14; normative
structure of, 299–305; participatory
parity relating to, 300–302; self and,
297–99, 302–7; solidarity and, 297,
302–5
relative decoupling, 241
religion, 7; *Bhagavad Gita*, 39–40;
Buddhist ethics, 42–47; creation
relating to, 8; divine mandate, 8;
ethics and, 25; Hinduism, 39–41;
Islamic economics and ethics, 47–51;
primitive, 8; on usury, 25, 29–30,
48–49
religious wars, 73
republicanism, 310*n*14
resilience: moral resiliency, 306; of
resources, 79; scale and, 79
resources: extraction rates of, 245–46;
resilience limits of, 79
respect, 84
Ricardo, David, 5, 27
right relationship: distribution relating
to, 161; indicators and, 157–58, 160;
safe operating space and, 157–62

Rio Principles, 150, 184n1
Rockström team: on biodiversity, 179; on indicators and planetary boundaries, 150–51, 153, 157–61, 164, 167–68, 184n4; on IPCC, 171; on nitrogen loading, 176–78
Rogers, Raymond, 327
Roland-Morris low back pain and disability questionnaire, 144–45
"The Role of Wildlife in a Liberal Education" (Leopold), 215
runoff, 191–92
rural population, 211–12
Ruskin, John, 38–39

safe operating space, 155, 157–62
Sahlins, Marshall, 36, 272, 311n15
Saracci, Rodolpho, 197
scale, 4–5; aggregate indicators relating to, 162–63; indicators relating to, 128–30, 150, 162–63; resilience limits and, 79
scarcity: abundance and, 33–34; governance relating to, 166; property and ownership relating to, 34–35; standard economics relating to, 34
Schmidt, Jeremy, 82
Schrödinger, Erwin, 67
Schumacher, E. F., 42; background on, 45–46; Buddhist economics and, 45–47; Good Work, 46; A Guide for the Perplexed, 46–47; Small Is Beautiful, 43, 45–46
Schumpeter, Joseph, 27, 30
Schweitzer, Albert, 9
science: ethics, knowledge, and, 71–75; Leopold on, 214
scientific model, of environment, 119–20
scientific revolution, 8–9
scope, 128
secularization, 73
self: consumer culture and, 331; crises relating to, 330–31; ecological public philosophy on, 277–79; embedded, 69–70; -interest and justice, 99; Nadeau on, 69; relational liberty and, 297–99, 302–7; -rule, 41; solidarity and, 297, 302–5

Sen, Amartya, 80
Shue, Henry, 80
Small Is Beautiful (Schumacher), 43, 45–46
Smith, Richard, 196–97
social cognition studies, 352n17
social justice model, 107–8
social reality: cultural change relating to, 323–25; media and, 333–34
Soddy, Frederick, 4
soft-line objection, 93–94
solar flow, 78
solidarity, 297, 302–5
Speth, James Gustave, 327–28
Spiritual Franciscans, 33
standard economics: appeal of, 22; challenges of, 22–23; dangers of, 23; disembedding, 24–25; distress signs in, 23–24; freedom and, 22, 24; justice in, 98–102; overlooked elements of, 1–2; quasi-science relating to, 2; scarcity relating to, 34; shift from, 107–9, 111–16. See also capitalism
standing up: as, 304–5; for, 303–4; with, 304; beside, 303
stationary state, 30
status, 74
stewardship, 114–16
St. Francis: background on, 31–33; ethics of abundance and, 31–35; Franciscan movement and, 32–35; poverty and, 31–33; Spiritual Franciscans and, 33
stock-flow consistent macroeconomic model, 254–57
sustainable community, 57–59
swarm intelligence, 94
system dynamics, 242–44
systems theory: on person, 69–70; world-, 11, 319–20

T21 model, 243–44
Tansley, Arthur, 216–17
Taylor, Charles, 318, 334
technical revolution, 8–9
technological advance, 239, 276, 286
temperature: climate models and, 140–41; global, 136–37; measurements of, 135–39, 137, 138

Thatcher, Margaret, 10, 326
thresholds, 159, 160–61
time, 129
Timmerman, Peter, 10, 16–17, 321–22
tipping points, 157–58
traditions: alternative Western, 27–31;
    embeddedness of, 24–26
transactions, 365
transitional justice, 111–16
trees, 223–24

UK. *See* United Kingdom
uncertainty: global indicators and,
    157–58; indicators and, 141, 157–58;
    planetary boundaries relating to,
    157–60, *169*; trends in, *169*
UNEP. *See* United Nations
    Environmental Outlook
UNFCCC. *See* United Nations
    Framework Convention on Climate
    Change
United Kingdom (UK): carbon emissions
    in, 253, *253*; CICs in, 269*n*2
United Nations Environmental Outlook
    (UNEP), 328–29
United Nations Framework Convention
    on Climate Change (UNFCCC), 113
universe, 67–68, 71–75
"Unto This Last" (Ruskin), 38–39
usury: Aristotle on, 29–30; Islam on,
    48–49; religion on, 25, 29–30, 48–49
utilitarianism, 311*n*17
utilitarians, 73
Utopia (More), 310*n*12

validity: of indicators, 143–46, 156–57;
    measurements relating to, 143–46;
    reality compared to, 145, *146*

valuation techniques, 328
value: Aristotle on, 28–31; Marx on, 28;
    plurality of, 362; theory of, 28
values, 352*n*17
Victor, Peter, 11, 233–34; on economy,
    307*n*2; on ecosystems, 307*n*2
violence, 40–42
virtues: atonement, 82–83; courage as,
    81–82; of ecological economics,
    81–84; epistemological humility, 82;
    fair share, 83–84; respect, 84

Wallerstein, Immanuel, 340–41
waste, 77
water runoff, 191–92
wealth, 29, 79
welfare state, 338
Western culture, 8–9; consumer culture
    of, 330–32; emancipation project
    relating to, 67–68; global hegemony
    of, 75–76; on progress, 85; shifts in,
    336–37; superiority of, 76
Whately, Richard, 39
"What is Life" (Schrödinger), 67
WHO. *See* World Health Organization
work, 276, 287, 293–95
World 3 model, 243
World Health Organization (WHO),
    197–99
world system models, 242–44; linkages
    in, 244–46; stock-flow consistent,
    254–57
world-systems theory, 11, 319–20
Wuthnow, Robert, 336–37, 338–39,
    341–42, 345–46

zakat, 48
Žižek, Slavoj, 103–4